Studies in Systems, Decision and Control

Volume 611

The series "Studies in Systems, Decision and Control" (SSDC) covers both new developments and advances, as well as the state of the art, in the various areas of broadly perceived systems, decision making and control–quickly, up to date and with a high quality. The intent is to cover the theory, applications, and perspectives on the state of the art and future developments relevant to systems, decision making, control, complex processes and related areas, as embedded in the fields of engineering, computer science, physics, economics, social and life sciences, as well as the paradigms and methodologies behind them. The series contains monographs, textbooks, lecture notes and edited volumes in systems, decision making and control spanning the areas of Cyber-Physical Systems, Autonomous Systems, Sensor Networks, Control Systems, Energy Systems, Automotive Systems, Biological Systems, Vehicular Networking and Connected Vehicles, Aerospace Systems, Automation, Manufacturing, Smart Grids, Nonlinear Systems, Power Systems, Robotics, Social Systems, Economic Systems and other. Of particular value to both the contributors and the readership are the short publication timeframe and the world-wide distribution and exposure which enable both a wide and rapid dissemination of research output.

Indexed by SCOPUS, DBLP, WTI Frankfurt eG, zbMATH, SCImago.

All books published in the series are submitted for consideration in Web of Science.

Rajni Mohana · Aman Sharma · Anand Nayyar ·
Poonam Saini
Editors

Federated Learning Applications in the Industrial Internet of Everything (IoE)

Springer

Editors
Rajni Mohana
Department of Computer Science
Amity School of Engineering
and Technology
AUMP in Amity University Punjab
Sahibzada Ajit Singh Nagar, Punjab, India

Aman Sharma
Department of Computer Science
and Engineering and Information
Technology
Jaypee University of Information
Technology
Solan, Himachal Pradesh, India

Anand Nayyar
School of Computer Science
Duy Tan University
Da Nang, Vietnam

Poonam Saini
Department of Computer Science
and Engineering
Punjab Engineering College
Chandigarh, Punjab, India

ISSN 2198-4182 ISSN 2198-4190 (electronic)
Studies in Systems, Decision and Control
ISBN 978-3-031-99269-8 ISBN 978-3-031-99270-4 (eBook)
https://doi.org/10.1007/978-3-031-99270-4

This Springer imprint is published by the registered company Springer Nature Switzerland AG
The registered company address is: Gewerbestrasse 11, 6330 Cham, Switzerland

This book is dedicated to all the editors and their family members. Without their support and cooperation, this book would not have become a reality.

Preface

The advent of the Industrial Internet of Everything (IoE) marks a pivotal moment in our technological evolution, weaving a complex digital tapestry that connects an unprecedented number of devices, systems, and critical infrastructures. This hyper-connected ecosystem generates a deluge of data, offering profound opportunities for optimization, automation, and intelligent decision-making. However, this data-rich environment simultaneously presents formidable challenges. The very act of centralizing vast, heterogeneous, and often sensitive data for traditional machine learning analysis runs afoul of escalating privacy regulations, such as the GDPR, and exposes organizations to significant security vulnerabilities. This creates a fundamental dichotomy: how can we harness the collective intelligence of distributed data without compromising the foundational principles of privacy and security?

This volume, Federated Learning Applications in the Industrial Internet of Everything (IoE), addresses this critical question by positing Federated Learning (FL) as a transformative paradigm. FL introduces a decentralized machine learning methodology wherein collaborative models are trained across multiple entities without the need to exchange or centralize raw data. Instead of moving data to a central model, the model is brought to the data. This approach not only inherently preserves data privacy and sovereignty but also mitigates the immense communication and storage costs associated with large-scale data aggregation. This collection of works provides a rigorous and comprehensive exploration of FL, moving beyond theoretical constructs to present its practical applications, architectural nuances, and the solutions it offers to the most pressing challenges within the IoE landscape.

The book comprises eleven chapters, each providing a deep dive into a specific facet of Federated Learning's role in the industrial and technological sphere. Chapter 1 titled "**Introduction to Federated Learning and Its Application in Industry,**" lays the foundational groundwork, defining the core concepts of FL and distinguishing its decentralized framework from traditional machine learning paradigms. Chapter 2, titled "**Federated Learning in the Healthcare Industry,**" explores the application of FL in one of the most data-sensitive domains, demonstrating how collaborative models can be built for diagnostics and treatment while upholding stringent patient privacy. Chapter 3, titled "**Federated Learning for**

Remote Health Monitoring—A Review," narrows the focus to the burgeoning field of remote patient care, reviewing how FL enables real-time health monitoring through distributed wearable and IoT devices. Chapter 4, titled "**Predicting Alzheimer's Disease: A Case Study Using Federated Learning Framework,**" provides a practical and compelling case study, illustrating the step-by-step implementation of an FL framework for early disease prediction. Chapter 5, titled "**Improving Healthcare Privacy and Efficiency with Federated Learning,**" further investigates the dual benefits of FL in healthcare, analyzing its impact on both enhancing data privacy and streamlining operational efficiency. Chapter 6, titled "**Securing Collaborative Model Training: Navigating Privacy Challenges in Federated Learning for Industrial IoT,**" delves into the specific security threats inherent in Industrial IoT environments and examines the robust cryptographic and security protocols that fortify FL systems. Chapter 7, titled "**Challenges and Opportunities in Federated Learning for the Internet of Everything (IoE),**" offers a high-level strategic overview of the broader FL landscape, systematically outlining the key hurdles and future prospects for its large-scale adoption. Chapter 8, titled "**Revolutionizing Industry 4.0: Multi-level Federated Learning for a Dynamic Ecosystem,**" introduces advanced architectural concepts, exploring how multi-level and hierarchical FL models can bring new levels of dynamism and adaptability to Industry 4.0. Chapter 9, titled "**Federated Learning for Energy Optimization in the Industrial Internet of Things: Applications and Challenges,**" addresses the critical goal of sustainability, detailing how FL can be deployed for intelligent energy management and resource optimization in industrial settings. Chapter 10, titled "**Revolutionizing Smart Transportation with Federated Learning: Applications, Challenges, and Future Directions,**" examines the impact of FL on creating safer and more efficient mobility solutions, from autonomous vehicle navigation to public transport optimization. Chapter 11, titled "**Federated Learning in Modern Education: Balancing Privacy, Scalability, and Effectiveness,**" explores the novel application of FL in education, discussing how it can personalize learning and generate institutional insights while protecting student data.

This book is intended for a broad and technically-oriented audience, including graduate students, academic researchers, AI and IoT engineers, and industry practitioners working at the intersection of industrial automation, machine learning, and distributed systems. It serves as an essential scholarly resource for anyone seeking to understand, design, and deploy Federated Learning solutions to unlock the full potential of the Industrial Internet of Everything in a secure, scalable, and privacy-preserving manner.

Sahibzada Ajit Singh Nagar, India	Rajni Mohana
Solan, India	Aman Sharma
Da Nang, Vietnam	Anand Nayyar
Chandigarh, India	Poonam Saini

Acknowledgments The editors would like to acknowledge all the authors for their contributions. We present our sincere gratitude to those who contributed their time and expertise to this book. We also want to thank the reviewers who took part in the review process and gave their valuable comments to the contributors to refine their work and adhere to the quality, coherence, and content of the chapters. Without their support, this book would not have become a reality.

Contents

Introduction to Federated Learning and Its Application in Industry

Monali Gulhane, Amit Chauhan, Nitin Rakesh, and Yash Tripathi

Abstract Federated learning (FL) represents a revolutionary approach to machine learning, particularly addressing data privacy and security challenges in industrial applications. The chapter the concept and implementation framework of FL and provides practical implications more precise than traditional centralized machine learning methods. FL allows collaborative models to be trained on nodes without data centralization, thus supporting data privacy and enhancing security. The chapter discusses different classifications of FL based on data classification and learning models and provides a mathematical framework to identify integrated learning models. Furthermore, it explores the importance of FL in data governance and addressing data silo issues, in particular under complex regulations such as the GDPR. FL has the potential to be applied in various areas such as health care, processing natural languages, computer vision, autonomous vehicles, and IoT. Integration of cryptographic solutions with customer security measures highlights FL's ability to maintain data integrity and privacy in data-sensitive environments paving the way for greater adoption.

Keywords Federal learning · Machine learning · Data privacy · Industry 4.0 · GDPR · Cryptography · Data governance · Artificial intelligence · Internet of Things (IoT) · Healthcare · NLP · Computer vision

---

M. Gulhane · N. Rakesh
Symbiosis Institute of Technology, Symbiosis International (Deemed University), Pune, India

A. Chauhan (✉)
School of Computer Science Engineering and Technology (SCSET), Bennett University, Greater Noida, UP, India
e-mail: chauhanamit37@gmail.com

Y. Tripathi
Department of Computer Science Engineering and Information Technology, Jaypee University of Information Technology (JUIT), Waknaghat, Solan, India

R. Mohana et al. (eds.), *Federated Learning Applications in the Industrial Internet of Everything (IoE)*, Studies in Systems, Decision and Control 611,
https://doi.org/10.1007/978-3-031-99270-4_1

1 Introduction

The increasing capacity for storage and processing power has made the significance of statistical analysis in the engineering field more evident. In the past few years, industrial engineering has experienced rapid and notable progress in the fields of machine learning, artificial intelligence, smart manufacturing, and deep learning [20]. Nevertheless, two considerable obstacles exist in this domain regarding advancing data science. First and foremost, data governance is crucial. Legal considerations privatize data the European Union implemented the regulation known as the General Data Protection Regulation (GDPR) in 2018 [11], users are granted full ownership rights over their data. No institutions or organizations can employ a user's data without their consent. Furthermore, the issue of data silos is a significant obstacle to the advancement of the modern industry, as the availability of additional training data would enhance the performance of training processes. For instance, when comparing it to the initial version of AlphaGo, which relied on 160,000 cases including information about human chess moves and strategies to defeat beginner-level professional players. As described by Holcomb et al. [15], Alpha Zero utilized an extensive dataset of 28.6 billion chess configurations, consisting of human and machine-generated data. This dataset enabled Alpha Zero to surpass professional players with ease. Furthermore, data annotation requires expertise from professionals in specialized fields, such as the medical industry, leading to a shortage of dependable data. The limited availability of identifiable information also hinders progress in the industrial sector. Nevertheless, the advent of FL (Federated Learning) has presented a remedy to these obstacles in this industry.

1.1 Definition and Concept

Federated learning, sometimes called collaborative learning, is a subset of machine learning focused on situations where multiple participants, often known as clients, collaborate to train a model while keeping their data decentralized. Federated learning aims to develop machine learning algorithms, like deep neural networks, by utilizing various local datasets stored on individual nodes without the need to directly share data samples. The fundamental concept involves training localized models using data samples specific to each node, then periodically swapping parameters (such as the biases and weights within deep neural networks) among these nodes to create a global model shared by all nodes. Due to their rapid advancements, technologies such as Artificial Intelligence (AI), including Machine Learning (ML) and Deep Learning (DL), are increasingly utilized in numerous IT applications. These applications extend across multiple fields, such as healthcare, engineering, IoT, marketing, and business intelligence systems. These technological advancements, powered by sophisticated algorithms and machine learning, are not only beneficial but also signify a substantial achievement towards the 4th Industrial Revolution. However, these algorithms necessitate considerable datasets for training and assessing the models to achieve high precision and dependability. Acquiring such data can be straightforward in certain scenarios, especially where digital records are abundant, but can pose difficulties in more data-scarce environments, necessitating innovative data collection and preprocessing techniques. However, privacy concerns become a significant challenge when the models need to be trained on user data. Multiple legislation and legal policies, such as the General Data Protection Regulation (GDPR) [11], are implemented to safeguard user privacy-health insurance. The Health Insurance Portability and Accountability Act (HIPAA) [1]. Furthermore, in light of the growing utilization of artificial intelligence (AI), the European Commission is considering establishing a regulatory framework to manage AI and its associated risks [7]. Numerous issues persist in the current training of machine learning models. Data accessibility is restricted: Gathering and consolidating data can be highly costly. User data confidentiality and protection: Individual repositories of sensitive data may be susceptible. High costs and centralized resources for training: Substantial computing power and energy are necessary. Because of growing user awareness of privacy concerns and related legislation, individuals are worried about the security and confidentiality of their data, making them reluctant to share their information. The scarcity of data negatively impacts the precision of machine learning models. Training ML/DL models in a centralized manner demands extensive data and robust computational resources. Securing sufficient energy and electricity to facilitate the training and evaluation of these models poses an additional obstacle to implementing these methods. Moreover, data collected from different sources is in different forms, necessitating extensive preprocessing and cleaning before its utilization for model training. Due to the aforementioned factors, machine learning cannot achieve its maximum capability for industrial applications in specific instances [19, 24]. Federated Learning (FL), or Federated Machine Learning (FedML), has been introduced as a potential remedy to address the aforementioned challenges.

Data is of utmost significance for every company, individual, and entity worldwide. Every individual and business desires to prevent their data from being compromised. However, a substantial amount of high-quality data is required to train machine learning methods effectively. Conventional machine learning techniques include storing data on a single server and training the model. This strategy carries the potential for personal data breaches. Federated Learning (FL) is a method in machine learning that enables algorithms to be trained on distributed datasets across multiple locations without sharing the actual data. This approach facilitates the creation of a global model without centralized data storage. Moreover, it ensures that personal data remains localized, thereby reducing the risk of privacy violations and data breaches. In federated learning, the movement occurs so that the model is transferred to the data rather than the data being sent to the model. It signifies that the training happens with the final gadgets.

1.2 Objectives of the Chapter

The objectives of the chapter are:

Of course. Based on a thorough review of Chap. 1, "Introduction to Federated Learning and Its Application in Industry," here are three primary objectives that uniformly cover its content, presented in a professional and research-oriented manner.

Objectives of the Chapter:

- To establish the foundational principles of Federated Learning (FL) as a decentralized machine learning paradigm, critically contrasting its architecture and data handling methodologies with conventional centralized and distributed learning approaches to address the challenges of data privacy, governance (e.g., GDPR), and data silos in modern industry.
- To systematically classify the diverse frameworks of Federated Learning, including Horizontal FL, Vertical FL, and Federated Transfer Learning, while examining the core mathematical formulations, operational parameters, and essential privacy-preserving mechanisms such as cryptographic solutions and differential privacy that underpin its secure implementation.
- And, to survey the practical applications and strategic impact of Federated Learning across key sectors such as healthcare, finance, manufacturing, and the Internet of Things (IoT), while simultaneously analyzing the principal implementation challenges related to communication overhead, system heterogeneity, scalability, and the ethical and legal complexities of data ownership.

1.3 Organization of Chapter

The rest of the chapter is organized as: Sect. 2 discusses the differences between federated learning and traditional machine learning. Section 3 enlightens mathematical formula for federated learning. Section 4 highlights parameters for federated learning. Sections 5 and 6 discusses federated learning in Finance and Manufacturing, while Sects. 7 and 8 focuses on ethical, legal and scalability issues. And, finally, Sect. 9 concludes the chapter.

2 Difference Between Federated Learning and Traditional Machine Learning

The primary distinction between federated learning (FL) and distributed learning (DL) is based on the assumptions made about the characteristics of the local datasets involved, as depicted in the Fig. 1. Distributed learning primarily aims to parallelize computational power, whereas federated learning concentrates on training with diverse datasets. In distributed learning, a single model is trained across multiple servers. However, it is assumed that the local datasets are independent and identically distributed (i.i.d.) and of similar size. Federated learning does not make these assumptions; instead, datasets often exhibit varied characteristics, and their sizes can differ substantially. Furthermore, clients participating in federated learning can be less reliable due to their greater susceptibility to failures or dropouts. This vulnerability arises because they frequently rely on less robust communication media, such as Wi-Fi, and battery-powered devices, including smartphones and IoT gadgets. In contrast, distributed learning typically involves nodes that are data centers with robust computational capabilities and fast network connections.

2.1 *Machine Learning Versus Centralized Approach*

Machine learning classifier utilizing a centralized approach individuals connect with a central server to transmit data in centralized machine learning. Specifically, the participants transfer the local information to the cloud server, where all the computational activities required to train the data are carried out. Centralized learning is computationally effective for participants because it relieves them of computational obligations that require more significant resources. However, participants are vulnerable to danger due to the potential evil nature of the cloud server or adversaries' ability to infer information. Additionally, large data upload can increase communication latency between participants and the cloud server.

2.2 *Classifier for Distributed Machine Learning*

Distributed machine learning methods are specifically developed to address the computing challenges posed by complicated algorithms when applied to large-scale datasets. Distributed machine learning methods provide superior efficiency and scalability compared to centralized techniques. The distributed models are taught using the same methods as centralized machine learning mathematical models, except they are trained individually on numerous participants [17]. During the training process of a distributed method, the participants autonomously train the models and transmit the weight updates to the centralized server. Simultaneously, the central server collects. Once a specific number of communication rounds have been completed, convergence testing is carried out on the centralized cloud servers.

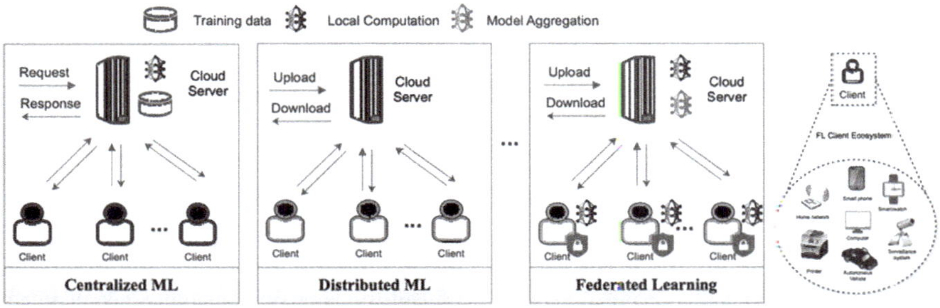

Fig. 1 Federated learning and traditional learning

2.3 Federated Learning Classifier

As decentralized machine learning, federated learning additionally develops the models autonomously. The only difference between decentralized machine learning and federated learning is that in federated learning, each participant begins the training process independently, without any other participants in the network at the outset. Federated learning involves collaborative and independent training undertaken by individual participants. Specifically, the learning parameters define local epochs, and each participant trains the data by performing these local epochs. At predetermined intervals, the local update is calculated, and participants send their updates to the cloud server. The cloud server receives updates from each participant, averages them, and aggregates the resulting global model. Participants use this global model to proceed with the training process for the next communication round. This method iterates until the desired level of convergence is achieved or the specified number of communication rounds is completed [8].

2.4 Categorization of Federated Learning

FL is categorized based on numerous characteristics, including data distribution across devices/clients, architecture, and model training.

Data Distribution Federated learning (FL) is a distributed machine-learning approach involving a central server and a group of clients. Clients are compute nodes that use their local data for local training. The central server transmits the standardized global patterns to the initial clients. Subsequently, clients employ their local data to train the entire model and communicate the revised local models back onto the server. The server combines the local model into a novel global one andÂ initiates a fresh training session. This approach can be iterated repeatedly till the entire model merges or reaches a specified threshold. The two primary distributions in federated learning are cross-device and cross-silo. Cross-device training involves deploying the model on many devices at the edge and training it using data that is accessible from each device. Moreover, Federated Learning (FL) is exceptionally efficient at disseminating data among several isolated systems, including data centers, and then consolidating it into a universal model at a single hub (Fig. 2).

Learning Models Types of foreign languages can also be classified according to the learning model. The model is trained with different methodologies, such as Vertical and horizontal, along with Federated Transfer Learning, all of which are illustrated in Fig. 3. Horizontal Federated Learning: All data sources employ the same feature space. The GBoard project by Google serves as a prime example of cross-device Horizontal Federated Learning that operates [10, 12]. Vertical Federated Learning, also known as Feature Space Federated Learning, is a model that uses different feature sets from many data resources, and its sample space may or may not be

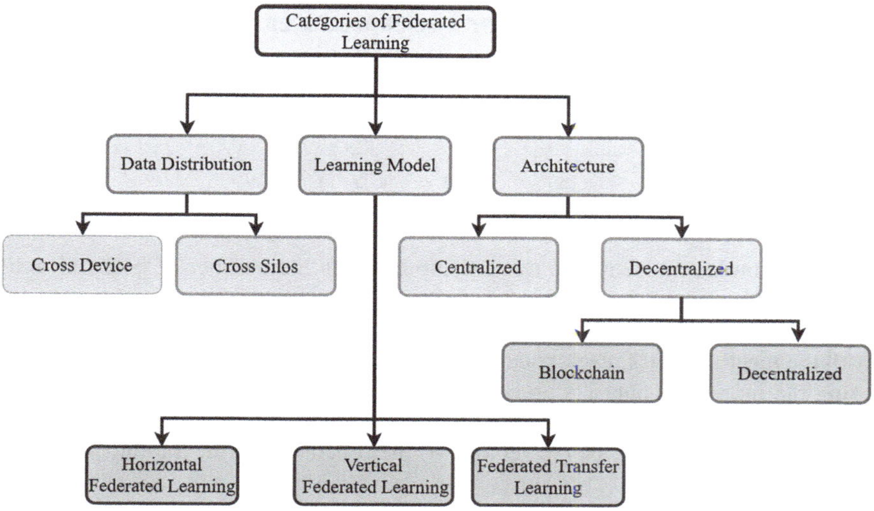

Fig. 2 Categories of federated learning

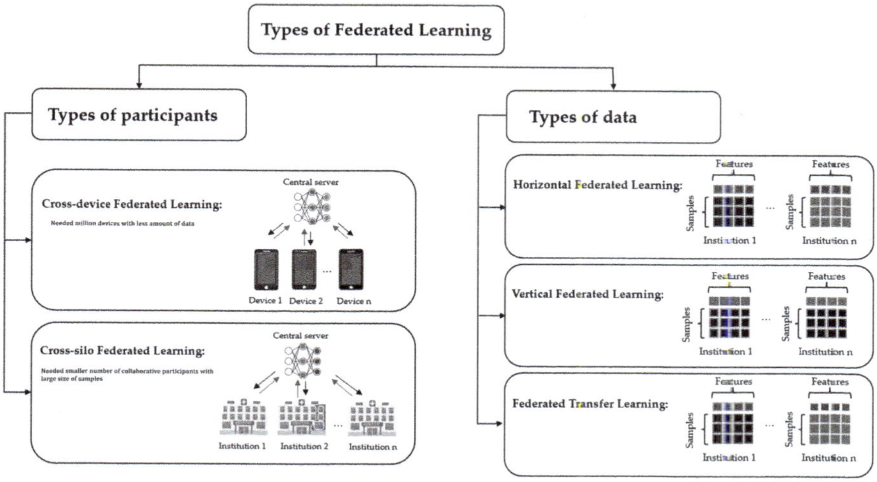

Fig. 3 Types of federated learning

identical. Federated Transfer Learning involves utilizing a pre-trained model that has been learned on a comparable dataset in conjunction using Vertical Federated Learning. This method addresses a distinct problem, namely generating customized movie suggestions for consumers by analyzing their past browsing behavior [23].

3 Mathematical Formula for Federated Learning

The purpose of federated learning can be defined mathematically as:

$$f(\mathbf{x}_1, \ldots, \mathbf{x}_K) = \frac{1}{K} \sum_{i=1}^{K} f_i(\mathbf{x}_i) \quad (1)$$

The formula represents the relationship between the number of nodes (K), the weights of the model as seen by each node ($\mathbf{x}_i$), and the local objective function (f_i) of each node. The local objective function describes how the model weights ($\mathbf{x}_i$) align with the local dataset of each node. Federated learning aims to train a shared model using the individual datasets on each node. The objective function $f(\mathbf{x}_1, \ldots, \mathbf{x}_K)$ is being optimized. Reaching agreement on the value of $\mathbf{x}_i$. Put simply, the vectors $\mathbf{x}_1, \ldots, \mathbf{x}_K$ converge to a shared vector $\mathbf{x}$ by the conclusion of the training procedure.

4 Parameters for Federated Learning

Following selecting the structure of the node network, the user can manipulate numerous variables of the federated training process, in addition to the hyperparameters of the machine learning model, to enhance learning:

The maximum number of consecutive federated learning iterations is denoted as T.
The total amount of nodes utilized in the process is denoted by K.
The proportion of nodes utilized in every iteration for each individual node:
The symbol C represents the mathematical concept of a set.
The local batch size utilized during each learning cycle is denoted as B.
Additional model-specific characteristics can also be adjusted, including:

- The number of iterations during local training preceding pooling is denoted as N.
- The local learning rate is denoted as η (eta).

The optimization of these parameters must be tailored to the restrictions of a machine learning application, such as the available processing power, memory, and bandwidth.

4.1 Federated Learning: A Security Solutions

The concept of Federated Machine Learning (FedML) was introduced to integrate privacy into machine learning models by offloading computations to edge devices and building trust among the communication entities. Given the heightened consciousness surrounding data privacy and the corresponding legislation, upholding data security throughout model training is imperative. Privacy is enhanced through several layers and by diverse means.

4.2 Cryptography Solutions

Data confidentiality is ensured by decentralizing and distributing data across multiple entities. Federated Learning (FL) does not require access to user data; instead, it transfers the learning structure to the information. Zhu et al. [33] stated that, information about users can still be extracted from the gradients, posing a privacy risk. One way to enhance security is by using Differential Privacy and Cryptographic techniques like Homomorphic Encryption and additionally secured Multiparty Computations (MPC). Differential privacy introduces random perturbations to the data before processing, effectively concealing the original data. However, this approach presents a balance between maintaining privacy and ensuring the model's accuracy. Introducing noise to the data can impact the model's precision. Cryptographic solutions, on the other hand, use encryption to protect user data before it is utilized to train the local model. Only the encrypted information is accessible, even after inverting the gradients, rendering it meaningless. Data decryption occurs solely on the server once the results from all clients are combined, thus ensuring privacy and security. This method does not compromise the model's correctness. However, the decryption and encryption of the data and results necessitate additional computations.

4.3 Security of Client

The selection of clients for training purposes of the federated learning models is of the utmost importance for ensuring data security. In FL, centralized data storage is not required. The clients retain the data and can delete it once the local model has been trained.

4.4 Federated Learning Applications

The primary objectives of Industry 4.0 are the utilization of intelligent gadgets and the advancement of automation in industrial solutions. FL's decentralized and privacy-preserving characteristics make it applicable in diverse sectors for intelligent solutions. A summary of some applications:

- **Medical care**: To stay abreast of the Industry 4.0 trends, various researchers across the world supported the use of Federated Learning (FL) in the fields of healthcare [27] and medical informatics, primarily because of its ability to maintain anonymity. For data-driven medicines and therapies, it is necessary to have access to patients' sensitive data. Therefore, FL is the optimal approach for such instances [25]. The statistical and system issues Biomedical Science are addressed by granting access to diverse clinical, pharmaceutical, and hospital data while ensuring data privacy, as highlighted by Xu et al. [29]. Sweden and other companies are utilizing Federated Learning (FL) to forecast the survival rate of emergency care patients in their project called "Federated Mortality Prediction" [16]. Another effort, called "FedHealth" [3], utilizes Federated Transfer Learning to train a model for tailored prediction using data received from wearable healthcare devices. This model was then tested for Parkinson's disease. Li et al. [18] implemented and assessed Federated Learning (FL) for Brain Tumor Segmentation on the BraTS dataset and employed differential privacy measures to ensure compliance with data privacy rules. Fed-BioMed is an open-source front-end platform for federated learning in medical field. This framework's practical application involves analyzing brain imaging data from multiple centres. Natural Language Processing (NLP) models are utilized to forecast medical conditions such as depression with the assistance of FL[2].

- **Natural Language Processing (NLP)**: NLP is a field that utilizes FL and has various applications within the realm of NLP. A renowned applications is Google's GBoard [30], where FL enhances question suggestions in the keyboard for Android phones. Another use is predicting the next word on mobile keyboards [13]. A word-level recurrent neural network with on-device training is utilized for emoji prediction, demonstrating that Federated Transfer Learning is a superior way of predicting emojis compared to the server-based approach. FedBERT proposes using pre-trained BERT models in a federated manner, restricting access to raw data and achieving improved outcomes.
- **Computer Vision**: AISweden, in collaboration with other organizations, is developing a test-bed for FL known as "FedBird". The study involved training an object detection model on two nodes with distinct datasets and sample spaces. Another use of FL in computer vision is a platform called "FedVision", utilized for identifying objects in films that are both expansive and costly to send [21].
- **Autonomous Systems and Internet of Things (IoT)**: Autonomous Systems and Internet of Things (IoT) are the major components of Industry 4.0 processes. These processes involve Cyber-Physical Systems (CPS) and edge Computing with AI and ML/DL solutions. As the use of IoT and CPS increases, it becomes necessary to include data privacy and smart processing in model training. To enhance the reliability of CPS at edge devices and enable the implementation of intelligent services, an FL framework "FengHuoLun" [31] is utilized for training machine learning models.
- **Big Data**: Within big data, individuals' details and behavioural information can be logged on mobile smart devices and edge servers. Data privacy protection is essential for ensuring the secure and confidential handling of personal information, which is crucial for advancing intelligent technologies. Consequently, multiple industries increasingly prioritise the security and confidentiality of sensitive data.

Through the collaborative efforts of numerous scholars, FL has played a significant role in various industries. The prevailing fields of application for FL are currently healthcare, finance, industry, and urban services [28].

5 Federated Learning in Finance

When financial firms conduct large-scale data analyses, they encounter numerous privacy-related challenges [22]. Federated Learning offers a solution to these obstacles by enabling the training of models without sharing raw data from devices. Instead, only updates are transmitted, ensuring privacy is preserved. In addition, Federated Learning aids the finance sector by improving fraud detection and enhancing risk management in financial models.

5.1 Risk Management and Fraud Detection

Federated learning improves risk management by avoiding the exposure of sensitive raw data. This helps create strong models that comply with regulations and reduce the risk of privacy breaches. By using federated learning, fraud can be easily detected because there are many edge devices, each with its own diverse personal dataset. This enables the detection of fraudulent activities while maintaining the privacy of the data.

6 Federated Learning in Manufacturing

Small manufacturers often face challenges with data collection and analysis, which are essential for implementing advanced machine learning methods. Federated Learning (FL) has been created to tackle these issues by promoting collaboration among small and medium-sized enterprises. It enables multiple enterprises to train models using their local data, improving accuracy and ensuring data privacy. This collaborative approach can [14] take place within a single company or across different companies, making it a versatile and secure solution for enhancing machine learning capabilities in the manufacturing sector.

Predictive maintenance involves predicting equipment failures and scheduling maintenance to prevent machinery downtime and prolong equipment life. Implementing Predictive Maintenance with FL improves accuracy, maintains data privacy, and significantly reduces costs by avoiding unexpected machine failures and downtime.

6.1 Quality Control

To maintain high product standards and ensure customer satisfaction, quality control plays a crucial role in manufacturing. FL can assist with product quality control by training a model to effectively detect defects. It also helps optimize the manufacturing process by analyzing data from various production lines, leading to improved product quality.

Federated Learning enables the collaboration of multiple devices to train ML models. However, this approach introduces [4, 14] a significant challenge of communication overhead on limited communication and computational resources, as devices need to frequently exchange updated learned models.

FL uses an iterative approach to exchanging its ML parameters. The time required depends on the ML parameters and the number of iterations. In realistic networks, such as wireless networks with limited bandwidth and power, parameter transmission may introduce a noticeable delay.

Another method to reduce communication overhead is proposed, where only a part of the model remains with the client while the other part is transferred to the server [32]. This technique is called split learning, and it mitigates the amount of communication data since only the activations of the split layer need to be communicated. A combination of federated learning and split learning named FedSL is introduced where the model is split at some specific layer, where the lower layers reside with the clients and upper layers with the server.

The data used in Federated Learning is held by different clients and is often non-IID. For example, one client's data may focus on a particular geographic region while another might cover a different one. Another challenge is imbalanced data and different features in different clients. Clients may have devices with significantly varying computational and communication capabilities [6]. A solution proposing the use of class prototypes to tackle data heterogeneity is introduced, wherein, by calculating class prototypes on each client device, the essence of each class from the local data only is captured. Class prototypes are samples or centroids of each class representing it in the feature space that can provide a summary of the class characteristics. Thus, to increase data uniformity and reduce communication overhead, only these prototypes are shared with the server rather than the entire model updates [26]. Another approach argues that in environments with heterogeneous data, local learning is crucial. In this approach, each client focuses on learning robust data and then contributes to the global model iteration. The research paper introduces local proximal optimization methods that compel the local updates to be close to the global server. This significantly mitigates data divergence among clients. Techniques such as weighted averaging, where more weight is given to devices with more representative data, can be used to balance the chaos created by heterogeneous data.

7 Ethical and Legal Issues

Data Ownership Various advantages of Federated Learning also come with a significant challenge of data ownership. Ideally, clients should have control over their data and the ability to consent to use it in training FL models. Therefore, the users should be fully transparent about the terms and conditions of how their data will be used. Determining the ownership of the data used in Federated Learning is a complex task. Various laws and regulations also grant individuals rights over their personal data.

Although federated learning enhances privacy by using localised data on the users' device itself, it also introduces new privacy concerns [5], such as the leakage of sensitive information while sharing model updates with the central server. Techniques like differential privacy and secure multiparty computation should be put into practice while abiding by the legal frameworks and agreements defining the data's ownership rights. Some frameworks imply that [6] data ownership cannot be held in the traditional sense, while others suggest some form of control over data. This lack of consensus over the existing framework is also a significant challenge. Therefore, a systematic and nuanced understanding of data ownership is required to take into account its attributes and implications.

Various regions have varying data privacy laws, like the General Data Protection Regulation (GDPR) in Europe or the California Consumer Privacy Act (CCPA) in California. Ensuring that Federated Learning systems comply with these regulations is a complex task. The training system must ensure that the data is not misused or exposed without the client's consent. Laws ensuring Data Sovereignty require data to be stored and processed within specific geographical boundaries, but with Federated Learning using multiple client data across different regions, complying with these laws becomes a complex task. In Federated Learning, data is not managed or stored centrally. This can pose a challenge for regulatory bodies that require transparency in data processing. Users must also fully understand the implications of their device's data being used for FL model training and, hence, consent to their data being used. The risks of model poisoning or data leakage through model updates must also be considered.

8 Scalability Issues

For the proper working of Federated Learning models, scalability becomes a crucial concern. The model should be able to scale up or down its resources according to the availability of devices for the training model. Out of the two types of Federated, Learning [9]. Namely, Cross-Device Federated Learning and Cross-Silo Federated Learning, the latter type provides better scalability and computation. Usually, the transition from the experimental implementation of the model to the [9] real-world implementation is a big step and can alter the model's efficiency. To decrease the gap, FEDn is a framework that has been proposed. It is created to bridge the gap between the experimental set-up and the real-world implementation by making it easy to go from pseudo-distributed mode to scalable distributed deployments, Ekmefjord et al. [9] made the scalability of the model more efficient during its implementation. Scalability hurdles can stem from resource constraints, communication overhead, client device heterogeneity, resource constraints, data distributions, resource allocation, and model convergence.

1. **Communication Overhead**: With the increase of devices in the model, the communication overhead of exchanging model updates also increases. To overcome this obstacle, researchers had put forward many techniques to minimize the size of the model updates while not compromising the model's accuracy. Some of them are gradient compression, quantization and sanctification.
2. **Resource Constraints**: The devices that partake in the training of the model are often resource-constrained, which leads to decreased performance and efficiency. To tackle resource constraints, small or light model architectures can be opted for, which would lead to a reduction of memory footprint and the resource utilization of edge devices.
3. **Resource allocation**: In FL models, its very crucial to effectively distribute computational resources in the edge devices, this can become difficult when the number of devices increase. To overcome this obstacle, the workload among the devices should be distributed, and no device should be overburdened. Also, for the training of the model, those devices should be chosen that match the capabilities for the computation in the model.
4. **Model Convergence**: It is difficult to achieve convergence of the model when the number of devices increases. Federated averaging can be implemented to tackle this obstacle of scalability. Federated averaging of the model updates when the devices increase can diminish the model's efficiency. To tackle this issue, decentralized or tree-based aggregation can opt for better scalability in the FL models.

9 Conclusion

The concept of Federated Learning (FL) underscores its potential to revolutionize machine learning, particularly in industrial settings where data privacy and security are of utmost importance. By decentralizing the training process and facilitating collaborative modelling across nodes, FL addresses significant challenges associated with traditional centralized machine learning approaches, such as governance, data silos, and compliance with strict privacy regulations like GDPR. This research showcases the implementation algorithm and FL classification, demonstrating its effectiveness in preserving data privacy without compromising the accuracy and efficiency of the machine learning model. The presented mathematical model elucidates the underlying FL mechanism and serves as a solid foundation for further research and applications. The real-world applications of Federated Learning (FL) extend across numerous industries, including healthcare, natural language processing (NLP), computer vision, autonomous systems, Internet of Things (IoT), and beyond. The advantages of FL are universally beneficial, as long as it guarantees the secure and efficient distribution of data. Furthermore, FL fosters trust and confidence by integrating advanced cryptographic techniques and customer integration security measures, addressing the pertinent solutions to today's technological challenges.

References

1. Accountability Act: Health insurance portability and accountability act of 1996. Public Law **104**, 191 (1996)
2. Basu, P., Roy, T.S., Naidu, R., Muftuoglu, Z., Singh, S., Mireshghallah, F.: Benchmarking differential privacy and federated learning for Bert models (2021). arXiv preprint arXiv:2106.13973
3. Chen, Y., Qin, X., Wang, J., Chaohui, Y., Gao, W.: Fedhealth: a federated transfer learning framework for wearable healthcare. IEEE Intell. Syst. **35**(4), 83–93 (2020)
4. Chen, M., Shlezinger, N., Vincent Poor, H., Eldar, Y.C., Cui, S.: Communication-efficient federated learning. Proc. Natl. Acad. Sci. **118**(17), e2024789118 (2021)
5. Chik, W., Gamper, F.: Ethical considerations and legal issues relating to federated learning. In: Federated Learning (pp. 369–391). Elsevier (2024)
6. Dai, Y., Chen, Z., Li, J., Heinecke, S., Sun, L., Ran, X.: Tackling data heterogeneity in federated learning with class prototypes. In: Proceedings of the AAAI Conference on Artificial Intelligence, vol. 37, pp. 7314–7322 (2023)
7. Di Noia, T., Tintarev, N., Fatourou, P., Schedl, M.: Recommender systems under European AI regulations. Commun. ACM **65**(4), 69–73 (2022)
8. Duan, M., Liu, D., Chen, X., Tan, Y., Ren, J., Qiao, L., Liang, L.: Astraea: self-balancing federated learning for improving classification accuracy of mobile deep learning applications. In: 2019 IEEE 37th International Conference on Computer Design (ICCD), pp. 246–254. IEEE (2019)
9. Ekmefjord, M., Ait-Mlouk, A., Alawadi, S., Åkesson, M., Singh, P., Spjuth, O., Toor, S., Hellander, A.: Scalable federated machine learning with fedn. In: 2022 22nd IEEE International Symposium on Cluster, Cloud and Internet Computing (CCGrid), pp. 555–564. IEEE (2022)
10. Federated Learning: Collaborative machine learning without centralized training data. Publication date: Thursday 6 (2017)
11. General Data Protection Regulation: Regulation (eu) 2018/1725 of the European parliament and of the council of 23 October 2018 on the protection of natural persons with regard to the processing of personal data by the union institutions, bodies, offices and agencies and on the free movement of such data, and repealing regulation (ec) no 45/2001 and decision no 1247/2002/ec. *data.europa.eu/eli/reg/2018/1725/oj* (2018)
12. Google: Federated learning: collaborative machine learning without centralized training data. https://ai.googleblog.com/2017/04/federated-learning-collaborative.html (2017). Accessed: 27 April 2024
13. Hard, A., Rao, K., Mathews, R., Ramaswamy, S., Beaufays, F., Augenstein, S., Eichner, H., Kiddon, C., Ramage, D.: Federated learning for mobile keyboard prediction. arXiv preprint arXiv:1811.03604 (2018)
14. Hegiste, V., Legler, T., Ruskowski, M.: Application of federated learning in manufacturing. arXiv preprint arXiv:2208.04664 (2022)
15. Holcomb, S.D., Porter, W.K., Ault, S.V., Mao, G., Wang, J.: Overview on deepmind and its Alphago zero ai. In: Proceedings of the 2018 International Conference on Big Data and Education, pp. 67–71 (2018)

16. Khan, M., Glavin, F.G., Nickles, M.: Federated learning as a privacy solution-an overview. Procedia Comput. Sci. **217**, 316–325 (2023)
17. Kraska, T., Talwalkar, A., Duchi, J.C., Griffith, R., Franklin, M.J., Jordan, M.I.: Mlbase: a distributed machine-learning system. CIDR **1**, 2–1 (2013)
18. Li, W., Milletarì, F., Xu, D., Rieke, N., Hancox, J., Zhu, W., Baust, M., Cheng, Y., Ourselin, S., Cardoso, M.J., et al.: Privacy-preserving federated brain tumour segmentation. In: Machine Learning in Medical Imaging: 10th International Workshop, MLMI 2019, Held in Conjunction with MICCAI 2019, Shenzhen, China, October 13, 2019, Proceedings 10. Springer (2019)
19. Li, T., Sahu, A.K., Talwalkar, A., Smith, V.: Federated learning: challenges, methods, and future directions. IEEE Signal Process. Mag. **37**(3), 50–60 (2020)
20. Li, L., Wang, Y., Lin, K.-Y.: Preventive maintenance scheduling optimization based on opportunistic production-maintenance synchronization. J. Intell. Manuf. **32**(2), 545–558 (2021)
21. Liu, Y., Huang, A., Luo, Y., Huang, H., Liu, Y., Chen, Y., Feng, L., Chen, T., Han, Y., Yang, Q.: Fedvision: an online visual object detection platform powered by federated learning. In: Proceedings of the AAAI Conference on Artificial Intelligence, vol. 34, pp. 13172–13179 (2020)
22. Liu, T., Wang, Z., He, H., Shi, W., Liangliang, L., Ran, A., Chenhao, L.: Efficient and secure federated learning for financial applications. Appl. Sci. **13**(10), 5877 (2023)
23. McMahan, B., Moore, E., D. Ramage, Hampson, S., Arcas, B.A.: Communication-efficient learning of deep networks from decentralized data. In: Artificial Intelligence and Statistics, pp. 1273–1282. PMLR (2017)
24. Mothukuri, V., Parizi, R.M., Pouriyeh, S., Huang, Y., Dehghantanha, A., Srivastava, G.: A survey on security and privacy of federated learning. Future Gener. Comput. Syst. **115**, 619–640 (2021)
25. Rieke, N., Hancox, J., Li, W., Milletari, F., Roth, H.R., Albarqouni, S., Bakas, S., Galtier, M.N., Landman, B.A., Maier-Hein, K., et al.: The future of digital health with federated learning. NPJ Digit. Med. **3**(1), 1–7 (2020)
26. Sharma, A., Dalmia, R., Saxena, A., Mohana, R.: A stacked deep learning approach for multiclass classification of plant diseases. Plant Soil 1–18 (2024)
27. Sharma, A., Dalmia, R., Saxena, A., Mohana, R.: A stacked deep learning approach for multiclass classification of plant diseases. Plant Soil 1–18 (2024)
28. Wen, J., Zhang, Z., Lan, Y., Cui, Z., Cai, J., Wensheng, Z.: A survey on federated learning: challenges and applications. Int. J. Mach. Learn. Cybern. **14**(2), 513–535 (2023)
29. Xu, J., Glicksberg, B.S., Su, C., Walker, P., Bian, J., Wang, F.: Federated learning for healthcare informatics. J. Healthc. Inf. Res. **5**, 1–19 (2021)
30. Yang, T., Andrew, G., Eichner, H., Sun, H., Li, W., Kong, N., Ramage, D., Beaufays, F.: Applied federated learning: improving google keyboard query suggestions. arXiv preprint arXiv:1812.02903 (2018)
31. Zhang, C., Liu, X., Zheng, X., Li, R., Liu, H.: Fenghuolun: a federated learning based edge computing platform for cyber-physical systems. In: 2020 IEEE International Conference on Pervasive Computing and Communications Workshops (PerCom Workshops), pp. 1–4. IEEE (2020)
32. Zhang, W., Zhou, T., Lu, Q., Yuan, Y., Tolba, A., Said, W.: Fedsl: a communication efficient federated learning with split layer aggregation. IEEE Internet Things J. (2024)
33. Zhu, L., Liu, Z., Han, S.: Deep leakage from gradients. Adv. Neural Inform. Process. Syst. **32** (2019)

Federated Learning in the Healthcare Industry

Ramkrishna Mondal

Abstract Advances in machine learning (ML), deep-learning (DL) algorithms, the Internet of Things (IoT), and artificial intelligence (AI) have enabled the development of reliable and scalable diagnostic and predictive models based on medical data. This has led to significant improvements in healthcare quality. However, adopting AI-based medical applications still faces challenges such as security, privacy, and quality of service standards. Several ML techniques have been incorporated into healthcare systems to develop stable data processing methods. However, applying ML to precise clinical data is challenging due to the lack of precise clinical data. Protecting patient privacy and data confidentiality is the main drawback of electronic medical record transmission. Federated learning (FL) offers a promising solution by allowing multiple clients to work together without transferring data to a single location. FL can lower communication and storage costs while maintaining high user privacy. Recent breakthroughs in FL have shown numerous success stories for smart healthcare applications that use data-driven insights to improve clinical care quality. The most recent innovations in intelligent healthcare include FL, AI, and Explainable Artificial Intelligence (XAI), which have the potential to reduce various limits and issues in the healthcare system. This chapter discusses the current and future state of FL technology in medical applications, research trends, and findings that illustrate the complexity of creating reliable and scalable FL models. However, obstacles remain in deploying a true FL system in IoT networks. Research is paving the way for breakthroughs in various healthcare domains to demonstrate that prototypes trained with FL methods can reach dependable performance, minimizing the impact on security and privacy. Decentralized networks are becoming increasingly important in today's healthcare business due to effective privacy issues.

Keywords Federated learning · Machine learning · Artificial intelligence · Deep learning · IoT · IoMT · AI · ML · DL

R. Mondal (✉)
Department of Hospital Administration, All India Institute of Medical Sciences Patna, Patna, Bihar, India
e-mail: dr.rkmondal@gmail.com

R. Mohana et al. (eds.), *Federated Learning Applications in the Industrial Internet of Everything (IoE)*, Studies in Systems, Decision and Control 611,
https://doi.org/10.1007/978-3-031-99270-4_2

1 Introduction

Modern technologies are advancing rapidly, particularly in the areas of Artificial Intelligence (AI) and the Internet of Things (IoT), which have led to significant advances in healthcare that have improved both the quality and length of human life [1]. A clear example of this revolutionary potential is the partnership with Google Inc. and researchers from relevant institutions who have demonstrated the amazing potential of AI [2]. AI-powered IoT devices are transforming healthcare by monitoring body motions, temperature, and sound waves, which are critical for COVID-19 treatment. Data-driven medical applications produce scalable prediction algorithms, but security, privacy, and Quality of Service (QoS) concerns impede the application [3]. Data owners, including individuals, patients, and hospitals, are wary of disclosing sensitive information, which has resulted in stringent rules such as General Data Protection Regulation (GDPR) and the Consumer Privacy Act [4, 5]. Traditional medical data processing on faraway cloud servers presents practical and privacy concerns, prompting a shift toward distributed data storage at the network edge for real-time applications [6, 7]. Edge computing-based AI convergence or Edge-AI has emerged as a powerful tactic in this environment [8, 9]. By leveraging the computing power of distributed devices, Edge-AI enables the localization of AI operations without violating data protection laws. Federated Learning (FL) goes one step further and allows models to be trained together on different medical datasets without requiring the disclosure of private patient information [10, 11]. The potential of FL is highlighted by its use in healthcare, which holds promise for significant improvements in public health and patient care systems [12, 13]. IoT devices generate significant amounts of data, which raises privacy and confidentiality concerns. FL solves these problems while maintaining device stability and communication throughput. Wearable sensors and AI algorithms make IoT technology useful for health monitoring [14–16]. Federated optimization was originally introduced by Google AI in the context of the discussion of collaborative ML without centralized training data and new distributed ML techniques for on-device intelligence [17]. Researchers are looking into the potential of FL in IoT, mobile applications, defense, and healthcare. FL facilitates early pathology identification, cardiac event hospitalizations, and patient data gathering. However, it may jeopardize privacy by allowing illegal access [18, 19]. The FL-based system leverages local model updates from the central server to reduce patient data loss while strengthening data protection, hence improving system security [20]. FL is a method for training models on decentralized datasets with various participants, such as mobile devices or entire enterprises, thereby overcoming privacy concerns and delivering accurate data for future technological improvement [21]. In addition to discussing the privacy and security concerns in exchanging health data, this chapter emphasizes the significance of FL in the healthcare industry. The primary objective of this chapter is to make the reader understand about the concept of Federated Learning (FL) technology and its uses in the Healthcare industry. The chapter organized initially with the basic concepts and frameworks of FL in medical field followed by the different applications of FL in healthcare industry and also, included different challenges faced and future research direction of FL in healthcare domain.

1.1 Organization of Chapter

The rest of the chapter is organized as: Sect. 2 illustrates the concept of Federated Learning (FL), Sect. 3 focuses on diverse applications of federated learning, followed by discussion on challenges of federated learning in Sect. 4. And, finally Sects. 5 and 6 concludes the chapter with future directions.

2 Concept of Federated Learning (FL)

2.1 Background

In recent years, FL has emerged as a promising distributed privacy-preserving solution for meeting the aforementioned urgent demand. In 2016, McMahan et al. used the term "federated learning" [22]. FL enables remote customers to work together on machine learning (ML) models for data-driven applications without centralizing client data. It protects user privacy, particularly in IoT, by not disclosing raw data with third parties, thereby increasing the efficiency of IoT applications [23, 24] FL provides privacy-preserving features, increased model performance, and flexible scalability in IoT networks. It limits the large-scale transmission of raw IoT data. FL training consists of three phases: task initialization, local model training with updating, and global model aggregation with updating. In 2018, the European Union proposed the GDPR to prevent data leaks and preserve privacy [14, 19]. The draft GDPR proposal intends to secure data security and user privacy. Traditional data processing methods have limits, which result in fragmentation and isolation concerns. FL provides a privacy-preserving solution based on Secure Multiparty Computing and homomorphic encryption [20]. FL is a paradigm that uses evaluation metrics to collaborate on privacy and data management issues without sharing samples [21]. FL, a method developed in another field, has acquired appeal in medicine due to its capacity to remove barriers to patient data collecting, enable e-health results interchange, and improve patient care and early-stage cognitive disease identification.

2.2 Federated Learning (FL) Basics

Google employed distributed mobile phones to conduct integrated ML, protect user data, and address privacy concerns via network model training [22].

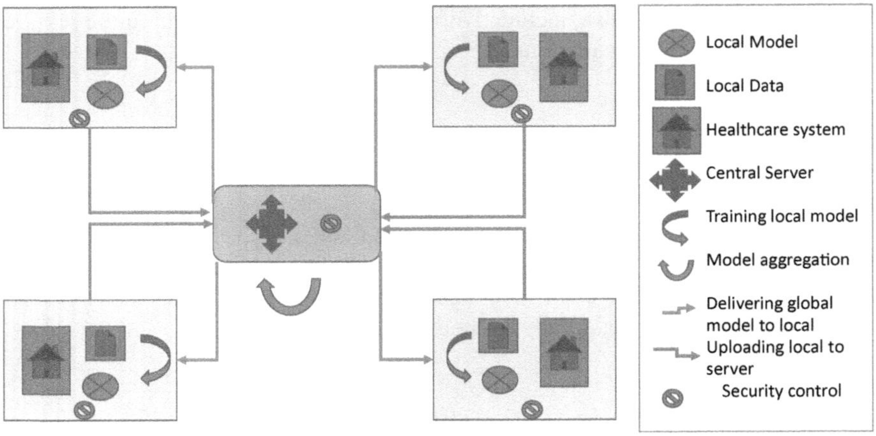

Fig. 1 FL for the Healthcare system

Computational Intelligence-based FL is a teaching methodology that employs raw data to generate a universal ML model while addressing issues such as user privacy, security, and exposure to large datasets (Fig. 1), FL is utilized in a variety of areas, including the military, IoTs, mobile communications, and healthcare. It entails sending local model parameters to a central controller, which produces a bigger reference model for continual improvement [23].

2.3 Different Aspects of the FL System

In the broad areas of ML, data protection, and distributed systems, a new approach called FL has emerged. The billing subjects in the FL system are the manager and the parties. The manager is the one who starts the conversation between the parties. The model is trained through computation and model parameters are exchanged through communication. Data partition, model, privacy layer, and communication architecture are four different aspects of FL [24]. Figure 2 shows the components of the FL system [15].

Federated Learning (FL) is used by healthcare companies because it saves data locally, decreasing data security and privacy concerns. FL facilitates access to huge datasets from many hospitals, hence addressing data governance concerns. FL collects enormous amounts of data and analyzes patients based on demographic characteristics, influencing stakeholders such as physicians, patients, and hospitals. FL is a promising approach for fair, accurate, and safe data models that gives AI developers access to a variety of data sources [14, 25, 26]. In essence, FL aggregation harmonizes the collective intelligence of individual data sources while preserving privacy,

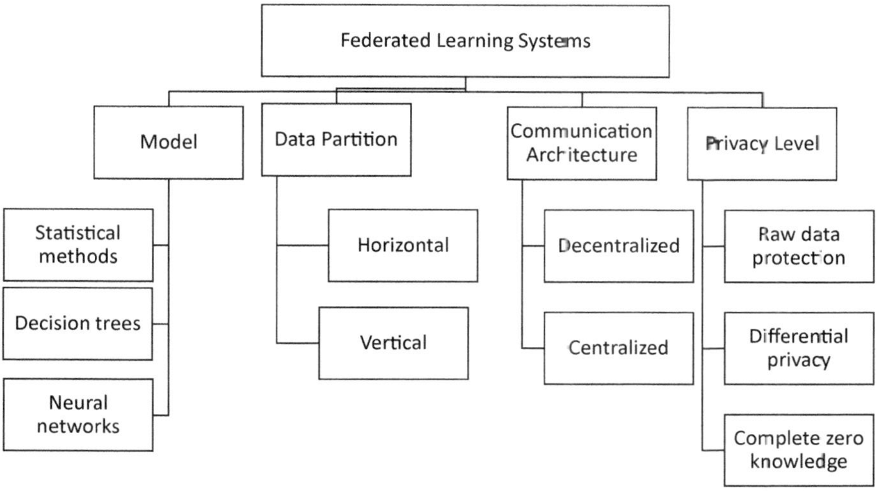

Fig. 2 Different component aspects of the FL system

ensuring security, and yielding a globally refined model. This unified model, generated collaboratively from diverse medical data points, holds significant potential to drive advancements in medical research, diagnostics, and treatment while upholding the ethical imperatives of data privacy and security. The categorization of FL frameworks within the medical domain is primarily determined by data distributions, as outlined below (Fig. 3).

2.3.1 Horizontal FL Framework

Horizontal FL is a data-split strategy that employs identical feature spaces across all devices, allowing hospitals to collaborate and train a single model that benefits from heterogeneous patient populations from various regions. This method handles low sample sizes in data training, as shown in an autism spectrum disorder prediction scenario involving four healthcare facilities in different regions [27, 28] (Fig. 3).

2.3.2 Structure of the Vertical FL

Vertical FL is a technique that integrates global models from many feature spaces with diverse data sets to increase feature dimensionality. It enables healthcare firms to aggregate data for holistic analysis while maintaining privacy. This strategy protects sensitive patient information, as illustrated by Chat et al. [36], who employed an auto-encoder FL model to translate client user attributes into latent dimensions for vertically partitioned medical data [29, 30] (Fig. 3).

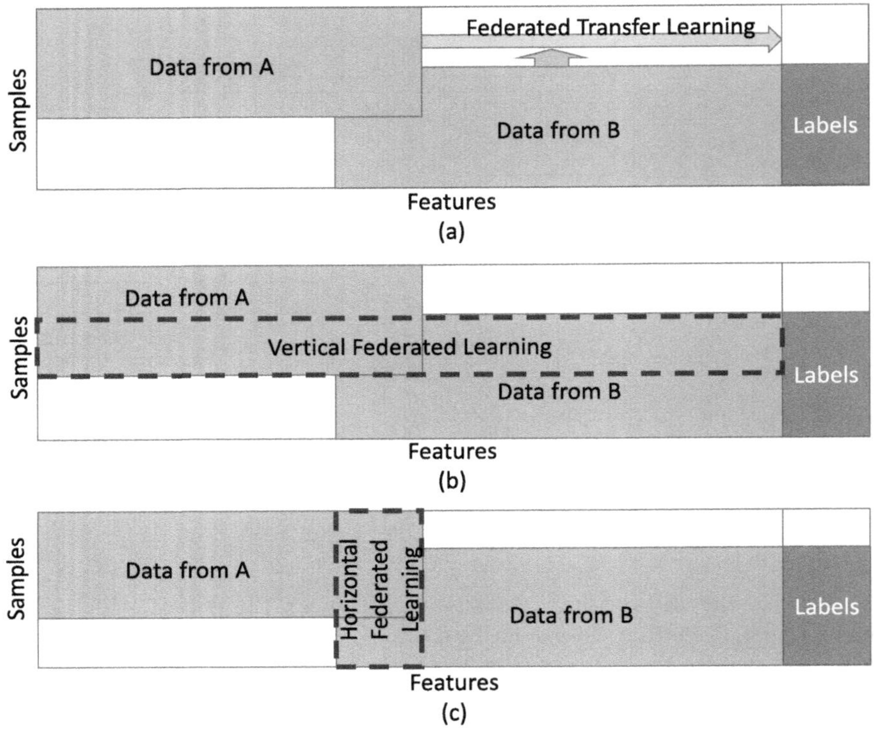

Fig. 3 Classification of FL

2.3.3 Federated Transfer Learning Framework

Federated transfer learning is a vertical learning strategy that employs pre-trained models to solve specific issues, especially when datasets differ in sample sizes and feature areas. This technique allows medical institutions to interact by exchanging model updates and knowledge collected from related tasks, hence improving model performance. A consortium of healthcare facilities, for example, can develop individualized cancer prediction models by training them for specific cancer types and sharing updates.

FedHealth employs federated transfer learning to develop tailored models for various companies, improving model performance and generalization while maintaining privacy. This approach applies human activity recognition tasks to Parkinson's disease classification [13, 31].

2.4 *FL Concepts and Framework in the Medical Field*

The application of AI in a dispersed healthcare context is predicated on trust, collaboration, efficiency, and scalability. This method ensures anonymity, lowers data transfer, and provides scalability. It addresses privacy and security concerns by localizing and aggregating data, providing differential data protection, and anonymization, and ensuring secure connection. The classic FL architecture includes a centralized global server, local model updates, and a summarized global model. This approach allows for collaborative model training while balancing healthcare insights with privacy and security considerations [32]. The FedAvg algorithm, developed by Google, is the fundamental algorithm in the field of FL. It aggregates local parameters from various clients at each iteration until the necessary convergence is reached. FedAvg performs well with non-IID data, but it lacks theoretical assurances in convex optimization contexts [33, 34]. Finally, home health care includes inpatient healthcare services, which are frequently delivered utilizing a variety of smart medical equipment. Remote patient monitoring is a component of the home care sector, offering patients tailored treatment in a familiar setting. This critical phase is essential for developing a better global model that improves accuracy while maintaining data privacy and security. Especially in the medical field, where patient confidentiality is essential.

3 Federated Learning (FL) Applications

3.1 *Application of FL to Different Industries*

Due to its many advantages, FL has enabled several important IoT applications in different industries. In this section, we will look at some of the most important ones.

3.1.1 Industry 4.0

The Industrial Internet of Things (IIoT) is driving Industry 4.0, the fourth industrial revolution, by emphasizing real-time data availability and connection. However, obstacles include insufficient data volume for reliable models and protecting privacy. The FL framework is intended to address these difficulties and improve supply chains.

3.1.2 Healthcare

IoT devices are changing people's lives and impacting privacy in personal healthcare data. FL approaches are critical for creating advanced machine-learning models in a heavily controlled data environment. Successful implementation could lead to precision medicine, which matches the correct treatment to the right patient.

3.1.3 Smart Home

Smart home systems driven by IoT devices are gaining popularity due to their increased comfort and quality of life. These wireless devices, which are controlled remotely via smartphones and microcontrollers, rely on sophisticated virtual assistants for hands-free operation.

3.1.4 Smart City

IoT-enabled smart cities improve efficiency and quality of life by monitoring physical assets in real time and giving intelligent information on road networks, transportation, public safety, healthcare, parking, and agriculture. Privacy concerns have shifted to local computing resources.

3.1.5 Autonomous Driving

Autonomous driving technology is being integrated into ordinary vehicles, necessitating real-time communication and sophisticated adaptability to environmental changes. FL in vehicular edge computing (FL) provides a possible approach by allowing vehicles to send limited data to the cloud.

3.1.6 Metaverse and Virtual Reality

The metaverse is a next-generation internet that provides immersive 3D virtual experiences via personal computers and augmented reality devices. It bridges the virtual and physical worlds with IoT data. FL promotes collaboration, improves performance, increases security, and allows users to access services without disclosing their personal information [14].

3.2 Applications of FL in Retail

When utilized in finance, sales, and numerous industries where data cannot be aggregated for the training of ML models due to considerations such as data security, privacy protection, and intellectual property rights, FL offers a lot of promise as a new modeling technique [35]. Machine learning (ML) approaches will be employed in the smart retail business to provide personalized services such as product recommendations and sales help. Data factors like as users' purchasing power, personal preferences, and product features are critical in this process. However, typical data storage from banks, social networking sites, and e-commerce sites generates data barriers, making model training challenging. Federated and transferable learning (FL) can help to overcome this issue.

FL enables the building of models for all parties without exporting company data, hence maintaining data confidentiality and privacy. Transferable learning addresses the issue of data heterogeneity and overcomes the constraints of traditional AI methodologies. The FL design enables many databases to be searched without revealing data. Federated transfer learning is projected to have a large impact on smart healthcare because health records are difficult to obtain and exist only in isolated facilities. Collaboration and data sharing can dramatically improve the effectiveness of ML algorithms trained on large volumes of medical data. A combination of transfer learning and FL may be required for intelligent healthcare development.

3.2.1 Wireless Communication

The complexity of wireless networks mandates the removal of older models. Deep learning networks are becoming more popular, resulting in the development of novel wireless network models. Researchers investigated the dependability and security of FL characteristics in edge computing and 5G networks. Tran and colleagues created a lightwave power-based FL model to achieve maximum transmission efficiency [34, 36]. Scientists created a reliable wireless noise reduction method that employs parallel optimization and sampling-based algorithms [37]. This resulted in higher forecast accuracy and lower loss. FL improved attack detection and communication efficiency in smart home systems. Savazz and colleagues suggested a serverless learning method for 5G wireless networks, as well as a hierarchical FL system for heterogeneous cellular networks [38].

3.2.2 Maintenance Recommendations

Google is creating an FL for Android mobile devices to improve keyboard input prediction accuracy while protecting user privacy and security [39]. FL has a wide range of applications, including recommendation systems and intelligent medical diagnosis. However, ensuring patient confidentiality is difficult due to the growing scarcity of medical data. Advances in AI have altered resource allocation and disease diagnosis. A multi-FL network based on the APOLLO network was created to overcome data shortages and labeling deficiencies, changing medical research data for future patient diagnoses [10, 40].

3.2.3 Medical Applications in Healthcare

FL often tries to be classified by Classification algorithms in ML and learn to classify or annotate a given set of phenomena using labels or classes. For example, classification tasks for COVID-19 detection [28, 41], cancer diagnosis [42, 43], and autism spectrum disorder (ASD) [30, 44] are considered in FL healthcare settings [68]. Other classification tasks studied include recognizing human emotions and activities [45, 46] and predicting patient hospitalizations [47, 48] (Table 1). Kose et al. [49] and Aledhari et al. [50] presented an overview of classification challenges in FL research for medical applications, with a focus on the role of electronic health record (EHR) data in biological research, including ML. EHRs provide useful patient data, but their inherent biases limit generalizability. Algorithms for specialty health programs may favor African Americans while attributing the same risk to Caucasian patients [51]. Misconfigurations in algorithms can occur as a result of insufficient training data or disparities in healthcare access. To avoid biased algorithms, learning from larger EHR datasets is required. However, discrepancies in data processing procedures and privacy considerations make it difficult to review everything at the same time [52]. Although data sharing allows for a more complete replication analysis, it remains a significant restriction. FL allows for the connection of EHR data in healthcare while retaining the data's privacy [53]. Improved iterative learning from various medical datasets can improve the effectiveness of machine-learning models in healthcare, such as predictive modeling, phenotyping, patient representation learning, and patient similarity training [54, 55]. Researchers employed tensor factorization models and federated techniques to extract useful phenotypes from massive EHRs, focusing on obesity-related comorbidities and patient presentation training in FL settings [56]. Researchers created SplitNN, a technique that allows healthcare institutions to train DL models jointly without exposing model parameters or sensitive raw data [57, 58]. Researchers developed FL to compare brain correlations across disorders and clinical cohorts. Scientist divide patients into comparable communities and train models for each. FL assists clinicians in understanding the benefits and dangers associated with early therapy [59–61]. Brisimi et al. 2018 [54] desired to solve an l1-regulated sparse SVM classifier in an FL framework to estimate the number of hospitalized patients with cardiovascular disease in the future utilizing EHR data collected across several Agents/data sources.

Oukin uses FL to predict treatment preferences and odds of survival. Pfohl investigates extended hospital stays and inpatient mortality. Privacy-preserving algorithms predict mortality in intensive care units, and training within FL produces comparable findings [64].

3.3 Application of FL in Healthcare

This section discusses some possible applications of FL in healthcare.

Table 1 Overview of current developments in areas of FL in Healthcare

ML method	Dataset	Problem statement	References
TF	MIMIC III, UCSD	Phenotyping	[54]
PCA	MIRIAD, PPMI, Biobank, UK ADNI	Representation learning	[59]
SVM	Boston Medical Center	Hospitalization prediction	[48]
LR, NN	Collaborative Research Dataset of eICU	Mortality prediction	[62]
CNN	UCI Smartphone	Activity recognition	[53]
NN	PhysioNet Dataset	Arrhythmia detection	[63]
LASSO, MLP, LRR	Mount Sinai COVID 19	Mortality prediction	[47]
NLP	MIMIC III	Phenotyping	[56]
Autoencoder	Collaborative Research Dataset eICU	Mortality prediction	[60]
RNN	Cerner Health Facts	Premature birth prediction	[61]
MLP, LR	MIMIC III	Mortality prediction	[17]

3.3.1 FL Applications for Medical Image Processing

Medical imaging has transformed healthcare by revealing insights into the human body. Deep learning algorithms have been shown to outperform MRI reconstruction. A FL-based system is proposed for combining MRI data from several institutions while maintaining patient privacy. FL is used to identify physician-associated persons, estimate hospitalization, mortality, and survival rates, and segment brain tumors. FL has the potential to analyze medical photos, safeguard patient privacy, and detect diseases such as COVID-19, skin disease, heart disease, and brain tumors. It is also used to segregate brain tumors from the BraTS dataset [65, 66].

3.3.2 FL in IoT-Enabled Smart Healthcare Applications

FL is an idea for developing intelligent IoT devices that allow physicians to use data from numerous medical institutions while not revealing patient information. It allows hospitals to develop their own AI models and set up shared care facilities for faster diagnostic tests. FL's utility in the smart healthcare sector has been demonstrated through studies, particularly in EHR management and health insurance participation. Traditional AI systems are unsuitable for modern health [67].

3.3.3 Manage Electronic Medical Records with FL

FL is used in the healthcare industry to provide flexible and secure EHR services. The EHR system is administered by the FL-based student-centered architecture, which connects several clinics to a cloud server. Google developed a data deletion technique to protect the learning experience from attackers. To provide accurate results, AlexNet NN modeling was used for the standard CIFAR 10 dataset. When investigating FL-based EHRs, differential privacy approaches can help to improve privacy. FL predicts undesired outcomes with comparable accuracy to centralized AI systems while preserving privacy [68, 69].

3.3.4 FL in Health Collaborative

FL technology can improve health coordination and service delivery by centralizing and securing data. A healthcare cooperation framework employs FL to create medical IoT devices. The FedHealth framework promotes portable device federation with FL technology, which allows AI algorithms to identify behavior and secure data. Chain-directed synchronous stochastic gradient descent reduces human latency in personal mobile sensing applications. FL has been demonstrated to help reduce COVID-19 infections [70, 71].

3.3.5 FL for Predicting a COVID Outbreak

FL is a novel approach that builds statistical models using local data samples rather than aggregated clinical data, which is problematic due to patient privacy concerns. The COVID-19 pandemic has resulted in the rapid spread of the virus, which may be detected using a variety of technologies, including computed tomography and X-rays. FL has shown promise in diagnosis, therapy, and prognosis prediction, with studies demonstrating increased detection sensitivity and performance above locally trained models. FL has also been shown to be more effective than typical machine-learning models for detecting COVID-19. Research testing FL's efficacy in predicting important outcomes for hospitalized COVID-19 patients is expected to generate a lot of interest. FL's ability to improve outcomes for COVID-19 patients is also being investigated [72, 73].

3.4 FL and Its Applications with Other Rising Advances

FL is a robust and dependable ML technique in healthcare, with prospective applications in areas such as NLP, transportation, and finance. Its therapeutic record protection is critical, and FL techniques increase productivity and efficiency in global healthcare systems when compared to typical data-driven therapeutic applications [74, 75].

3.4.1 Blockchain-Based FL in Healthcare

Blockchain technology is being used to improve the security, scalability, and performance of FL in a variety of industries, including healthcare. Blockchain is a decentralized, open record breakthrough that allows for collaborative learning across devices without a central aggregator. It has helped to reshape trade, transportation, coordination, and healthcare. The integration of blockchain and FL creates a secure data management system and efficient delivery of FL applications, increasing their robustness and security. Clinical trials, precision medicine, and illness classification are all potential applications of blockchain technology. The use of blockchain-based FL in healthcare provides various advantages, including improved information security, transparency, interoperability, consent administration, monetization, and adherence to data security legislation [76–78].

3.4.2 FL Empowered IoT-Based Healthcare

The integration of IoT and FL in healthcare has had a substantial impact on many areas, including real-time information management (FL). The use of IoT and FL in healthcare entails merging real-time data from IoT devices with privacy-preserving FL to assure security. The IoMT worldview, which blends FL with centralized ML methodologies, seeks to strike a compromise between information utility and privacy. The convergence of IoT and FL has enormous potential for real-time data preparation, accessible silent monitoring, infection forecasting, early detection, lower healthcare costs, flexibility, and continuous engagement. This combination is predicted to transform information assessment methodologies while ensuring security and progress in the healthcare setting [79–81].

3.4.3 Edge Computing Assisted FL for Healthcare

Edge computing and FL are gaining importance in healthcare. Edge computing brings computing resources closer to data sources, resulting in shorter response times and less information overload. FedHome, a leading FL system, combines cloud and edge resources to tailor home health monitoring. It achieves 95.41% precision by reducing communication costs with a class-balanced dataset and a lightweight display. Other novel approaches include fog computing models, BodyEdge, and personalized edge-assisted FL. Blockchain and Multipurpose Edge Computing (MEC) are also emerging as tools for healthcare transformation, with decentralized designs that use blockchain and MEC to securely share EHRs among integrated clinics. These innovations save energy on IoT devices while maintaining quality of care [82–85]. The combination of edge computing and edge computing in healthcare has the potential to transform the business by minimizing inactivity, maintaining transfer speed, and ensuring data security. This lightweight protection mechanism, which uses shared insights and weight coverings, keeps edges secure while improving performance and reducing hardware dropouts and malicious attacks. This strategy is very useful in resource-constrained circumstances [80].

3.5 *Viable Application of FL in Therapeutic Space*

FL is a groundbreaking AI technique that blends ML with restorative information to transform healthcare institutions. It promotes global collaboration among restorative educators by assuring data security and redefining global care, research, and healthcare frameworks, thus boosting collective efforts and patient protection.

3.5.1 FL-Enabled COVID-19 Discovery

Healthcare has revolutionized thanks to breakthroughs in smart healthcare and IoMT, particularly in response to global health emergencies like as COVID-19. AI has been utilized to enable rapid collaborative responses, particularly in the setting of FL. FL models have demonstrated potential for forecasting mortality, determining oxygen requirements, and predicting serious kidney injury. Collaborative FL systems were used in X-ray preparation and arrangement experiments, and a secure worldwide informatics FL framework was developed [86–89].

3.5.2 FL-Enabled Movement Acknowledgment for Upgraded Elderly Care

Computer vision-driven action recognition (FL) can change human exercise, including walking and running, to provide individualized treatment. FL can revolutionize aged care by offering individualized attention to older populations while maintaining their safety. Studies have demonstrated that FL can give effective models for human movement recognition tasks. Innovative systems such as Combined Multi Task Multiple Leveled Considerations demonstrate how FL may increase precision in multi-sensor data. A unified machine learning framework has been created to address the potential of deep neural networks. FL's potential extends to the elderly with Alzheimer's and dementia, improving care and understanding of cognitive disorders [90, 91].

3.5.3 Cases of FL in Therapeutic Devices

FL is used in therapeutic imaging to generate deep learning models from big datasets, minimizing the need for annotation exchange. FL has been used to detect malignancies in the breast, lungs, colorectum, prostate, stomach, liver, and pancreas. Researchers at the College of Pennsylvania, Penn Medicine, and Intel Labs have developed a technology that can distinguish glioblastoma tumors in Attractive Reverberation Imaging MRI images with 33 times greater accuracy than traditional methods. FL can also be used to build a brain cancer segmentation model from treatment imaging data, increasing accuracy and generality across different types [92–94].

TensorFlow

Unified TFF is an open-source solution for ML and computation on localized data that enables engineers to simulate integrated learning calculations on models and data. It is composed of two layers: Combine Center FC and Combine Learning FL APIs. The Unified Center API is a robust, practical programming environment that blends new unified learning calculations with TensorFlow, whereas the Combined Learning API enables developers to incorporate current FL calculations into their models. TFF supports FedAvg and can run on a single computer [94, 95].

Unified AI Innovation Enabler Destiny

Webank's AI Office has released Destiny, an open-source computing system that supports the AI ecosystem and employs privacy-preserving techniques such as homomorphic encryption and Secured Multi-Party Computation. It supports ML models such as NNs, GBDTs, and computed relapse and has six main elements. Destiny also includes rich documentation and algorithm-level interfacing, but non-expert users will need to make expert modifications [96].

PySyft

PySyft is a secure and private deep learning package that employs privacy-preserving techniques such as Differential Security and Homomorphic Encryption. It is compatible with deep learning systems such as Pytorch and TensorFlow but lacks thorough documentation and support for a broader computing viewpoint, such as IoT [97].

Leaf

LEAF is a benchmarking system for unified settings that includes programs such as unified learning, multitasking, meta-learning, and on-tool learning. It has six databases that address topics such as picture recognition, feeling investigation, and character expectation. However, it may be limited in certain areas and does not support decentralized unified learning, part learning, or vertical FL [98].

Paddle FL

Paddle FL is an open-source technology for transmitting combined learning frameworks in the form of distributed clusters, which offers substantial benefits to designers. It includes FL strategies, user-defined models, and an FL assignment generator. However, it is still in its early phases, with insufficient documentation and tests, and it requires established standards such as demonstrating DNN and Vertical Unified Learning [99, 100].

Clara Preparing Framework

The Clara Train SDK is an AI-assisted software system built for engineers that enables AI-powered applications such as Exchange Learning, Combined Learning, and AutoML. It enables designers to set up MMAR Restorative Show Chronicle work and monitor GPU performance.

Encouraged ML FedML

FedML is a two-layered solution that combines learning and FL benchmarking capabilities. It enables on-device preparation for IoT, adaptable computers, distributed computing, and single-machine imitation. FedML provides a wide range of calculations, models, and databases for maximum flexibility [101].

IBM Combined Learning

IBM Unified Learning is a Python system library for distributing ML forms in projects, with a focus on secure deployment, disappointment resistance, and rapid demonstration determination. Separate from neural systems and decision trees, IBM Combined Learning enables enterprise experts to study and apply differentiated protection to diverse ML models [102, 103].

3.6 Security and Privacy Issues in the FL

Security definition for federated learning in most cases is a combined exchange learning framework involving two partners. Because its protocols are comparable to vertical combined learning, the security definition for vertical unified learning can be enlarged here, as detailed below.

3.6.1 Vertical Unified Learning (VFL)

ML strategies for vertically separated information, such as slope plummet, classification, secure direct relapse, association run show mining, and factual investigation, have been offered for security, with a VFL strategy displaying person protection [104, 105]. VFL, also known as feature-based FL (Fig. 3b), can be used when two datasets share the same test ID space but have different highlight spaces. In the same location, e-commerce and banks have distinct clientele and target areas. Both seek to establish a product purchase expectation based on client information. In a unified learning framework, each party has the same personality and place, and everyone creates a shared wealth arrangement. In a two-party situation, two parties may not cooperate or be captured by a competitor. A semi-honest third-party STP is occasionally used to enable secure calculations between parties, which is supported by SMC's formal protection certification [105]. Following the learning preparation, each participant has a fair understanding of the parameters of associated models, as well as their claim to interesting traits. As a result, the two parties must collaborate during the induction process to deliver a result.

3.6.2 Horizontal Federated Learning (HFL)

Figure 3c depicts sample-based combined learning or level unified learning for datasets that share a common area but differ in the number of tests. There is extremely little overlap between the client bases of two territorial banks, each with a distinct clientele. Overall, the included areas are indistinguishable due to their similar industries. The authors advocated collaboratively profound learning [106]. In 2017, Google advocated an FL technique for Android phone models, which allows for local parameter changes that are then transferred to the centralized Android cloud. This safe aggregation method protects client upgrades and stores data with additively homomorphic encryption. Multitask learning can handle issues such as high fault tolerance and communication costs. A secure client–server topology, as well as model-building methodologies, can avoid data loss and lower transmission costs for centrally placed models [107]. The transmission capacity of large-scale preparation has been reduced, and the security of HFL (Hybrid Learning Framework) is widely accepted, as servers can interfere with information members' security. Modern security demonstrations, which include malicious clients, have created privacy issues [107, 108].

Privacy concerns in healthcare are a major concern. A new framework strives for a centralized server, yet sensitive data is stored locally. Combining engineering solves privacy concerns by integrating privacy, judgment, availability, non-repudiation, and accountability. The subjective evaluation of security mechanisms in healthcare systems reveals potential for security [109]. Traditional solutions for integrating EHR systems can jeopardize data security and long-term information. Cryptographic processes are employed to safeguard transmission, but they have little control over access control tools. Discretionary Access Control (DAC), Mandatory Access Control (MAC), and Role Based Access Control (RBAC) are all important access control components in healthcare systems. Federated Health Data Frameworks (FHIS) must facilitate the secure sharing of healthcare information by integrating pseudonymization and encryption to produce a hybrid model [110]. A unified information storage system for evidence-based medications is required to meet privacy concerns. Depersonalization and pseudonymization are employed to protect privacy, while object-oriented reference data display Edge provides a single-league storage solution for the medical community. A secure EHR system is presented to improve healthcare quality through cryptographic measures. The Therapeutic Cyber-Physical Framework (MCPS) is used to hold sensitive patient data that requires strict security. A new Interruption Discovery Framework (IDS) is presented to overcome the limitations of existing IDSs and achieve reliable attack localization. FL creates a worldwide exhibition by averaging weights from each device across communication channels [111].

4 Challenges of Federated Learning

4.1 Challenges Within the Community

To realize the full potential of FL in the aforementioned applications, the following problems are identified, that serve as critical barriers to enabling FL on potentially billions of IoT devices. The following depicts key issues, followed by openings with a high likelihood of addressing those challenges.

4.1.1 Constrained On-Device Assets

The deployment of FL at the edge is hampered by restricted IoT resources, making conventional ML models computationally intensive. To boost performance, analysts are focused on lightweight and hardware-friendly model techniques, as well as condensing current models to make them smaller, more computationally and energy efficient. FedMask is offered as an FL system that is both computationally and communicationally efficient, allowing each device to learn a heterogeneous and organized inadequate double veil while using less computing, memory, and energy [112].

4.1.2 Constrained Arrange Transmission Capacity

Communication bottlenecks pose significant issues in FL-based IoT setups, as most devices connect via remote systems with slower transfer speeds than cable transmission capacity in data centers. As more devices join, the problem grows more complicated due to the various asset allocations. Strategies such as slope compression have been investigated to reduce transmission capacity demand, but also degrade preparation quality. Mercury offered a sampling-based technique for effective on-device preparation while maintaining quality. Researchers identified transfer speed asset assignment and client determination as optimization difficulties in training FL models [113, 114].

4.1.3 Irregular Network and Accessibility

The intermittent network of IoT devices can lead to dropouts, especially in large-scale IoT systems. The present FL system is based on synchronous server updates, which might be difficult due to varying transmission capacity and data dispersion. In real-world scenarios, some devices may disconnect during training, complicating synchronous upgrades. To solve this, several experts developed FedBuff, an unconventional accumulation procedure designed to control stragglers and permit secure aggregatement. However, IoT devices are heterogeneous and have varying computational rates, making it impossible to estimate training progress. A nonconcurrent method that takes into account the variety of IoT devices may be a more effective and promising option [115].

4.1.4 System Heterogeneity

The FL framework confronts issues in cross-device contexts due to the variety of equipment, software, and data collected by different devices. FL approaches must deal with small groups, devices, and periodic disconnections. Multitasking and meta-learning views can help with personalization. FL research focuses on a single non-IID feature, but there is no thorough medical dataset analysis. Creating an automated approach for picking hyperparameters is critical [116].

4.1.5 Transient Flow and Ceaseless Learning

IoT sensing devices are continually collecting useless data, which can be used for deep learning. However, IoT devices with limited memory capacity struggle with model inference and preparation, resulting in disastrous overlooks. To overcome this, lightweight ML motors such as FedGKT can be utilized to reduce memory consumption during on-device learning. FedGKT enables IoT devices to exchange data from numerous CNN models to a larger CNN on a cloud server, hence decreasing memory utilization for successful on-device training [117].

4.1.6 Trustworthiness

The increase in cyber threats, including phishing and DDoS, has highlighted the importance of IoT risk assessment and security, particularly in healthcare. To overcome this, lightweight security protocols like distributed learning (FL) can be included in the IoT architecture. However, FL continues to face security threats as a result of data exchange [118, 119].

4.1.7 Standardization and Framework Improvement

When developing a combined learning framework for IoT systems, analysts must consider communication APIs, dataflow models, and device characteristics. Standardization is required to accelerate the transmission of FL frameworks via IoT networks and foster an open environment for content sharing. Existing FL-creation and benchmarking tools, such as TensorFlow Combined, PySyft, and FedML, should be improved to support IoT applications. A user-friendly simulation environment is required for planning and evaluating the full FL framework on massive IoT networks. Researchers should investigate extending present training methods to edge FL scenarios, as edge computing systems prioritize edge computation above learning [120].

4.1.8 Privacy

Federated Learning (FL) raises privacy concerns because of the revealing of personal information during training. Differential privacy and Secure Multiparty Computing (SMC) can be used to protect privacy. However, the protocol is subject to assault. An approach has been presented for preserving client differential privacy in FL models. Fingerprint scanners and signature recognition are examples of reliable identifying systems used for secure data sharing and encryption. FL shares model gradients to secure sensitive data, but it frequently compromises system efficiency or model performance [78, 121, 122].

4.1.9 Federated Fairness

Federated learning (FL) creates new opportunities and problems for fairness researchers. FL enables the utilization of partitioned data, which facilitates advanced fair modeling theory and practice. However, it raises new concerns about fairness, as contextual factors like connection quality, device kind, geographic location, usage patterns, and dataset size can all influence sample selection. Frameworks such as agnostic FL can help to avoid bias in training objectives. Fairness improvements in federated training algorithms are critical as other FL system components, like as model compression, reach technology limits. Adapting and enhancing these methods for the Florida environment remains a problem [123, 124].

4.2 Challenges in Future Directions and Possible Solutions

While FL has some advantages, it also presents drawbacks that must be considered when planning collaborative learning initiatives.

4.2.1 Privacy and Security

Healthcare data protection is critical, and FL provides problems in terms of alternatives, techniques, and outcomes. FL largely maintains privacy by sharing method changes, but it also poses some risks. Trust levels in FL partnerships can be classified into two categories: trustworthy and untrustworthy consortiums. Collaborations can be challenging to form because the technology is deployed in multiple contexts. Threats are mitigated via security measures such as cryptography, participant permission, and transaction tracking. Data breaches can arise as a result of the exchange of patients' medical information between FL devices and hospitals, and standard training methods involve many parties in the training process [125].

4.2.2 Data Heterogeneity

Health data differs depending on gathering methods, enterprise biomedical applications, and area demographics. Traditional approaches distribute information independently, however FL can be learned using non-IID data. However, approaches such as FedAvg frequently fail, limiting active learning. Researchers investigated this using FedProx and partitioning techniques. Data heterogeneity can result in local solutions that are less optimal than global solutions. As a result, agreement on optimal model training is essential before training begins [126].

4.2.3 Traceability and Accountability

FL in healthcare necessitates reproducibility due to safety–critical applications. It demands traceability of network connections, configuration changes, and hyperparameter adjustment throughout the training phase. This ensures that training and testing datasets do not overlap. Execution integrity is critical for accountability in nontrusted federations. FL does not allow researchers to view images of completed designs, but organizations can give a central viewer or explain optimization techniques.

4.2.4 System Architecture

Large-scale FL in healthcare is better suited to computational resources and networks than consumer devices because it enables more product training and data distribution. However, FL brings both potential and challenges, such as security prevention, effective encryption solutions, and downtime reduction via independent system and network settings.

4.2.5 Client Management

Client management is critical in FL, as opposed to ML's centralized architecture. It entails assisting patients in developing a strategy for coordinating critical support services for the best possible outcomes. Customer management includes several components, such as patient identification, eligibility determination, assessment, care planning, goal setting, planning, implementation, and tracking. The American Society of Case Management and the Canadian National Case Management Network define case management as consisting of six fundamental parts [127].

4.2.6 Health Dataset Issues

The healthcare industry can manage a wide range of data and content, including text, photos, audio, time series, blood kinds, heart rates, facial images, and body temperatures. Although privacy-preserving FL-based healthcare services have been developed, the majority of them have only been evaluated on one dataset. A central server navigates different FL techniques via private ensemble learning, allowing participants to employ a collection of heterogeneous models without explicitly summarizing the data in one location [127].

4.3 Challenges of FL in Healthcare

Integrating FL into real-world healthcare environments poses numerous significant challenges that must be solved. The primary problems include communication, statistical, computational, security, and privacy concerns. In these subsections, we will delve into the subtleties of these topics and emphasize their importance in the medical sector [128].

4.3.1 Communication Challenges

Effective communication is critical in centralized or distributed FL systems, but in healthcare, geographically dispersed institutions might slow down the process due to faulty or unpredictable network connections. Bandwidth limitations, geographical variations, latency challenges, and data confidentiality concerns can all impede the collaborative training of robust ML models.

4.3.2 Statistical Problems

The sharing of health data, particularly from Electronic Health Records (EHR), is a major concern in the topic of Financial Liability (FL) in healthcare applications. This distribution can cause problems such as data heterogeneity, bias, generalizability, data economy, and non-I.I.D., complicating the optimization process and delaying client risk [129].

4.3.3 Computing Challenges

By 2030, more than 80% of edge device data will be handled at the network's edge, sparking renewed interest in AI deployment. The challenges of edge computing in healthcare include peripheral expansion, equipment dependability, lightweight models, data inconsistency, and scalability. These reasons make effective model training and deployment essential for modern healthcare organizations with limited resources [130, 131].

4.3.4 Security and Privacy Concerns

Because medical information is so delicate, security and privacy are critical in Florida healthcare. Compliance with regulations such as HIPAA and GDPR makes applying these safeguards difficult. However, communicating gradient updates with a central server can result in reverse engineering attacks that compromise the underlying FL system. Encrypting global gradients in a distributed network can help guard against backdoor assaults, but it reduces model performance significantly. Existing technical measures, such as additive homomorphic encryption, can help protect against data breaches and leaks [132, 133].

In conclusion, the problems of applying FL in healthcare are multifaceted and necessitate novel solutions that meet communication, statistical, computational, security, and privacy concerns. Addressing these problems is crucial to reaching the full promise of FL while maintaining the security and confidentiality of healthcare data.

4.4 Implementation Issues and Challenges

The deployment of an FL system can result in a variety of problems, including confidentiality and system assaults, data constraints, biases and distortions, and communication concerns between servers, peripheral devices, and numerous servers.

4.4.1 Data Protection and Attacks

Threats to the ML DL model include adversarial assaults, evasion/scout attacks, insufficient model preparation, and gradient inversion attacks. Adversary attacks can be exploited because of improper hyperparameters, and FL systems lack privacy-preserving algorithms. Evasion/scout attacks are widespread and do not affect the training data. Inadequate model preparation can result in confidentiality breaches, model poisoning, and theft. Gradient inversion attacks can recover input data using model knowledge, however, genomic data privacy remains an issue due to the lack of privacy protection in sequencing technology. These challenges underscore the necessity for better security measures in FL systems [48].

4.4.2 Challenges Related to Data

The healthcare industry has data gathering issues such as noise, incorrect annotation, distortion, false positives, and false negatives. Data gathering in EHRs, medical pictures, and radiology reports involves human intervention, resulting in unbalanced datasets and class imbalance. Data bias has an impact on ML models, specifically training data, cognitive sampling, reporting, and confirmation bias. Data quality is critical for FL systems because data points that are independently and identically distributed improve accuracy. Addressing patient variability can help to improve patient safety, lower healthcare costs, and boost individual and system performance metrics [30, 54].

4.4.3 Communication Problems

Distributed learning relies on identically distributed IID datasets, yet data sets might be diverse and vast, resulting in communication failures and significant costs. Federation learning confronts computational and communication overhead difficulties as wireless networks and end-user internet connections become more expensive and unreliable. Federated averaging and scarification have significantly reduced communication costs, but secure MPC experiments are more expensive. Another concern is energy usage while communicating, as IoT relies on mobile devices for processing and sensing, resulting in significant battery consumption on local devices and reduced user engagement [103].

4.5 *Challenges Faced and Lessons Learned*

This section focuses on the use of federated learning (FL) in medicine, namely in medical image processing, IoT-enabled smart healthcare applications, and COVID-19 epidemic prediction. FL uses electronic medical data to estimate hospitalization, survival, and mortality rates for patients who are connected to clinicians. It also uses fMRI classification to identify sickness risk variables and has possible answers for COVID-19. FL also improves user privacy and reduces latency in IoT applications by enabling doctors to view patient data from numerous institutions. It also helps predict COVID outbreaks through decentralized research, which improves study design and data processing. However, there are still unresolved difficulties in real-world health data environments that can be addressed using real-world medical machine-learning environments [132, 133].

5 Future Research Directions

This section examines the root causes of this problem, examines how FL can support the future of digital health, and identifies issues and challenges that need to be considered [134]. AI research has transformed fields such as radiology, pathology, and genetics. Deep learning models must learn millions of parameters from big datasets while being secure, fair, and portable. FL, a learning paradigm, addresses data governance and privacy concerns by training algorithms collaboratively without disclosing data. It is increasingly popular in healthcare applications, surpassing models trained on isolated single-institution data. FL can be used in EHRs to define and identify patients, forecast hospital admissions, mortality, and ICU stays, and develop AI models for mammography assessment. It can also have a direct impact on patient treatment by fostering collaboration between healthcare and research institutions [134–136].

5.1 *Effects on Interested Parties*

The paradigm shift from centralized data lakes to FL is substantial, and the implications for the various stakeholders in an FL ecosystem must be understood.

5.1.1 Clinicians

Clinicians can employ ML-based systems to improve their knowledge and maintain consistency in diagnosis, but they must follow data structure, annotation, and report procedure to ensure that collaborators have a common understanding of the information presented.

5.1.2 Patients

Globally established FL could enhance clinical decisions for patients in rural places and with rare diseases. It may also lessen the barrier to data contribution, as patients may rely on their institution's data, and access can be removed.

5.1.3 Hospitals and Practices

Hospitals can maintain control over patient data through traceability, reducing third-party misuse. This necessitates investing in computing infrastructure and adhering to established data formats. Even modest institutions can profit from the collaborative models produced by ML models.

5.1.4 Researchers and AI Developers

Researchers and AI developers might benefit from access to large amounts of real-world data, particularly at smaller labs and start-ups. This permits resources to be directed toward clinical demands and technical difficulties while exploring algorithmic strategies for FL is critical. FL-based development limits the capacity to analyze and see all learned data [136].

5.1.5 Healthcare Provider

Healthcare providers are transitioning from volume-based to focused treatment, prioritizing improved outcomes and cost savings. FL has the potential to improve healthcare AI accuracy, resulting in precision medicine.

5.1.6 Manufacturers

FL can assist healthcare software and hardware producers by combining learning from various devices to improve ML-based systems, but it necessitates major improvements in local processing, data storage, networking, and software.

A "honesty broker system" can be used to administer a federation by allowing data access through a trusted third party while maintaining stringent confidentiality. This approach necessitates an independent organization, which incurs additional expenditures and complexity. However, it provides flexibility and fast updates [135].

5.2 New Upcoming Future Issues

The research of FL is ongoing, with progress made in addressing difficulties. However, certain questions remain unsolved. This overview quickly examines historical difficulties as well as new issues in the federal framework, such as production and benchmarking.

5.2.1 Extreme Communication Schemes

FL necessitates communication, whereas ML optimization can tolerate imperfection for generalization. Classic data center communication strategies such as one shot and divide and conquer are less well recognized in heterogeneous networks [83, 137].

5.2.2 Communication Reduction and the Pareto Frontier

Federated training can benefit from local updates and model compression, but it is critical to understand how they interact and balance accuracy and communication. Effective approaches improve at the Pareto frontier, resulting in higher accuracy across several profiles [138].

5.2.3 Novel Models of Asynchrony

Distributed optimization intensively investigates asynchronous and bulk synchronous strategies. Worker nodes in data centers are dedicated to the workload and ready to receive tasks from the central node. In federated networks, devices are inactive, necessitating a device-centric communication strategy.

5.2.4 Heterogeneity Diagnostics

Local dissimilarity metrics are used to quantify statistical heterogeneity in federated networks, although they are difficult to compute before training. The usefulness of these indicators is debatable, as forecasting heterogeneity before deployment is challenging. Using existing or new definitions may improve convergence in federated optimization methods [139].

5.2.5 Detailed Privacy Restrictions

The Privacy in FL section defines both local and global privacy for all network devices. However, a more thorough definition is required due to differing privacy restrictions. Recent work samples provide reduced privacy for more accurate models.

5.2.6 Beyond Supervised Learning

The present approaches are geared toward supervised learning in federated networks, assuming that all data is labeled. Realistic networks may provide unlabeled or incorrectly tagged data. Scalability, heterogeneity, and privacy are all important considerations for advanced tasks such as reinforcement learning [140].

5.2.7 Productionizing Federated Learning

The work focuses on practical issues in FL production systems, such as idea drift, diurnal variations, and cold-start concerns, and addresses various real-world system issues for readers [141].

5.2.8 Benchmarks

Real-world scenarios, assumptions, and data sets have the potential to impact the future of FL, and researchers should develop implementations and testing tools such as TensorFlow Federated [142].

5.3 Future Research Directions for FL in the Medical Domain

Future healthcare activities can be enhanced by implementing collaborative tactics across many institutions, which enable the integration of heterogeneous data and multi-expert annotation. However, FL in healthcare is still in its infancy, and more study is needed to create a solid foundation for its use. Future research should bridge the research gap and point in promising directions.

5.3.1 FL for Healthcare 5.0

In a standard ML paradigm, the data utilized to train models creates security and privacy concerns. In this context, FL has been found to improve scalability, accuracy, training time, performance, privacy, and security, among others [143].

5.3.2 FL for Digital Twin in Healthcare

Future healthcare will be automated, emphasizing proactive health empowerment, prevention, and clinical decision-making. Digital twin technology improves patient care by employing AI to make life-saving decisions based on real-time and historical data. Rules and decentralized learning models (FL) are utilized to protect users' privacy and localize their data within FL models, resulting in better data analytics and patient outcomes [144].

5.3.3 FL and Blockchain for Healthcare

The FL-based technique can help healthcare practitioners improve AI model accuracy and resilience, making it more applicable for real-time application. It also benefits software and hardware vendors by allowing for collaborative learning and continual enhancement of AI models. Researchers and developers profit from the abundance of genuine data available [145].

5.3.4 FL for Collaborative Robotics in Healthcare

Cobots are collaborative robot technologies used in healthcare to increase productivity, streamline operations, and optimize safety protocols. These automated systems handle administrative work, laboratory tests, patient care, and surgical instruments. They fill the gap between employee shortages and industry hardship. The market is growing due to increased demand for automation and technology improvements, which reduce infection risks and provide health personnel more time to focus on essential concerns [146].

5.3.5 FL and Explainable AI for Healthcare

IoT has transformed healthcare by adding IoMT; yet, problems include distributed environment analysis, data creation, and device security. Explainable AI is a promising tool for detecting compromised data during cyberattacks, as well as enhancing communication, user acceptance, and task effectiveness in FL-based wearable health monitoring systems [147].

5.3.6 FL for Integration with 6G in Healthcare

The growing use of IoMT devices in daily activities causes an increase in data traffic and ML. 6G networks are changing wireless communication by providing sophisticated AI applications, security, and privacy protection in healthcare. Implementing FL-based edge intelligence in 6G boosts performance, low latency, and system privacy [28, 148].

5.3.7 Hyperparameter Optimization for Efficient FL Systems

Hyperparameter optimization is difficult in FL research since it requires an AI model with optimal hyperparameters to train numerous customers using their sensitive medical data. Despite this, optimizing hyperparameters on a federated health dataset is difficult due to customer opposition [149].

5.3.8 Security and Confidentiality of FL Systems

FL systems, particularly in medical applications, raise security concerns because of their collaborative nature with various clients, which increases vulnerability to attacks such as model poisoning. While there are ways to distinguish between benign and malignant models, they have limitations that must be addressed in actual practice [150].

5.3.9 Effective Communication Paradigm for FL Systems

Research should emphasize creating lightweight AI models to improve communication efficiency in FL systems, particularly smartwatches and healthcare gadgets. Sparsification, subsampling, and quantization are examples of strategies that can reduce message size while optimizing bandwidth consumption [151].

5.3.10 Addressing the Heterogeneity of Healthcare

FL systems encounter issues owing to heterogeneous data dissemination across healthcare institutions, resulting in decreased performance. Combining horizontal and longitudinal health data from many places remains a challenging task. Further research is required to generalize the findings and investigate the intricacies of medical data variability across numerous institutions [152].

5.3.11 Developing Intelligent Incentive Algorithms for the FL System

To assist in the creation of robust models, federal healthcare customers must be motivated to actively participate in FL processes. Current research focuses on developing incentive systems to promote equitable participation and reward high-quality data [153]. Developing clever FL algorithms that can recognize and remove consumers with low-quality data can also increase the model's generalizability [153].

5.3.12 Contrast Learning as a Solution to Unlabelled Dissemination of Medical Data

FL research primarily uses labeled models, but unlabeled datasets may grow. Semi-supervised and unsupervised learning methods struggle with medical imaging datasets. Self-learning is increasingly popular for learning unlabeled data representations.

5.3.13 Benchmarking FL

Because of FL's expanding popularity in medical areas, comprehensive benchmarking tools, open health datasets, and access to the research community are required to assure repeatability, generalizability, and creativity in FL solutions, as well as comparative analysis of medical applications.

5.3.14 Regulatory Framework for FL-Enabled Health Apps

The establishment of rules for reviewing health data using the FL framework is critical due to GDPR laws. Collaboration between computer scientists and law researchers is required to ensure compliance and improve the explanatory power of FL models in medical settings [13].

5.3.15 FL and the Three Most Innovative Technologies

Innovative technologies in smart healthcare include FL, AI, and explainable artificial intelligence (XAI). They enable decentralized communication while also addressing security, privacy, stability, and dependability concerns. The combination of FL, AI, and XAI has the potential to solve medical problems and lead future research themes in FL-based AI for healthcare management [84].

6 Conclusion

Federated learning (FL) is a promising technique for developing dependable, accurate, secure, fault-tolerant, and impartial digital health models. It eliminates issues associated with sensitive medical data by allowing various parties to train together without centralization. FL is suitable for the IoTs, which can benefit research, economic development, and health care. FL has a substantial impact on stakeholders and treatment cycles by increasing medical image analysis, precision medicine, and drug discovery. It saves costs and time to market, hence improving precision medicine and medical care [134]. FL is an emerging AI technology in healthcare that enables algorithms to train on servers or decentralized devices without exchanging data. It improves efficiency by sharing global model parameters. Currently, FL is used to diagnose diseases and track patient health. However, studies frequently overlook its limitations, and countermeasures are required to reduce attacks. FL is also considered feasible for data privacy and security, particularly when minimal data is available. The FL methodology is a powerful, accurate, safe, robust, and unbiased tool for simulating healthcare outcomes. It facilitates communication without data centralization, hence addressing crucial health information issues. FL has the potential to generate new economic opportunities and enhance healthcare around the world. However, it confronts several problems, including system unpredictability, communication limitations, and security and privacy concerns. To advance FL in medical applications, a collaborative effort from several diverse research fields is required. Rapid practical use of FL in medical applications may result in more advanced protocols, as well as increased security and privacy. According to Jayaram et al., major R&D efforts are centered on future 6G wireless technologies, such as incorporating FL features into medical devices, leveraging 6G for large-scale FL-based healthcare, and providing new services such as AI and FL capabilities to improve patient quality of life and reduce hospitalizations [132].

References

1. Dimitrov, D.V.: Medical internet of things and big data in healthcare. Healthc. Inform. Res. **22**, 156 (2016). https://doi.org/10.4258/hir.2016.22.3.156
2. Gulshan, V., Peng, L., Coram, M., Stumpe, M.C., Wu, D., Narayanaswamy, A., Venugopalan, S., Widner, K., Madams, T., Cuadros, J., Kim, R., Raman, R., Nelson, P.C., Mega, J.L., Webster, D.R.: Development and validation of a deep learning algorithm for detection of diabetic retinopathy in retinal fundus photographs. JAMA **316**, 2402 (2016). https://doi.org/10.1001/jama.2016.17216
3. Cascini, F., Santaroni, F., Lanzetti, R., Failla, G., Gentili, A., Ricciardi, W.: Developing a data-driven approach in order to improve the safety and quality of patient care. Front. Public Health **9** (2021). https://doi.org/10.3389/fpubh.2021.667819
4. Hoofnagle, C.J., Van Der Sloot, B., Borgesius, F.Z.: The European union general data protection regulation: what it is and what it means. Inf. Commun. Technol. Law **28**, 65–98 (2019). https://doi.org/10.1080/13600834.2019.1573501

5. John, M.M., Olsson, H.H., Bosch, J.: Developing ML/DL Models (2020). https://doi.org/10.1145/3379177.3388892
6. Xu, D., Li, T., Li, Y., Su, X., Tarkoma, S., Jiang, T., Crowcroft, J., Hui, P.: Edge Intelligence: Architectures, Challenges, and Applications. https://arxiv.org/pdf/2003.12172
7. Zhang, P., Jiang, C., Pang, X., Qian, Y.: STEC-IoT: a security tactic by virtualizing edge computing on IoT. IEEE Internet Things J. **8**, 2459–2467 (2020). https://doi.org/10.1109/jiot.2020.3017742
8. Li, E., Zeng, L., Zhou, Z., Chen, X.: Edge AI: on-demand accelerating deep neural network inference via edge computing. IEEE Trans. Wireless Commun. **19**, 447–457 (2019). https://doi.org/10.1109/twc.2019.2946140
9. Ray, P.P., Dash, D., De, D.: Edge computing for Internet of Things: a survey, e-healthcare case study and future direction. J. Netw. Comput. Appl. **140**, 1–22 (2019). https://doi.org/10.1016/j.jnca.2019.05.005
10. Rieke, N., Hancox, J., Li, W., Milletarì, F., Roth, H.R., Albarqouni, S., Bakas, S., Galtier, M.N., Landman, B.A., Maier-Hein, K., Ourselin, S., Sheller, M., Summers, R.M., Trask, A., Xu, D., Baust, M., Cardoso, M.J.: The future of digital health with federated learning. NPJ Digit. Med. **3** (2020). https://doi.org/10.1038/s41746-020-00323-1
11. Xu, J., Glicksberg, B.S., Su, C., Walker, P., Bian, J., Wang, F.: Federated learning for healthcare informatics. J. Healthc. Inform. Res. **5**, 1–19 (2020). https://doi.org/10.1007/s41666-020-00082-4
12. Rasouli, J.J., Shao, J., Neifert, S., Gibbs, W.N., Habboub, G., Steinmetz, M.P., Benzel, E., Mroz, T.E.: Artificial intelligence and robotics in spine surgery. Global Spine J. **11**, 556–564 (2020). https://doi.org/10.1177/2192568220915718
13. Rauniyar, A., Hagos, D.H., Jha, D., Håkegård, J.E., Bagci, U., Rawat, D.B., Vlassov, V.: Federated learning for medical applications: a taxonomy, current trends, challenges, and future research directions. IEEE Internet Things J. **11**, 7374–7398 (2023). https://doi.org/10.1109/jiot.2023.3329061
14. Zhang, T., Gao, L., He, C., Zhang, M., Krishnamachari, B., Avestimehr, A.S.: Federated learning for the internet of things: applications, challenges, and opportunities. IEEE Internet Things Mag. **5**, 24–29 (2022). https://doi.org/10.1109/iotm.004.2100182
15. Amalraj, J.R., Lourdusamy, R.: Security and privacy issues in federated healthcare—an overview. Open Comput. Sci. **12**, 57–65 (2022). https://doi.org/10.1515/comp-2022-0230
16. Ahmed, I.H., Djebbari, A., Kachenoura, A., Senhadji, L.: Telemedical transport layer security based platform for cardiac arrhythmia classification using quadratic time–frequency analysis of HRV signal. J. Supercomput. **78**, 13680–13709 (2022). https://doi.org/10.1007/s11227-022-04387-6
17. Li, T., Sahu, A.K., Talwalkar, A., Smith, V.: Federated learning: challenges, methods, and future directions. IEEE Signal Process. Mag. **37**, 50–60 (2020). https://doi.org/10.1109/msp.2020.2975749
18. Yu, H., Liu, Z., Liu, Y., Chen, T., Cong, M., Weng, X., Niyato, D., Yang, Q.: A Fairness-aware Incentive Scheme for Federated Learning (2020). https://doi.org/10.1145/3375627.3375840
19. Marulli, F., Verde, L., Marrone, S., Barone, R., De Biase, M.S.: Evaluating Efficiency and Effectiveness of Federated Learning Approaches in Knowledge Extraction Tasks (2021). https://doi.org/10.1109/ijcnn52387.2021.9533946
20. Campanile, L., Marrone, S., Marulli, F., Verde, L.: Challenges and trends in federated learning for well-being and healthcare. Procedia Comput. Sci. **207**, 1144–1153 (2022). https://doi.org/10.1016/j.procs.2022.09.170
21. Rana, N., Marwaha, H.: Role of federated learning in healthcare systems: a survey. Math. Found. Comput. **7**, 459–484 (2024). https://doi.org/10.3934/mfc.2023023
22. McMahan, B., Moore, E., Ramage, D., Hampson, S., Arcas, B.A.Y.: Communication-Efficient Learning of Deep Networks from Decentralized Data. https://proceedings.mlr.press/v54/mcmahan17a.html
23. Lim, W.Y.B., Luong, N.C., Hoang, D.T., Jiao, Y., Liang, Y.-C., Yang, Q., Niyato, D., Miao, C.: Federated learning in mobile edge networks: a comprehensive survey. IEEE Commun. Surv. Tutor. **22**, 2031–2063 (2020). https://doi.org/10.1109/comst.2020.2986024

24. Imteaj, A., Thakker, U., Wang, S., Li, J., Amini, M.H.: A survey on federated learning for resource-constrained IoT devices. IEEE Internet Things J. **9**, 1–24 (2021). https://doi.org/10.1109/jiot.2021.3095077
25. Kędzior, M.: GDPR and beyond—a year of changes in the data protection landscape of the European Union. ERA Forum. **19**, 505–509 (2019). https://doi.org/10.1007/s12027-019-00549-x
26. Rivest, R.L., Adleman, L., Dertouzos, M.L.: Massachusetts Institute of Technology: On Data Banks and Privacy Homomorphisms (1978)
27. Arikumar, K.S., Prathiba, S.B., Alazab, M., Gadekallu, T.R., Pandya, S., Khan, J.M., Moorthy, R.S.: FL-PMI: federated learning-based person movement identification through wearable devices in smart healthcare systems. Sensors **22**, 1377 (2022). https://doi.org/10.3390/s22041377
28. Konečný, J., McMahan, H.B., Ramage, D., Richtarik, P.: Federated Optimization: Distributed Machine Learning for On-Device Intelligence. arXiv (Cornell University) (2016). https://doi.org/10.48550/arxiv.1610.02527
29. Alazab, M., Rm, S.P., M, P., Maddikunta, P.K.R., Gadekallu, T.R., Pham, Q.-V.: Federated learning for cybersecurity: concepts, challenges, and future directions. IEEE Trans. Indus. Inform. **18**, 3501–3509 (2021). https://doi.org/10.1109/tii.2021.3119038
30. Li, Q., Wen, Z., Wu, Z., Hu, S., Wang, N., Li, Y., Liu, X., He, B.: A survey on federated learning systems: vision, hype and reality for data privacy and protection. IEEE Trans. Knowl. Data Eng. **35**, 3347–3366 (2021). https://doi.org/10.1109/tkde.2021.3124599
31. Song, J., Wang, W., Gadekallu, T.R., Cao, J., Liu, Y.: EPPDA: an efficient privacy-preserving data aggregation federated learning scheme. IEEE Trans. Netw. Sci. Eng. **10**, 3047–3057 (2022). https://doi.org/10.1109/tnse.2022.3153519
32. Reddy, K.D., Gadekallu, T.R.: A comprehensive survey on federated learning techniques for healthcare informatics. Comput. Intell. Neurosci. **2023**, 1–19 (2023). https://doi.org/10.1155/2023/8393990
33. Lu, W., Wang, J., Chen, Y., Qin, X., Xu, R., Dimitriadis, D., Qin, T.: Personalized Federated Learning with Adaptive Batchnorm for Healthcare. https://arxiv.org/abs/2112.00734
34. Zhang, W., Zhou, T., Lu, Q., Wang, X., Zhu, C., Sun, H., Wang, Z., Lo, S.K., Wang, F.-Y.: Dynamic-fusion-based federated learning for COVID-19 detection. IEEE Internet Things J. **8**, 15884–15891 (2021). https://doi.org/10.1109/jiot.2021.3056185
35. Romanini, D., Hall, A.J., Papadopoulos, P., Titcombe, T., Ismail, A., Cebere, T., Sandmann, R., Roehm, R., Hoeh, M.A.: PyVertical: A Vertical Federated Learning Framework for Multi-Headed SplitNN. arXiv (Cornell University) (2021). https://doi.org/10.48550/arxiv.2104.00489
36. Cha, D., Sung, M., Park, Y.-R.: Implementing vertical federated learning using autoencoders: practical application, generalizability, and utility study. JMIR Med. Inform. **9**, e26598 (2021). https://doi.org/10.2196/26598
37. Chen, Y., Qin, X., Wang, J., Yu, C., Gao, W.: FedHealth: a federated transfer learning framework for wearable healthcare. IEEE Intell. Syst. **35**, 83–93 (2020). https://doi.org/10.1109/mis.2020.2988604
38. Yu, H., Liu, Z., Liu, Y., Chen, T., Cong, M., Weng, X., Niyato, D., Yang, Q.: A sustainable incentive scheme for federated learning. IEEE Intell. Syst. **35**, 58–69 (2020). https://doi.org/10.1109/mis.2020.2987774
39. Xia, Y., Yang, D., Li, W., Myronenko, A., Xu, D., Obinata, H., Mori, H., An, P., Harmon, S.A., Turkbey, E., Turkbey, B., Wood, B.J., Patella, F., Stellato, E., Carrafiello, G., Ierardi, A., Yuille, A.L., Roth, H.: Auto-FedAvg: Learnable Federated Averaging for Multi-Institutional Medical Image Segmentation. arXiv (Cornell University) (2021). https://doi.org/10.48550/arxiv.2104.10195
40. Li, X., Huang, K., Yang, W., Wang, S., Zhang, Z.: On the Convergence of FedAvg on Non-IID Data. https://arxiv.org/abs/1907.02189
41. Yang, Q., Liu, Y., Chen, T., Tong, Y.: Federated Machine Learning: Concept and Applications. arXiv (Cornell University) (2019). https://doi.org/10.48550/arxiv.1902.04885

42. Niknam, S., Dhillon, H.S., Reed, J.H.: Federated learning for wireless communications: motivation, opportunities, and challenges. IEEE Commun. Mag. **58**, 46–51 (2020). https://doi.org/10.1109/mcom.001.1900461
43. Ang, F., Chen, L., Zhao, N., Chen, Y., Wang, W., Yu, F.R.: Robust federated learning with noisy communication. IEEE Trans. Commun. **68**, 3452–3464 (2020). https://doi.org/10.1109/tcomm.2020.2979149
44. Savazzi, S., Nicoli, M., Rampa, V.: Federated learning with cooperating devices: a consensus approach for massive IoT networks. IEEE Internet Things J. **7**, 4641–4654 (2020). https://doi.org/10.1109/jiot.2020.2964162
45. Mansour, Y., Mohri, M., Ro, J., Suresh, A.T.: Three Approaches for Personalization with Applications to Federated Learning. https://arxiv.org/abs/2002.10619
46. Bharati, S., Podder, P., Mondal, M.R.H., Prasath, V.B.S., Gandhi, N.: Ensemble learning for data-driven diagnosis of polycystic ovary syndrome. In: Lecture Notes in Networks and Systems, pp. 1250–1259 (2022). https://doi.org/10.1007/978-3-030-96308-8_116
47. Bharati, S., Mondal, M.R.H.: 12 Applications and challenges of AI-driven IoHT for combating pandemics: a review. De Gruyter eBooks 213–230 (2021). https://doi.org/10.1515/9783110712254-012
48. Sarma, K.V., Harmon, S., Sanford, T., Roth, H.R., Xu, Z., Tetreault, J., Xu, D., Flores, M.G., Raman, A.G., Kulkarni, R., Wood, B.J., Choyke, P.L., Priester, A.M., Marks, L.S., Raman, S.S., Enzmann, D., Turkbey, B., Speier, W., Arnold, C.W.: Federated learning improves site performance in multicenter deep learning without data sharing. J. Am. Med. Inform. Assoc. **28**, 1259–1264 (2020). https://doi.org/10.1093/jamia/ocaa341
49. Kose, U., Deperlioglu, O., Hemanth, D.J.: Deep Learning for Biomedical Applications (2021). https://doi.org/10.1201/9780367855611
50. Aledhari, M., Razzak, R., Parizi, R.M., Saeed, F.: Federated learning: a survey on enabling technologies, protocols, and applications. IEEE Access. **8**, 140699–140725 (2020). https://doi.org/10.1109/access.2020.3013541
51. Xiao, Z., Xu, X., Xing, H., Song, F., Wang, X., Zhao, B.: A federated learning system with enhanced feature extraction for human activity recognition. Knowl.-Based Syst. **229**, 107338 (2021). https://doi.org/10.1016/j.knosys.2021.107338
52. Chhikara, P., Singh, P., Tekchandani, R., Kumar, N., Guizani, M.: Federated learning meets human emotions: a decentralized framework for human-computer interaction for IoT applications. IEEE Internet Things J. **8**, 6949–6962 (2020). https://doi.org/10.1109/jiot.2020.3037207
53. Vaid, A., Jaladanki, S.K., Xu, J., Teng, S., Kumar, A., Lee, S., Somani, S., Paranjpe, I., De Freitas, J.K., Wanyan, T., Johnson, K.W., Bicak, M., Klang, E., Kwon, Y.J., Costa, A., Zhao, S., Miotto, R., Charney, A.W., Böttinger, E., Fayad, Z.A., Nadkarni, G.N., Wang, F., Glicksberg, B.S.: Federated learning of electronic health records to improve mortality prediction in hospitalized patients With COVID-19: machine learning approach. JMIR Med. Inform. **9**, e24207 (2021). https://doi.org/10.2196/24207
54. Brisimi, T.S., Chen, R., Mela, T., Olshevsky, A., Paschalidis, I.C., Shi, W.: Federated learning of predictive models from federated electronic health records. Int. J. Med. Informatics **112**, 59–67 (2018). https://doi.org/10.1016/j.ijmedinf.2018.01.007
55. Glicksberg, B.S., Johnson, K.W., Dudley, J.T.: The next generation of precision medicine: observational studies, electronic health records, biobanks and continuous monitoring. Hum. Mol. Genet. **27**, R56–R62 (2018). https://doi.org/10.1093/hmg/ddy114
56. Miotto, R., Wang, F., Wang, S., Jiang, X., Dudley, J.T.: Deep learning for healthcare: review, opportunities and challenges. Brief. Bioinform. **19**, 1236–1246 (2017). https://doi.org/10.1093/bib/bbx044
57. Obermeyer, Z., Powers, B., Vogeli, C., Mullainathan, S.: Dissecting racial bias in an algorithm used to manage the health of populations. Science **366**, 447–453 (2019). https://doi.org/10.1126/science.aax2342
58. Hripcsak, G., Duke, J.D., Shah, N.H., Reich, C.G., Huser, V., Schuemie, M.J., Suchard, M.A., Park, R.W., Wong, I.C.K., Rijnbeek, P.R., Van Der Lei, J., Pratt, N., Norén, G.N., Li, Y.-C.,

Stang, P.E., Madigan, D., Ryan, P.B.: Observational health data sciences and informatics (OHDSI): opportunities for observational researchers. PubMed. **216**, 574–578 (2015)
59. Yang, D., Xu, Z., Li, W., Myronenko, A., Roth, H.R., Harmon, S., Xu, S., Turkbey, B., Turkbey, E., Wang, X., Zhu, W., Carrafiello, G., Patella, F., Cariati, M., Obinata, H., Mori, H., Tamura, K., An, P., Wood, B.J., Xu, D.: Federated semi-supervised learning for COVID region segmentation in chest CT using multi-national data from China, Italy, Japan. Med. Image Anal. **70**, 101992 (2021). https://doi.org/10.1016/j.media.2021.101992
60. Kim, Y., Sun, J., Yu, H., Jiang, X.: Federated Tensor Factorization for Computational Phenotyping (2017). https://doi.org/10.1145/3097983.3098118
61. Lee, J., Sun, J., Wang, F., Wang, S., Jun, C.-H., Jiang, X.: Privacy-preserving patient similarity learning in a federated environment: development and analysis. JMIR Med. Inform. **6**, e20 (2018). https://doi.org/10.2196/medinform.7744
62. Liu, D., Dligach, D., Miller, T.: Two-Stage Federated Phenotyping and Patient Representation Learning (2019). https://doi.org/10.18653/v1/w19-5030
63. Gupta, O., Raskar, R.: Distributed learning of deep neural network over multiple agents. J. Netw. Comput. Appl. **116**, 1–8 (2018). https://doi.org/10.1016/j.jnca.2018.05.003
64. Vepakomma, P., Gupta, O., Swedish, T., Raskar, R.: Split Learning for Health: Distributed Deep Learning Without Sharing Raw Patient Data. https://arxiv.org/abs/1812.00564
65. Federated Learning in Distributed Medical Databases: Meta-Analysis of Large-Scale Subcortical Brain Data. https://ieeexplore.ieee.org/document/8759317
66. Boughorbel, S., Jarray, F., Venugopal, N., Moosa, S., Elhadi, H., Makhlouf, M.: Federated Uncertainty-Aware Learning for Distributed Hospital EHR Data. arXiv (Cornell University) (2019). https://doi.org/10.48550/arxiv.1910.12191
67. Sharma, P., Shamout, F.E., Clifton, D.A.: Preserving Patient Privacy while Training a Predictive Model of In-Hospital Mortality. https://arxiv.org/abs/1912.00354
68. Pfohl, S.R., Dai, A.M., Heller, K.A.: Federated and Differentially Private Learning for Electronic Health Records. arXiv (Cornell University) (2019). https://doi.org/10.48550/arxiv.1911.05861
69. Yuan, B., Ge, S., Xing, W.: A Federated Learning Framework for Healthcare IoT devices. arXiv (Cornell University) (2020). https://doi.org/10.48550/arxiv.2005.05083
70. Bharati, S., Mondal, M.R.H., Podder, P., Prasath, V.B.S.: Federated learning: applications, challenges and future directions. Int. J. Hybrid Intell. Syst. **18**, 19–35 (2022). https://doi.org/10.3233/his-220006
71. Guo, P., Wang, P., Zhou, J., Jiang, S., Patel, V.M.: Multi-institutional Collaborations for Improving Deep Learning-based Magnetic Resonance Image Reconstruction Using Federated Learning (2021). https://doi.org/10.1109/cvpr46437.2021.00245
72. Linardos, A., Kushibar, K., Walsh, S., Gkontra, P., Lekadir, K.: Federated learning for multicenter imaging diagnostics: a simulation study in cardiovascular disease. Sci. Rep. **12** (2022). https://doi.org/10.1038/s41598-022-07186-4
73. Hao, M., Li, H., Xu, G., Liu, Z., Chen, Z.: Privacy-Aware and Resource-saving Collaborative Learning for Healthcare in Cloud Computing (2020). https://doi.org/10.1109/icc40277.2020.9148979
74. Kaissis, G.A., Makowski, M.R., Rückert, D., Braren, R.F.: Secure, privacy-preserving and federated machine learning in medical imaging. Nat. Mach. Intell. **2**, 305–311 (2020). https://doi.org/10.1038/s42256-020-0186-1
75. Wu, Q., He, K., Chen, X.: Personalized federated learning for intelligent IoT applications: a cloud-edge based framework. IEEE Open J. Comput. Soc. **1**, 35–44 (2020). https://doi.org/10.1109/ojcs.2020.2993259
76. Nguyen, D.C., Ding, M., Pathirana, P.N., Seneviratne, A.: Blockchain and AI-based solutions to combat coronavirus (COVID-19)-like epidemics: a survey. IEEE Access **9**, 95730–95753 (2021). https://doi.org/10.1109/access.2021.3093633
77. Hui, D.S., Azhar, E.I., Madani, T.A., Ntoumi, F., Kock, R., Dar, O., Ippolito, G., Mchugh, T.D., Memish, Z.A., Drosten, C., Zumla, A., Petersen, E.: The continuing 2019-nCoV epidemic threat of novel coronaviruses to global health—the latest 2019 novel coronavirus outbreak

in Wuhan, China. Int. J. Infect. Dis. **91**, 264–266 (2020). https://doi.org/10.1016/j.ijid.2020.01.009
78. Kanne, J.P., Little, B.P., Chung, J.H., Elicker, B.M., Ketai, L.H.: Essentials for radiologists on COVID-19: an update—radiology scientific expert panel. Radiology **296**, E113–E114 (2020). https://doi.org/10.1148/radiol.2020200527
79. Yan, B., Wang, J., Cheng, J., Zhou, Y., Zhang, Y., Yang, Y., Liu, L., Zhao, H., Wang, C., Liu, B.: Experiments of federated learning for COVID-19 chest X-ray images. In: Communications in Computer and Information Science, pp. 41–53 (2021). https://doi.org/10.1007/978-3-030-78618-2_4
80. Cheng, Y., Liu, Y., Chen, T., Yang, Q.: Federated learning for privacy-preserving AI. Commun. ACM **63**, 33–36 (2020). https://doi.org/10.1145/3387107
81. Fang, M., Cao, X., Jia, J., Gong, N.Z.: Local Model Poisoning Attacks to Byzantine-Robust Federated Learning. arXiv (Cornell University) (2019). https://doi.org/10.48550/arxiv.1911.11815
82. Ramanan, P., Nakayama, K.: BAFFLE : Blockchain Based Aggregator Free Federated Learning (2020). https://doi.org/10.1109/blockchain50366.2020.00017
83. Vazirani, A.A., O'Donoghue, O., Brindley, D., Meinert, E.: Blockchain vehicles for efficient Medical Record management. NPJ Digit. Med. **3** (2020). https://doi.org/10.1038/s41746-019-0211-0
84. Połap, D., Srivastava, G., Yu, K.: Agent architecture of an intelligent medical system based on federated learning and blockchain technology. J. Inform. Secur. Appl. **58**, 102748 (2021). https://doi.org/10.1016/j.jisa.2021.102748
85. Wang, R., Lai, J., Zhang, Z., Li, X., Vijayakumar, P., Karuppiah, M.: Privacy-preserving federated learning for internet of medical things under edge computing. IEEE J. Biomed. Health Inform. **27**, 854–865 (2022). https://doi.org/10.1109/jbhi.2022.3157725
86. Zhao, Y., Zhao, J., Jiang, L., Tan, R., Niyato, D., Li, Z., Lyu, L., Liu, Y.: Privacy-preserving blockchain-based federated learning for IoT devices. IEEE Internet Things J. **8**, 1817–1829 (2020). https://doi.org/10.1109/jiot.2020.3017377
87. Abreha, H.G., Bernardos, C.J., De La Oliva, A., Cominardi, L., Azcorra, A.: Monitoring in fog computing: state-of-the-art and research challenges. Int. J. Ad Hoc Ubiquitous Comput. **36**, 114 (2021). https://doi.org/10.1504/ijahuc.2021.113384
88. García-Valls, M., Calva-Urrego, C., García-Fornes, A.: Accelerating smart eHealth services execution at the fog computing infrastructure. Futur. Gener. Comput. Syst. **108**, 882–893 (2018). https://doi.org/10.1016/j.future.2018.07.001
89. Rahman, Md.A., Hossain, M.S., Loukas, G., Hassanain, E., Rahman, S.S., Alhamid, M.F., Guizani, M.: Blockchain-based mobile edge computing framework for secure therapy applications. IEEE Access **6**, 72469–72478 (2018). https://doi.org/10.1109/access.2018.2881246
90. Asif-Ur-Rahman, M., Afsana, F., Mahmud, M., Kaiser, M.S., Ahmed, M.R., Kaiwartya, O., James-Taylor, A.: Toward a heterogeneous mist, fog, and cloud-based framework for the internet of healthcare things. IEEE Internet Things J. **6**, 4049–4062 (2018). https://doi.org/10.1109/jiot.2018.2876088
91. Dayan, I., Roth, H.R., Zhong, A., Harouni, A., Gentili, A., Abidin, A.Z., Liu, A., Costa, A.B., Wood, B.J., Tsai, C.-S., Wang, C.-H., Hsu, C.-N., Lee, C.K., Ruan, P., Xu, D., Wu, D., Huang, E., Kitamura, F.C., Lacey, G., De Antônio Corradi, G.C., Nino, G., Shin, H.-H., Obinata, H., Ren, H., Crane, J.C., Tetreault, J., Guan, J., Garrett, J.W., Kaggie, J.D., Park, J.G., Dreyer, K., Juluru, K., Kersten, K., Rockenbach, M.A.B.C., Linguraru, M.G., Haider, M.A., AbdelMaseeh, M., Rieke, N., Damasceno, P.F., Silva, P.M.C.E., Wang, P., Xu, S., Kawano, S., Sriswasdi, S., Park, S.Y., Grist, T.M., Buch, V., Jantarabenjakul, W., Wang, W., Tak, W.Y., Li, X., Lin, X., Kwon, Y.J., Quraini, A., Feng, A., Priest, A.N., Turkbey, B., Glicksberg, B., Bizzo, B., Kim, B.S., Tor-Díez, C., Lee, C.-C., Hsu, C.-J., Lin, C., Lai, C.-L., Hess, C.P., Compas, C., Bhatia, D., Oermann, E.K., Leibovitz, E., Sasaki, H., Mori, H., Yang, I., Sohn, J.H., Murthy, K.N.K., Fu, L.-C., De Mendonça, M.R.F., Fralick, M., Kang, M.K., Adil, M., Gangai, N., Vateekul, P., Elnajjar, P., Hickman, S., Majumdar, S., McLeod, S.L., Reed, S.,

Gräf, S., Harmon, S., Kodama, T., Puthanakit, T., Mazzulli, T., De Lavor, V.L., Rakvongthai, Y., Lee, Y.R., Wen, Y., Gilbert, F.J., Flores, M.G., Li, Q.: Federated learning for predicting clinical outcomes in patients with COVID-19. Nat. Med. **27**, 1735–1743 (2021). https://doi.org/10.1038/s41591-021-01506-3
92. Naz, S., Phan, K.T., Chen, Y.P.: A comprehensive review of federated learning for COVID-19 detection. Int. J. Intell. Syst. **37**, 2371–2392 (2021). https://doi.org/10.1002/int.22777
93. Cao, Y.: Near real-time federated machine learning approach over chest computed tomography for COVID-19 diagnosis. In: Communications in Computer and Information Science, pp. 21–36 (2022). https://doi.org/10.1007/978-981-19-1166-8_3
94. Sakib, S., Fouda, M.M., Fadlullah, Z.M., Nasser, N.: On COVID-19 Prediction Using Asynchronous Federated Learning-Based Agile Radiograph Screening Booths (2021). https://doi.org/10.1109/icc42927.2021.9500351
95. Brenčič, N.S., Dragoi, M., Mocanu, I., Winiarski, T.: Intuitive and intelligent solutions for elderly care. In: Advances in Predictive, Preventive and Personalised Medicine, pp. 101–108 (2020). https://doi.org/10.1007/978-3-030-49815-3_12
96. Oppedal, K., Álvaro, F.Q.: Federated Learning for Dementia Classification in a European Multicentre Dementia Study. https://uis.brage.unit.no/uis-xmlui/handle/11250/2687294
97. Roth, H.R., Chang, K., Singh, P., Neumark, N., Li, W., Gupta, V., Gupta, S., Qu, L., Ihsani, A., Bizzo, B.C., Wen, Y., Buch, V., Shah, M., Kitamura, F., Mendonça, M., Lavor, V., Harouni, A., Compas, C., Tetreault, J., Dogra, P., Cheng, Y., Erdal, S., White, R., Hashemian, B., Schultz, T., Zhang, M., McCarthy, A., Yun, B.M., Sharaf, E., Hoebel, K.V., Patel, J.B., Chen, B., Ko, S., Leibovitz, E., Pisano, E.D., Coombs, L., Xu, D., Dreyer, K.J., Dayan, I., Naidu, R.C., Flores, M., Rubin, D., Kalpathy-Cramer, J.: Federated learning for breast density classification: a real-world implementation. In: Lecture Notes in Computer Science, pp. 181–191 (2020). https://doi.org/10.1007/978-3-030-60548-3_18
98. Pati, S., Baid, U., Edwards, B., Sheller, M., Wang, S.-H., Reina, G.A., Foley, P., Gruzdev, A., Karkada, D., Davatzikos, C., Sako, C., Ghodasara, S., Bilello, M., Mohan, S., Vollmuth, P., Brugnara, G., Preetha, C.J., Sahm, F., Maier-Hein, K., Zenk, M., Bendszus, M., Wick, W., Calabrese, E., Rudie, J., Villanueva-Meyer, J., Cha, S., Ingalhalikar, M., Jadhav, M., Pandey, U., Saini, J., Garrett, J., Larson, M., Jeraj, R., Currie, S., Frood, R., Fatania, K., Huang, R.Y., Chang, K., Balaña, C., Capellades, J., Puig, J., Trenkler, J., Pichler, J., Necker, G., Haunschmidt, A., Meckel, S., Shukla, G., Liem, S., Alexander, G.S., Lombardo, J., Palmer, J.D., Flanders, A.E., Dicker, A.P., Sair, H.I., Jones, C.K., Venkataraman, A., Jiang, M., So, T.Y., Chen, C., Heng, P.A., Dou, Q., Kozubek, M., Lux, F., Michálek, J., Matula, P., Keřkovský, M., Kopřivová, T., Dostál, M., Vybíhal, V., Vogelbaum, M.A., Mitchell, J.R., Farinhas, J., Maldjian, J.A., Yogananda, C.G.B., Pinho, M.C., Reddy, D., Holcomb, J., Wagner, B.C., Ellingson, B.M., Cloughesy, T.F., Raymond, C., Oughourlian, T., Hagiwara, A., Wang, C., To, M.-S., Bhardwaj, S., Chong, C., Agzarian, M., Falcão, A.X., Martins, S.B., Teixeira, B.C.A., Sprenger, F., Menotti, D., Lucio, D.R., LaMontagne, P., Marcus, D., Wiestler, B., Kofler, F., Ezhov, I., Metz, M., Jain, R., Lee, M., Lui, Y.W., McKinley, R., Slotboom, J., Radojewski, P., Meier, R., Wiest, R., Murcia, D., Fu, E., Haas, R., Thompson, J., Ormond, D.R., Badve, C., Sloan, A.E., Vadmal, V., Waite, K., Colen, R.R., Pei, L., Ak, M., Srinivasan, A., Bapuraj, J.R., Rao, A., Wang, N., Yoshiaki, O., Moritani, T., Turk, S., Lee, J., Prabhudesai, S., Morón, F., Mandel, J., Kamnitsas, K., Glocker, B., Dixon, L.V.M., Williams, M., Zampakis, P., Panagiotopoulos, V., Tsiganos, P., Alexiou, S., Haliassos, I., Zacharaki, E.I., Moustakas, K., Kalogeropoulou, C., Kardamakis, D.M., Choi, Y.S., Lee, S.-K., Chang, J.H., Ahn, S.S., Luo, B., Poisson, L., Wen, N., Tiwari, P., Verma, R., Bareja, R., Yadav, I., Chen, J., Kumar, N., Smits, M., Van Der Voort, S.R., Alafandi, A., Incekara, F., Wijnenga, M.M.J., Kapsas, G., Gahrmann, R., Schouten, J.W., Dubbink, H.J., Vincent, A.J.P.E., Van Den Bent, M.J., French, P.J., Klein, S., Yuan, Y., Sharma, S., Tseng, T.-C., Adabi, S., Niclou, S.P., Keunen, O., Hau, A.-C., Vallières, M., Fortin, D., Lepage, M., Landman, B., Ramadass, K., Xu, K., Chotai, S., Chambless, L.B., Mistry, A., Thompson, R.C., Gusev, Y., Bhuvaneshwar, K., Sayah, A., Bencheqroun, C., Belouali, A., Madhavan, S., Booth, T.C., Chelliah, A., Modat, M., Shuaib, H., Dragos, C., Abayazeed, A., Kolodziej, K., Hill, M., Abbassy, A., Gamal, S., Mekhaimar,

M., Qayati, M., Reyes, M., Park, J.E., Yun, J., Kim, H.S., Mahajan, A., Muzi, M., Benson, S., Beets-Tan, R.G.H., Teuwen, J., Herrera-Trujillo, A., Trujillo, M., Escobar, W., Abello, A., Bernal, J., Gómez, J., Choi, J., Baek, S., Kim, Y., Ismael, H., Allen, B., Buatti, J.M., Kotrotsou, A., Li, H., Weiss, T., Weller, M., Bink, A., Pouymayou, B., Shaykh, H.F., Saltz, J., Prasanna, P., Shrestha, S., Mani, K.M., Payne, D., Kurc, T., Pelaez, E., Franco-Maldonado, H., Loayza, F., Quevedo, S., Guevara, P., Torche, E., Mendoza, C., Vera, F., Ríos, E., López, E., Velastin, S.A., Ogbole, G., Soneye, M., Oyekunle, D., Odafe-Oyibotha, O., Osobu, B., Shu'aibu, M., Dorcas, A., Dako, F., Simpson, A.L., Hamghalam, M., Peoples, J.J., Hu, R., Tran, A., Cutler, D., Moraes, F.Y., Boss, M.A., Gimpel, J., Veettil, D.K., Schmidt, K., Bialecki, B., Marella, S., Price, C., Cimino, L., Apgar, C., Shah, P., Menze, B., Barnholtz-Sloan, J.S., Martin, J., Bakas, S.: Federated learning enables big data for rare cancer boundary detection. Nat. Commun. **13** (2022). https://doi.org/10.1038/s41467-022-33407-5

99. Kennedy, G.T., Azari, F.S., Bernstein, E., Nadeem, B., Chang, A., Segil, A., Carlin, S., Sullivan, N.T., Encarnado, E., Desphande, C., Kularatne, S., Gagare, P., Thomas, M., Kucharczuk, J.C., Christien, G., Lacombe, F., Leonard, K., Low, P.S., Criton, A., Singhal, S.: Targeted detection of cancer at the cellular level during biopsy by near-infrared confocal laser endomicroscopy. Nat. Commun. **13** (2022). https://doi.org/10.1038/s41467-022-30265-z
100. TensorFlow Federated. https://www.tensorflow.org/federated
101. FederatedAI: GitHub—FederatedAI/FATE: An Industrial Grade Federated Learning Framework. https://github.com/FederatedAI/FATE
102. Ryffel, T., Trask, A., Dahl, M., Wagner, B., Mancuso, J., Rueckert, D., Passerat-Palmbach, J.: A Generic Framework for Privacy Preserving Deep Learning. arXiv (Cornell University) (2018). https://doi.org/10.48550/arxiv.1811.04017
103. OpenMined: GitHub—OpenMined/PySyft: Perform Data Science on Data that Remains in Someone Else's Server. https://github.com/OpenMined/PySyft
104. PaddleFL—PaddleFL 0.1.0.Beta Documentation. https://paddlefl.readthedocs.io/en/latest/introduction.html
105. PaddlePaddle: GitHub—PaddlePaddle/PaddleFL: Federated Deep Learning in PaddlePaddle. https://github.com/PaddlePaddle/PaddleFL
106. He, C., Li, S., So, J., Zhang, M., Wang, H., Wang, X., Vepakomma, P., Singh, A., Qiu, H., Shen, L., Zhao, P., Kang, Y., Liu, Y., Raskar, R., Yang, Q., Annavaram, M., Avestimehr, S.: FedML: A Research Library and Benchmark for Federated Machine Learning. arXiv (Cornell University) (2020). https://doi.org/10.48550/arxiv.2007.13518
107. Ibm: federated-learning-lib/README.md at main IBM/federated-learning-lib. https://github.com/IBM/federated-learning-lib/blob/main/README.md
108. Joshi, M., Pal, A., Sankarasubbu, M.: Federated learning for healthcare domain—pipeline, applications and challenges. ACM Trans. Comput. Healthc. **3**, 1–36 (2022). https://doi.org/10.1145/3533708
109. Wan, L., Ng, W.K., Han, S., Lee, V.C.S.: Privacy-Preservation for Gradient Descent Methods (2007). https://doi.org/10.1145/1281192.1281275
110. Goldreich, O., Micali, S., Wigderson, A.: How to Play ANY Mental Game (1987). https://doi.org/10.1145/28395.28420
111. Shokri, R., Shmatikov, V.: Privacy-Preserving Deep Learning (2015). https://doi.org/10.1145/2810103.2813687
112. Lin, Y., Han, S., Mao, H., Wang, Y., Dally, W.J.: Deep Gradient Compression: Reducing the Communication Bandwidth for Distributed Training. arXiv (Cornell University) (2017). https://doi.org/10.48550/arxiv.1712.01887
113. Ma, X., Ma, J., Li, H., Jiang, Q., Gao, S.: PDLM: privacy-preserving deep learning model on cloud with multiple keys. IEEE Trans. Serv. Comput. **14**, 1251–1263 (2018). https://doi.org/10.1109/tsc.2018.2868750
114. Xu, J., Glicksberg, B.S., Su, C., Walker, P., Bian, J., Wang, F.: Federated Learning for Healthcare Informatics. arXiv (Cornell University) (2019). https://doi.org/10.48550/arxiv.1911.06270

115. Sicuranza, M., Ciampi, M., De Pietro, G., Esposito, C.: Secure healthcare data sharing among federated health information systems. Int. J. Crit. Comput.-Based Syst. **4**, 349 (2013). https://doi.org/10.1504/ijccbs.2013.059023
116. Heurix, J., Karlinger, M., Schrefl, M., Neubauer, T.: A hybrid approach integrating encryption and pseudonymization for protecting electronic health records. Biomed. Eng. (2010). https://doi.org/10.2316/p.2010.723-117
117. Khan, I.A., Razzak, I., Pi, D., Khan, N., Hussain, Y., Li, B., Kousar, T.: Fed-inforce-fusion: a federated reinforcement-based fusion model for security and privacy protection of IoMT networks against cyber-attacks. Inform. Fusion. **101**, 102002 (2023). https://doi.org/10.1016/j.inffus.2023.102002
118. Zeng, X., Yan, M., Zhang, M.: Mercury: Efficient On-Device Distributed DNN Training Via Stochastic Importance Sampling (2021). https://doi.org/10.1145/3485730.3485930
119. Chen, M., Yang, Z., Saad, W., Yin, C., Poor, H.V., Cui, S.: A joint learning and communications framework for federated learning over wireless networks. IEEE Trans. Wireless Commun. **20**, 269–283 (2020). https://doi.org/10.1109/twc.2020.3024629
120. Nguyen, J., Malik, K., Zhan, H., Yousefpour, A., Rabbat, M., Malek, M., Huba, D.: Federated Learning with Buffered Asynchronous Aggregation. arXiv (Cornell University) (2021). https://doi.org/10.48550/arxiv.2106.06639
121. Diao, E., Ding, J., Tarokh, V.: HeteroFL: Computation and Communication Efficient Federated Learning for Heterogeneous Clients. arXiv (Cornell University) (2020). https://doi.org/10.48550/arxiv.2010.01264
122. Van Berkel: Multi-Core for Mobile Phones. IEEE Xplore (2009). https://ieeexplore.ieee.org/document/5090858
123. Nguyen, T.D., Marchal, S., Miettinen, M., Fereidooni, H., Asokan, N., Sadeghi, A.-R.: DÏoT: A Federated Self-learning Anomaly Detection System for IoT (2019). https://doi.org/10.1109/icdcs.2019.00080
124. Geiping, J., Bauermeister, H., Dröge, H., Moeller, M.: Inverting Gradients—How Easy Is It to break Privacy in Federated Learning? arXiv (Cornell University) (2020). https://doi.org/10.48550/arxiv.2003.14053
125. Zhang, T., He, C., Ma, T., Gao, L., Ma, M., Avestimehr, S.: Federated Learning for Internet of Things (2021). https://doi.org/10.1145/3485730.3493444
126. Bhowmick, A., Duchi, J., Freudiger, J., Kapoor, G., Rogers, R.: Protection Against Reconstruction and Its Applications in Private Federated Learning, https://arxiv.org/abs/1812.00984
127. Mandapati, S., Kadry, S., Kumar, R.L., Sutham, K., Thinnukool, O.: Deep learning model construction for a semi-supervised classification with feature learning. Complex Intell. Syst. **9**, 3011–3021 (2022). https://doi.org/10.1007/s40747-022-00641-9
128. Khan, F., Prasad, B.V.V.S., Syed, S.A., Ashraf, I., Ramasamy, L.K.: An efficient, ensemble-based classification framework for big medical data. Big Data **10**, 151–160 (2021). https://doi.org/10.1089/big.2021.0132
129. Pillutla, K., Laguel, Y., Malick, J., Harchaoui, Z.: Federated learning with superquantile aggregation for heterogeneous data. Mach. Learn. **113**, 2955–3022 (2023). https://doi.org/10.1007/s10994-023-06332-x
130. Dinh, C.T., Tran, N.H., Nguyen, T.D.: Personalized Federated Learning with Moreau Envelopes. arXiv (Cornell University) (2020). https://doi.org/10.48550/arxiv.2006.08848
131. Mothukuri, V., Parizi, R.M., Pouriyeh, S., Huang, Y., Dehghantanha, A., Srivastava, G.: A survey on security and privacy of federated learning. Futur. Gener. Comput. Syst. **115**, 619–640 (2020). https://doi.org/10.1016/j.future.2020.10.007
132. Jayaram, K.R., Verma, A., Verma, A., Thomas, G., Sutcher-Shepard, C.: MYSTIKO: Cloud-Mediated, Private, Federated Gradient Descent (2020). https://doi.org/10.1109/cloud49709.2020.00039
133. Yoo, J.H., Jeong, H., Lee, J., Chung, T.-M.: Open problems in medical federated learning. Int. J. Web Inform. Syst. **18**, 77–99 (2022). https://doi.org/10.1108/ijwis-04-2022-0080

134. Rocher, L., Hendrickx, J.M., De Montjoye, Y.-A.: Estimating the success of re-identifications in incomplete datasets using generative models. Nat. Commun. **10** (2019). https://doi.org/10.1038/s41467-019-10933-3
135. Agarwal, N., Suresh, A.T., Yu, F., Kumar, S., Mcmahan, H.B.: cpSGD: Communication-Efficient and Differentially-Private Distributed SGD. https://arxiv.org/abs/1805.10559
136. Eichner, H., Koren, T., McMahan, H.B., Srebro, N., Talwar, K.: Semi-Cyclic Stochastic Gradient Descent. https://arxiv.org/abs/1904.10120
137. Ross, D.: Digital twinning [virtual reality avatars]. Eng. Technol. **11**, 44–45 (2016). https://doi.org/10.1049/et.2016.0403
138. Ramu, S.P., Boopalan, P., Pham, Q.-V., Maddikunta, P.K.R., Huynh-The, T., Alazab, M., Nguyen, T.T., Gadekallu, T.R.: Federated learning enabled digital twins for smart cities: concepts, recent advances, and future directions. Sustain. Cities Soc. **79**, 103663 (2022). https://doi.org/10.1016/j.scs.2021.103663
139. Aich, S., Sinai, N.K., Kumar, S., Ali, M., Choi, Y.R., Joo, M.-I., Kim H.-C.: Protecting Personal Healthcare Record Using Blockchain & Federated Learning Technologies (2021). https://doi.org/10.23919/icact51234.2021.9370566
140. Gadekallu, T.R., Pham, Q.-V., Nguyen, D.C., Maddikunta, P.K.R., Deepa, N., Prabadevi, B., Pathirana, P.N., Zhao, J., Hwang, W.-J.: Blockchain for edge of things: applications, opportunities, and challenges. IEEE Internet Things J. **9**, 964–988 (2021). https://doi.org/10.1109/jiot.2021.3119639
141. Gadekallu, T.R., Huynh-The, T., Wang, W., Yenduri, G., Ranaweera, P., Pham, Q.-V., Benevides, D.C.D., Liyanage, M.: Blockchain for the Metaverse: A Review. arXiv (Cornell University) (2022). https://doi.org/10.48550/arxiv.2203.09738
142. Maddikunta, P.K.R., Pham, Q.-V., B, P., Deepa, N., Dev, K., Gadekallu, T.R., Ruby, R., Liyanage, M.: Industry 5.0: a survey on enabling technologies and potential applications. J. Indus. Inform. Integr. **26**, 100257 (2021). https://doi.org/10.1016/j.jii.2021.100257
143. Nayak, S., Patgiri, R.: 6G communication technology: a vision on intelligent healthcare. In: Studies in Computational Intelligence, pp. 1–18 (2021). https://doi.org/10.1007/978-981-15-9735-0_1
144. Deng, Y., Lyu, F., Ren, J., Wu, H., Zhou, Y., Zhang, Y., Shen, X.: AUCTION: automated and quality-aware client selection framework for efficient federated learning. IEEE Trans. Parallel Distrib. Syst. **33**, 1996–2009 (2021). https://doi.org/10.1109/tpds.2021.3134647
145. Paragliola, G., Coronato, A.: Definition of a novel federated learning approach to reduce communication costs. Expert Syst. Appl. **189**, 116109 (2021). https://doi.org/10.1016/j.eswa.2021.116109
146. Kang, J., Xiong, Z., Niyato, D., Xie, S., Zhang, J.: Incentive mechanism for reliable federated learning: a joint optimization approach to combining reputation and contract theory. IEEE Internet Things J. **6**, 10700–10714 (2019). https://doi.org/10.1109/jiot.2019.2940820
147. Fallah, A., Mokhtari, A., Ozdaglar, A.: Personalized Federated Learning: A Meta-Learning Approach. https://arxiv.org/abs/2002.07948
148. Kassem, H., Alapatt, D., Mascagni, P., Karargyris, A., Padoy, N.: Federated cycling (FedCy): semi-supervised federated learning of surgical phases. IEEE Trans. Med. Imaging **42**, 1920–1931 (2022). https://doi.org/10.1109/tmi.2022.3222126
149. Zhu, H., Xu, J., Liu, S., Jin, Y.: Federated learning on non-IID data: a survey. Neurocomputing **465**, 371–390 (2021). https://doi.org/10.1016/j.neucom.2021.07.098
150. Li, Q., He, B., Song, D.: Model-Contrastive Federated Learning (2021). https://doi.org/10.1109/cvpr46437.2021.01057
151. Chen, T., Kornblith, S., Norouzi, M., Hinton, G.: A Simple Framework for Contrastive Learning of Visual Representations. arXiv (Cornell University) (2020). https://doi.org/10.48550/arxiv.2002.05709
152. Samek, W., Wiegand, T., Müller, K.-R.: Explainable Artificial Intelligence: Understanding, Visualizing and Interpreting Deep Learning Models. https://arxiv.org/abs/1708.08296v1
153. Chen, P., Du, X., Lu, Z., Wu, J., Hung, P.C.K.: EVFL: An explainable vertical federated learning for data-oriented artificial intelligence systems. J. Syst. Architect. **126**, 102474 (2022). https://doi.org/10.1016/j.sysarc.2022.102474

Federated Learning for Remote Health Monitoring—A Review

Venkatesh Upadrista, Sajid Nazir, and Huaglory Tianfield

Abstract Remote Health Monitoring (RHM) use cases produce large amounts of sensitive data from a variety of sources, requiring effective storage, management, and analysis to extract valuable insights. While the cloud is often used to store and analyse substantial health data, directly connecting diverse medical devices to the cloud can be inefficient, negatively impact performance, and pose data privacy risks. Edge computing provides a solution by enabling devices to process data at the network's edge, either through a nearby local data center or directly on the device itself. This approach satisfies the low-latency and data security needs of healthcare applications. Despite these advantages, edge computing faces limitations in the amount of data it can handle, and training machine learning (ML) algorithms solely on local data can reduce model effectiveness. Federated learning has emerged as a promising solution to these challenges. This chapter includes a comprehensive review of Federated learning applications for RHM to evaluate how well these solutions address the limitations of edge computing. A systematic search was conducted across databases such as Arxiv, IEEE, Medrxiv, and ScienceDirect, focusing on Federated Learning, Edge Computing, and Remote Health Monitoring. Out of 5,015 references identified, 233 papers were selected for full-text analysis after removing duplicates and irrelevant entries. Ultimately, 87 relevant studies were included in our analysis. The review indicates that federated edge solutions are being developed to overcome edge computing constraints. However, these solutions require further exploration, particularly concerning security, as they are not yet fully ready for widespread adoption. Federated learning remains in its early implementation phase and finding solutions to these challenges.

V. Upadrista (✉) · S. Nazir · H. Tianfield
Department of Computing, Glasgow Caledonian University, Glasgow, Scotland
e-mail: vupadr200@caledonian.ac.uk

S. Nazir
e-mail: Sajid.Nazir@gcu.ac.uk

H. Tianfield
e-mail: H.Tianfield@gcu.ac.uk

R. Mohana et al. (eds.), *Federated Learning Applications in the Industrial Internet of Everything (IoE)*, Studies in Systems, Decision and Control 611,
https://doi.org/10.1007/978-3-031-99270-4_3

Keywords Federated learning · Edge computing · Remote health monitoring · Cloud computing · Fog computing · Healthcare

1 Introduction

There is an increasing trend on the usage of wearable devices especially in areas of fitness and activity awareness [1, 2]. In addition to these niche recreational fitness devices, Remote Health Monitoring (RHM) use cases are using wearable devices for critical use cases such as patient care [3, 4], detecting critical illness [5, 6] and clinical access to patient's physiological information [7, 8].

RHM use cases often rely on cloud services for storing, analyzing, and processing data [9–13]. The cloud has transformed the healthcare industry by enabling innovative applications [14, 15], but it also brings challenges related to security, privacy [11, 16–18], latency and availability [19–21]. Healthcare data is extremely sensitive, demanding strong safeguards to prevent unauthorized access, data breaches, and manipulation. Strict regulations, such as the General Data Protection Regulation (GDPR) and the Health Insurance Portability and Accountability Act (HIPAA), enforce these protections and set guidelines for securing, communicating, and storing healthcare data [22]. Cloud-based deployments carry risks of regulatory breaches and identity theft [23, 24]. Although various authors have proposed solutions to enhance data security and privacy in cloud environments [23, 24] these issues continue to be significant challenges for the healthcare sector. Additionally, latency has become a critical concern in healthcare implementations due to the time-sensitive nature of many use cases, with cloud architectures often exacerbating latency-related problems [25–31]. Edge computing is revolutionizing how data is collected, stored, and processed, aiming to address issues associated with cloud computing, such as latency, security, and privacy. The core principle of edge computing lies in processing of the data at the point of generation and collection. This approach brings the tasks of processing, analytics, and storage closer to the originating devices. For instance, a smartwatch worn by an individual can monitor heart rhythms and immediately alert a doctor if any irregularities are detected. This is in contrast to a cloud-based system, where heartbeat data must be sent to the cloud for processing, which could result in delays. Edge computing enhances health systems by optimizing how data is collected, stored, and analyzed. Several studies have discussed the benefits of edge computing and outlined strategies to tackle the cloud computing challenges related to latency, data security, and privacy [32–42]. Additionally, reliability is a fundamental aspect of healthcare applications and is a primary factor driving the increased popularity of edge computing [37–42]. Based on these discussions it is quite obvious that edge computing is trying to address multiple gaps arising from cloud computing, which is quite encouraging. However, edge computing comes with its own challenges as it has limited storage and processing power [43, 44].

RHM use cases involves complex data processing tasks as most of the uses cases rely on machine learning to predict patient health. Because edge computing has limited processing power and storage capabilities, it is difficult to run machine learning algorithms at the edge [43, 44]. While increasing processing power at the edge can help mitigate some limitations, machine learning (ML) on the edge remains constrained by the availability of local data, making the ML model's performance dependent on the quality and scope of this data [45]. Privacy is another challenge with edge computing since ML model trained at the edge can be reverse engineered to regenerate individual patient data [46]. On the other hand, data plays a crucial role in the outcome of these edge computing models and they are susceptible to several security attacks [46, 47]. To address these limitation, federated edge is now coming to limelight.

Federated edge computing enables patient data to be stored locally at the edge, where a machine learning (ML) model is trained on the localized data. Once training is complete, only the model parameters are sent to a central server. This server then combines the local models to develop a comprehensive global or master model. Essentially, federated learning facilitates the development of a unified model by utilizing data from various edge locations, all without the need to share any individual datasets. This means that a model can be trained on data that belongs to different patients, or hospitals without sharing the data, which enforces data privacy in RHM use cases. With federated edge even if hackers reverse engineered the trained ML model to recreate data, it would mostly be corrupted data and on the other hand it is very difficult to identify the data owners thereby ensuring data privacy [48]. In 2016, Google researchers introduced federated learning as a promising approach to tackle challenges related to communication costs, data privacy, and regulatory compliance [44, 49–51]. Federated learning is a decentralized machine learning (ML) method where models are trained on local edge devices. After training, the models are sent to a central server, where they are combined with other locally trained models, all without the need to share the original datasets. This process ensures that data privacy is maintained throughout the training process.

This chapter presents a literature review to explore the use of Federated Learning across various Remote Health Monitoring (RHM) use cases, including critical illness, COVID-19, elderly care, and general RHM applications. Unlike other studies, this chapter focuses on how Federated Learning addresses challenges related to security, privacy, and latency. It also examines how Federated Learning tackles resource limitations inherent in edge computing, which is crucial since most healthcare edge devices have limited capacity and processing power. Given that most RHM use cases generate large volumes of data, edge computing poses a significant challenge. The literature was reviewed to identify Federated Learning solutions that address this issue. Additionally, the chapter discusses other open challenges associated with Federated Learning that could serve as the basis for future research.

1.1 Organization of Chapter

The rest of the chapter is organized as: Sect. 2 shares the Federated learning working model. Section 3 covers the existing literature of Federated learning under various RHM application areas. Section 4 shares the open research challenges, Sect. 5 provides a discussion and, finally, Sect. 6 concludes the chapter with discussion.

2 Federated Learning Overview

There is an increasing trend on the usage of wearable devices especially in the area of personal health management and fitness [1, 2]. In addition to these niche rec Federated Learning (FL) is a framework designed for secure, distributed machine learning, enabling the creation of a collective model without necessitating the sharing of sensitive data among dispersed clients [52]. This approach facilitates the development of a comprehensive global model by combining data contributions from all participants, where each one updates their local models based on the collective insights. FL mirrors the efficacy of conventional centralized training approaches, where data from various sources are amalgamated on a single server for analysis, but it does so without compromising data privacy or integrity [53]. In FL, it is commonly accepted that all participants are equal stakeholders in establishing common data management policies, ensuring data privacy through non-direct data exchange.

Initially, the concept of FL involves N participants aiming to collaboratively develop a global model by leveraging their collective data, without revealing individual datasets. This method contrasts with traditional models that rely on aggregating all data for training, thereby preserving privacy. FL's objective is to minimize performance discrepancies between the federated global model and centralized models through iterative optimization and secure model updates. This process involves participants training local models on their data, encrypting model updates, and sharing these with a central server that aggregates these updates to enhance the global model continuously.

To formulate the concept described for Federated Learning (FL) mathematically, let's define the key components:

- N: The number of participants in the FL framework.
- $\{D_i\}_{i=1}^N$ The dataset of each participant, where i indexes the participants.
- M: The global model trained collaboratively through FL.
- Performance (M): A metric that assesses the efficacy of the collective model M.
- Δ: A non-negative metric that measures the difference in efficacy between the federated model and an ideal model trained on a centralized dataset encompassing all participants' data.

The aim within this framework is $\{D_i\}_{i=1}^N$ to construct the collective model M by utilizing the datasets, without necessitating direct data sharing among participants.

The efficacy of the collective model M is evaluated through the metric Performance (M), and the FL approach seeks to optimize this metric while ensuring the privacy of the data involved.

The discrepancy in efficacy, denoted by Δ, is conceptualized as the variance in performance between the federated model and a hypothetical centralized model. The specifics of Δ's calculation are dependent on the performance metrics employed and the structural intricacies of the FL setup. Without detailed information on how Δ is calculated, a general expression can be posited as follows:

$$\Delta = g(\text{Performance}(M), N, \{D_i\}) \quad (1)$$

In this expression, g denotes a theoretical function that calculates the performance discrepancy, considering the model's efficacy, the number of participants, and their datasets. The exact nature of g would be determined by the performance evaluation criteria and the architectural specifics of the federated learning model.

2.1 Federated Learning Categorization

Federated Learning (FL) is classified into three primary types based on the similarity of data features or the extent of user overlap among participants: Horizontal Federated Learning (HFL), Vertical Federated Learning (VFL), and Federated Transfer Learning (FTL). HFL is ideal when participants have common features across different users, focusing on enhancing model accuracy and generalization by integrating similar data characteristics. VFL applies when participant datasets share common users but differ significantly in data features, focusing on aligning and training on these overlapping user segments. FTL addresses scenarios where participant data varies greatly, facilitating model learning across distinct data domains with minimal shared features or labels, thus expanding FL's applicability while adhering to privacy and security standards.

2.1.1 Horizontal Federated Learning (HFL)

In horizontal federated learning (HFL), the convergence of data attributes across various clients' datasets is more substantial than the overlap of user entities. This means that while there may be many common attributes among the datasets, the users associated with these attributes vary [54]. HFL focuses on the attributes' dimensions, pooling together the similar attributes from different users for combined learning. The merging of these attributes from across participants leads to a larger pool of training samples, which in turn enhances the precision and the broader applicability of the trained model.

In HFL, the datasets across different clients share common features but differ in terms of the data samples. This is well-suited for situations where different clients collect information in the same feature space but for different users.

Given a set of clients C = {c1,c2,...,cN}, each with a dataset D_i, the global learning objective can be described as:

$$\min_{\theta} = \sum_{i=1}^{N} \frac{|D_i|}{D} L_i(\theta) \tag{2}$$

Here, θ represents the parameters of the global model, $|D_i|$ is the size of the dataset of client i, D is the total size of all datasets combined, and L_i is the loss function computed on client i's dataset.

2.1.2 Vertical Federated Learning (VFL)

In contrast, vertical federated learning (VFL) presents a different scenario where the datasets from each client share a higher number of common users than they do attributes [55]. The attributes among these datasets are rarely identical. Unlike HFL, which aligns data based on attribute dimensions, VFL aligns data based on the users, selecting the attributes that are common among users for joint learning. This process requires aligning the datasets of each participant to identify common user data before proceeding with the training on the aligned datasets.

Mathematical Formulation:

VFL is used when clients have different feature sets for the same set of users. This means that each client can contribute additional dimensions of data for the same individuals.

In VFL, let client C_i have a feature set F_i for the same set of users U. The combined feature set from all clients is $F = \bigcup_i = 1NF_i$.

The objective function in VFL typically tries to minimize:

$$\min_{\theta} \sum_{u \in U} L\left(\theta, \oplus_{i=1}^{N} x_{u,i}\right) \tag{3}$$

Here, θ again are the global model parameters, $x_{u,i}$ represents the feature vector of user u from client i, and ⊕ denotes a feature vector concatenation operation across different clients for the same user. L is the loss function over the combined feature space.

2.1.3 Federated Transfer Learning (FTL)

Federated transfer learning (FTL) emerges as a novel approach within the FL paradigm to accommodate the divergent nature of datasets that participants may hold in real-world scenarios—where the overlap in attributes or users between datasets is minimal or non-existent [56]. FTL aims to construct a robust and precise global model, even when dealing with a limited quantity of data that has little to no overlap in features or labels, all while adhering to stringent data privacy and security norms. FTL overcomes the limitations of HFL and VFL by enabling models to leverage similarities in data or model structures across different domains, allowing for the application of insights from one domain (source) to enhance predictions in another (target). Commonly, in such applications, labels from the source domain aid in predicting the label accuracy within the target domain.

Mathematical Formulation:

In FTL, let D_s be the source domain dataset and D_t be the target domain dataset. The objective is to transfer knowledge from the source to the target domain:

$$\min_{\theta_s,\theta_t} L_s(\theta_s, D_s) + L_t(\theta_t, D_t) + \lambda\Omega(\theta_s, \theta_t) \tag{4}$$

where L_s and L_t are the loss functions for the source and target domains, respectively, θ_s and θ_t are the parameters for the source and target domain models, and Ω represents a regularization term that measures the discrepancy between the domains, encouraging them to learn from each other. λ is a hyperparameter that balances the contribution of the regularization term.

It has to be noted that the mathematical formulas described above are abstract representations and would require specific definitions of the loss functions and the data structures used for a particular federated learning task. The exact algorithms and optimization techniques employed can vary greatly depending on the application and the data.

The taxonomy presented in the chapter for federated learning (FL) in remote health monitoring (RHM) is comprehensive, covering various dimensions:

Application Areas:

- **Critical Illness (e.g., heart attacks, strokes)**: FL is utilized for early diagnosis, disease progression detection, and proactive management, thus potentially reducing hospital usage and saving lives.
- **COVID-19**: FL is employed for predicting mortality, detecting lung abnormalities in CT scans, generating realistic images for privacy-enhanced detection, and diagnosing COVID-19 through clustered FL.
- **Generic Health Monitoring**: FL is applied to areas such as contactless monitoring, movement detection, and IoMT device data processing, enhancing model accuracy while preserving privacy.
- **Elderly Care**: FL is utilized for human activity recognition, enabling personalized care while maintaining data privacy and addressing latency issues.

FL Architectures:

- **Standard FL Frameworks**: Standard FL frameworks are proposed and evaluated by diverse researchers for specific health conditions (e.g., epileptic seizure detection, arterial blood pressure inference).
- **Customized FL Approaches**: Tailored FL approaches are developed to address specific challenges (e.g., arrhythmia classification, Alzheimer's disease detection).
- **Hierarchical FL**: Hierarchical FL architectures are proposed to optimize computation and communication, particularly for edge devices with limited resources.

Challenges and Solutions:

- **Privacy and Security**: FL addresses privacy concerns by training models locally on edge devices, utilizing differential privacy, and employing secure multi-party computation.
- **Communication Efficiency**: Strategies like federated averaging and hierarchical FL are employed to reduce communication costs and bandwidth consumption.
- **Systems Heterogeneity**: Approaches such as robust aggregation algorithms and model auditing are developed to handle variations in device capabilities and connectivity.
- **Model Corruption and Integrity**: Countermeasures against malicious attacks, such as data and model poisoning, Byzantine faults, and non-IID data distribution, are explored to ensure the integrity of the global model.

Open Research Challenges:

- **Model Corruption**: Addressing issues like data and model poisoning, Byzantine faults, and non-IID data distribution.
- **Increased Cost of Communication**: Finding efficient communication strategies to reduce bandwidth consumption and latency.
- **Systems Heterogeneity**: Developing techniques to synchronize and train models efficiently across diverse systems.
- **Security and Privacy**: Enhancing FL's resistance to security breaches and privacy concerns, particularly regarding model inference and reverse engineering.

Overall, the taxonomy delineates FL applications, architectures, challenges, and solutions across various domains of RHM, providing a comprehensive framework for understanding and advancing FL in healthcare.

3 Federated Learning Working Model

The use of wearable devices has been on the rise in recent years, particularly in personal healthcare, fitness, and activity monitoring [1, 2]. Alongside these specialized applications, the healthcare industry is increasingly adopting advanced technologies like artificial intelligence and machine learning for the Internet of Medical Things

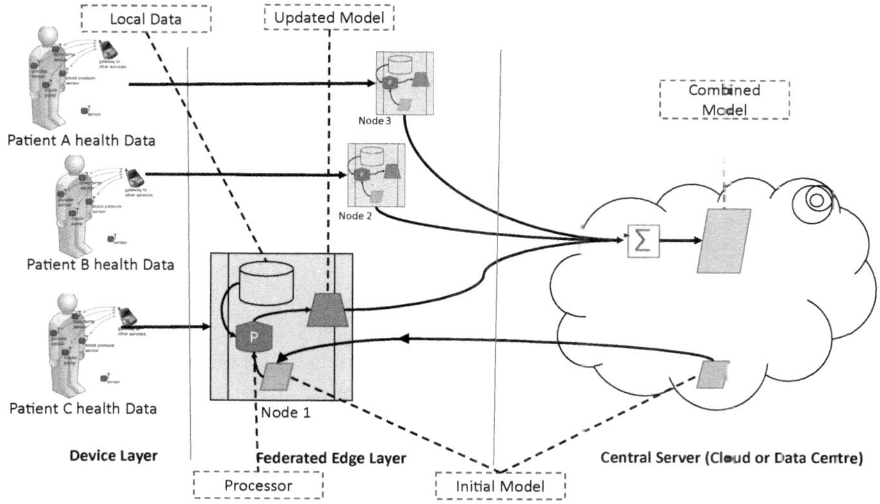

Fig. 1 Federated learning data flow

(IoMT), with the goal of revolutionizing traditional healthcare systems by making them more efficient, convenient, and personalized. Applying Federated Learning in this context represents a significant technological leap, potentially transforming the healthcare sector, especially in the domain of Remote Health Monitoring (RHM), which is closely tied to human well-being. On one hand, edge computing can overcome challenges related to latency and availability [37–39] while on the other hand, Federated learning can address privacy concerns by keeping patient data local and by enabling to create ML models that cannot tampered to trace data back to the original owner [44, 49, 50]. In addition, Federated learning aggregates learnings from multiple patient datasets to create a holistic and superior trained model.

A scalable Federated learning architecture comprises of three layers as shown in Fig. 1. The foundation of federated learning is quite simple. Every client that has data with them, including smartphones, hospitals, patients, and medical facilities, trains their unique models using their own data. They then send the model (not the data) to a centralized server, which combines them and distributes the updated combined model to each client for additional updating cycles.

3.1 Device Layer

Connected healthcare devices play a crucial role in the Internet of Medical Things (IoMT) by collecting data. Various sensors capture physical parameters from the real world, such as patient health data. These healthcare devices consist of both hardware and software, with many modern devices featuring onboard operating systems, processing capabilities, storage, and connectivity [57]. Examples of IoMT devices include blood pressure monitors, heart rate sensors, telemetry equipment, and activity trackers [4]. Additionally, devices like insulin pumps and infusion systems are widely used for remote wellness management and monitoring chronic diseases. Although many devices can store and process data, the vast amount of data generated by IoMT presents challenges.

3.2 Federated Edge Layer

The processing of data collected by the IoMT devices takes place at this layer. The local model can improve in performance by learning from multiple datasets, and at the same time ensuring that data of one entity is not shared with others, thereby preserving privacy.

As the first step a base machine learning model termed as the "initial model" is stored on the central server. The participating devices (e.g., Node 1, as shown in Fig. 1) contact the central sever to download a copy of the model and use local data to train the model. This helps to improve the model on the device using local data. After the model is trained, model's parameters are sent back to the central server. The other devices (e.g., Node 2 and Node 3, as shown in Fig. 1) can similarly pick up the global model and perform similar tasks of updating the model with their local data. The key advantage of this model is that the local data is not shared with the other devices, ensuring data privacy, and the model is also better through training on multiple data sets.

In summary based on the above steps, a global model is trained on data belonging to hundreds and thousands of patients, without them ever having to share data.

3.3 Centralized Server (Aggregator)

The aggregation of all the models that are trained individually at respective edge locations takes place at the central server. A central server takes the responsibility to coordinating the training of all edge devices (nodes) that are part of the training process. The central server creates a global model by receiving model parameters from individuals edge devices (nodes) and consolidating all these models. In addition, the central server also defines the architecture of the global model. Some implementations deploy central servers on the cloud, while others use local data centers to deploy these servers. The centralized server maintains and shares the updated model.

The information for each node remains private since only model parameters are being transferred between nodes. This is crucial for RHM use cases where protecting patient privacy comes first.

4 Review on Federated Learning for RHM

In recent years, RHM use cases have predominantly depended on cloud services for data storage, analysis, and processing [9–13]. However, the industry is now adopting a new data management approach that strategically integrates edge computing for data processing. For instance, it is more efficient to reduce the transmission of patient wearable device readings to the cloud by sending only processed results at specific intervals. Through Federated Learning, instead of transmitting data to central servers, a machine learning model is deployed directly on individual RHM devices to train on local data [58, 59]. The parameters from these locally trained models are then sent to a central ML model for global training. This approach helps protect patient privacy by minimizing the exposure of sensitive information to potential intruders or hackers. Furthermore, by keeping personal or sensitive patient data on-site at the edge, RHM systems can meet stringent data handling and privacy regulations such as HIPAA. In response to the growing use of digital technology, congress mandated federal privacy rules for individually identifiable health information, which are now part of HIPAA [60].

It is clear that Federated Learning will play a crucial role in the next phase of RHM innovation. However, the lack of understanding regarding the intersection between Federated Learning and edge computing may pose a challenge. Therefore, investing in both technologies is essential to advancing RHM use cases.

4.1 Federated Learning for Critical Illness

Critical illnesses, such as heart attacks, brain tumors and strokes are some of the deadly diseases with a high mortality rate [61]. Early diagnosis of these diseases results in a better survival rate for patients and federated learning is a technology that can help to achieve this vision [61]. RHM systems using Federated learning technology can detect disease progression and facilitate proactive management, possibly reducing expensive hospital usage, and in many cases saving lives of individuals. This section provides a review of Federated learning for critical illness and a summary is depicted in Table 1. Based on the review of these papers it is apparent that Federated learning has been used to address issues related to security and data privacy that are commonly found when cloud computing is used. Most of the papers have supported the argument that Federated learning can enhance privacy and security which they have supported with a working prototype.

Baghersalimi et al. [58] highlighted that sending raw patient data to centralize cloud servers poses significant privacy risks and consumes substantial energy. To address this issue, authors designed and evaluated a standard Federated Learning framework for epileptic seizure detection, utilizing a deep learning-based approach operated across multiple machines. This framework enabled each patient to train a personalized deep neural network (DNN) in collaboration with other patients,

Table 1 Comparative summary of federated learning for critical illness

References	Device used for data processing	Main challenges	Use cases	Prototype	Processing
[58]	Mobile devices	Privacy, energy consumption	Epileptic seizure detection	Yes	Data processing and storage happen at the edge, and the final trained model is aggregated on a cloud centralized server
[59]	Data collected from Smartwatch and processed on the local edge server	Cost and privacy	Ischemic heart disease	No	All processing and storage happen at the edge
[62]	Edge server	Security, privacy	Detect arrhythmia by monitoring single lead electrocard	Yes	Processing happens on the edge servers
[63]	Edge server		Cardiac activity monitoring	Yes	Processing happens on the edge servers. No mention of using cloud
[64]	IoT devices at home		Detection of Alzheimer's disease (AD)	Yes	Processing happens at the edge and models are centrally stored on the cloud
[65]	Smart bands/ mobile devices		Cardiac activity data to track stress levels	No	Learning happens on the edge/updated trained models are sent to the cloud for storage and retrieval

while also reducing the energy consumption associated with communication between local battery-powered mobile platforms and the server. The model's accuracy and performance were assessed, resulting in a sensitivity of 81.25%, a specificity of 82.00%, and a geometric mean of 81.62%. Authors also explored a customized Federated Learning approach, where each machine independently trained a DNN to identify unique electrocardiography (ECG) patterns associated with seizures, using local data. This approach demonstrated that the DNN could benefit from a well-trained model without the need to share raw patient data with a server or centralized cloud repository. The personalized Federated Learning method further improved key performance metrics, achieving a sensitivity of 90.24%, a specificity of 91.58%, and a geometric mean of 90.90%.

In another study, Brophy et al. [59] developed a machine learning framework designed to infer arterial blood pressure (ABP) from a single optical photoplethysmogram (PPG) sensor. The training process was distributed across multiple local models and data sources to simulate a large-scale collaborative learning experiment, which could be applied to low-cost wearable devices. The framework demonstrated a mean error of 2.95 mmHg and a standard deviation of 19.33 mmHg. Overall, this approach laid the groundwork for continuous ABP measurement on a large scale, offering a sustainable and private real-world application. The method also resulted in a more power-efficient learning framework, which supports the development of smaller, more affordable wearable devices without compromising the accuracy of ABP measurements or patient privacy.

Data privacy and security pose significant challenges in edge computing, particularly given the limited computational and communication capacities of wearable healthcare devices, which constrain the implementation of standard Federated Learning models [62]. To address this, Yuan et al. [62] proposed an advanced Federated Learning framework designed to train deep neural networks for arrhythmia detection by monitoring single-lead electrocardiograms. This approach involved segmenting the network and distributing it across multiple IoMT devices for Federated Learning, while part of the training was managed by a centralized server. By handling the majority of the computational workload on more powerful servers, rather than on the constrained edge devices, this model reduced accuracy loss while requiring just 0.2% of the synchronization traffic needed for traditional Federated Learning.

In a seperate study, Sakib et al. [63] explored Federated Learning architectures for arrhythmia classification, utilizing private electrocardiogram (ECG) data collected by smart sensors deployed on edge nodes. This architecture not only preserved data privacy but also facilitated online knowledge sharing through localized and distributed learning in a lightweight manner. The proposed Federated Learning framework for ECG analysis was further enhanced by asynchronously updating both shallow and deep model parameters in a custom Convolutional Neural Network (CNN)-based lightweight Artificial Intelligence (AI) model. This customization minimized valuable bandwidth and memory consumption, while still achieving significant accuracy in arrhythmia detection. The authors noted that as the number of edge nodes increased, the model led to reduced network overhead, including bandwidth, time, and memory usage.

Li et al. [64] introduced a privacy-preserving system called ADDetector, designed for Alzheimer's disease (AD) detection using IoT devices and advanced security mechanisms. The system captured audio data from users through IoT devices in a smart home environment and utilized innovative topic-based linguistic features to enhance detection accuracy. ADDetector employed a Federated Learning framework combined with differential privacy (DP) techniques to prevent the leakage of raw data and model details during transmission. The system achieved AD detection with robust privacy protection through its unique three-level Federated Learning architecture, consisting of the user's home, detection client, and data analysis cloud. When all privacy-preserving techniques—Federated Learning, DP, and cryptography-based aggregation—were applied, the model achieved an accuracy of 81.9% with a time overhead of only 0.7 s.

Wassan et al. [65] employed the authors employed Federated Learning to collaboratively and iteratively train a deep learning model. Each participant trained the model on their own confidential data and sent updates to the model based on local learning to a centralized cloud system for storage. This iterative process continued until the model was fully trained. Authors also presented a differential privacy Federated Learning method based on an adaptive Gradient Boosted Trees model (GBTM), which introduced noise to model parameters depending on the features and gradients of the training data. Using patient information, the GBTM model effectively identified medical fraud and demonstrated superior results in terms of data privacy.

4.2 Federated Learning for COVID-19

Utilizing machine learning for medical diagnostics has proven to be an effective and precise method for detecting COVID-19 infections. However, the sharing of patient health data between medical institutions is often restricted due to privacy concerns, leading to challenges in obtaining sufficiently large datasets needed for training machine learning models. Federated learning is a technology that can produces an unbiased global model from multiple data sets without exchanging clients' local data.

This section offers a literature review on the application of Federated Learning in COVID-19 research. A summary of the key findings is presented in Table 2. Creating Federated learning models for COVID-19 is a complex task, requiring multiple hospitals to collaborate with each other to create the Federated learning models [66, 67]. The increase in the number of participating hospitals has an associated improvement in the model learning.

Vaid et al. [66] assessed the effectiveness of applying Federated Learning to predict COVID-19 mortality rates using clinical data from Electronic Health Records (EHR). Data was gathered from five hospitals, each with diverse patient characteristics in terms of demographics, outcomes, sample sizes, and lab values. This study mirrored real-world conditions, where Federated Learning was used across hospitals with varied patient populations. The primary outcomes indicated that the federated Multilayer Perceptron (MLP) and Logistic Regression with L1 regularization (LASSO) models outperformed the locally trained models. The results suggested that Federated Learning can potentially mitigate the limitations associated with fragmented local models.

In another study by Dou et al. [67], Federated Learning strategies were explored to develop an AI model for detecting COVID-19 lung abnormalities in CT scans through a multinational validation study. Authors successful demonstrated that a federated deep learning model based on convolutional neural networks could accurately identify chest CT abnormalities in COVID-19 patients. Remarkably, the AI model trained on a specific dataset (from Hong Kong) performed well not only on internal tests but also on external independent datasets from hospitals in other regions, such as

Table 2 Comparative summary of federated learning for COVID-19

References	Device used for data processing	Main challenges	Prototype	Processing
[66]	Edge severs placed at hospital	Privacy	No	Models were trained with local data at hospitals and aggregated centrally on a server
[67]	Edge servers placed at hospital		Yes	Processing and storage happen on the edge servers
[68]	Edge servers		No	Models were trained on edge and aggregated centrally on a cloud central server
[69]	Edge servers	Security and privacy, latency	Yes	Models were trained on edge and aggregated centrally on a cloud central server
[70]	Mobile phones and fog computing	Computational complexity and patient mobility	No	Low computations are performed by mobile devices and high computations are performed on fog located near the patient's premises
[71]	Edge servers	Communication costs	No	Models were trained with local data at hospitals and aggregated centrally on a central server (no cloud)

Asia and Europe. The diversity of data in this multicenter study highlighted the feasibility of developing robust and generalizable AI tools for combating COVID-19 through image analysis in varied clinical environments. The model showed potential for real-time use, as it took approximately 40 ms to process a single CT volume.

A novel Federated Learning scheme, termed FedGAN, was proposed by Nguyen et al. [68] to generate realistic COVID-19 images and facilitate privacy-enhanced detection of the virus using generative adversarial networks (GANs) within an edge-cloud computing framework. This model was highly effective in generating realistic COVID-19 X-ray images, aiding in automatic COVID-19 detection. The proposed collaborative data augmentation approach involved alternating the training of GAN components (discriminator and generator) across different edge-based institutions, with trained parameters being updated to a cloud server without sharing actual image samples. Theoretical analysis and simulations revealed that this approach enhanced the performance of COVID-19 detection, delivering high accuracy and low processing time compared to other state-of-the-art methods.

To address challenges related to security, privacy, and service quality (such as low latency), clustered Federated Learning (CFL) was employed for the automatic diagnosis of COVID-19 by Qayyum et al. [69]. The primary goal of CFL was to create a unified model using data from multiple modalities. The proposed CFL-based collaborative learning framework efficiently processed visual data at the edge by training a multi-modal ML model capable of diagnosing COVID-19 using both X-ray and ultrasound imagery. Authors further evaluated evaluated the framework's performance using two benchmark datasets and various experimental setups. The specialized models, each focused on a specific image modality, were trained using centralized data, and promising results were obtained on both datasets, with improvements of 16 and 11% in overall F1-Scores on X-ray and ultrasound datasets, respectively, compared to the baseline model.

Real-time data processing is crucial for Remote Health Monitoring (RHM), which presents multiple challenges, including computational complexity and patient mobility [70]. Combining Federated Learning with fog and cloud computing, alongside machine learning (ML) and deep learning (DL), can address these challenges. Authors discussed combating the COVID-19 pandemic using a Federated Learning-based system. Authors proposed an architecture called Fog-based Smart Healthcare System (FSHS), which embedded all preprocessing tasks and ML/DL collaborative training models at the fog layer. Mobile phones carried by patients served as the fog layer, handling lightweight preprocessing tasks. For more complex processing functions (like DL models), a dedicated computing and storage node was placed near the COVID-19 patients. In emergencies, notifications and alerts were transmitted directly from the fog layer to healthcare providers. In routine scenarios, the collected data was sent to the knowledge and preprocessing layer for advanced analysis in the cloud computing nodes, enhancing healthcare services.

Image analysis (CT scan or X-Ray's) using Artificial Intelligence (AI) and ML has become the cornerstone to detect COVID-19 infections, but the individual patient reports or scans cannot be shared across medical institutions due to privacy issues, resulting in insufficient data sets for training [71]. Although Federated Learning can help address this issue, the model introduces significant communication costs for transferring updates and struggles to maintain performance when faced with severe data heterogeneity among clients. To overcome this, a dynamic fusion-based Federated Learning approach was proposed for CT scan image analysis to diagnose COVID-19 infections. This approach involved dynamically selecting participating clients based on the model fusion process, with aggregation scheduled according to each client's training time. The central server initiated the machine learning tasks and coordinated the Federated Learning process, while clients trained local models using their own data and computational resources. The results indicated that this approach outperformed standard Federated Learning in terms of model accuracy, fault tolerance, robustness, and communication efficiency.

4.3 Federated Learning for Generic Health Monitoring

This section discusses the use of Federated learning for most common RHM use cases in areas of health monitoring, contactless monitoring, and movement detection, with a summary provided in Table 3. Most of the papers highlight that IoMT devices have limited computations and storage capacity, due to which it is always recommended to use edge severs placed near the devices for data processing [72–77]. Most of the papers are also recommending using cloud for model aggregation and sending long term data for data preservation.

To allow the flexible deployment of IoMT services, a Federated learning driven IoMT (FLDIoMT) framework was developed by Fan et al. [77]. Additionally, a workflow for IoMT services was proposed, showcasing an efficient data processing and analysis scheme tailored for specific medical applications. A novel layered framework was designed to ensure the secure and dynamic deployment of IoMT services within a Federated Learning context. The FLDIoMT was tested on a sleep monitoring system known as iSmile, achieving a model accuracy of 74.19%, which was comparable to the accuracy of centralized models.

Hakak et al. [78] proposed an edge-assisted data analytics framework that utilized Federated Learning to train local machine learning models using user-generated data at the edge. The framework stored basic patient data and the trained models in the cloud. This conceptual framework employed pretrained models to generate personalized insights for users while maintaining data privacy. The proposed model could assist healthcare practitioners by providing data-driven insights for diagnosing or predicting diseases, analyzing mobility levels, and monitoring behaviors using data from wearable devices. Empowering and involving patients in disease prevention and monitoring is crucial for the sustainability of modern healthcare systems, and the proposed model could facilitate the effective integration of user-generated wellness and behavioral data on a large scale.

A Deep Federated Learning framework was developed for decentralized healthcare systems to preserve user privacy within a distributed architecture by Elayan et al. [79]. Similar to other models, this approach kept users' healthcare data on their IoT devices, training machine learning models locally and limiting data access to the owners. After completing local model training, a global model was collaboratively trained using the local models. A skin-disease detection system was created using images captured by the devices, and the results showed a higher Area Under the Curve (AUC) percentage after several rounds of Federated Learning. Additional classification metrics, such as accuracy, recall, and F1-Score, demonstrated strong performance during the Federated Learning rounds.

Wu et al. [80] introduced a model called FedHome, a cloud-edge-based Federated Learning framework for in-home health monitoring. This model utilized local data to train a global model, followed by personalized model learning through knowledge transfer. After training, the model was placed on the network edge to achieve data privacy by keeping user data locally. The challenges related to statistics and communication typically found in Federated learning, were managed by training the model as a generative convolutional autoencoder, effectively managing imbalanced and non-IID (non-independent and identically distributed) data. Experiments based on realistic human activity recognition data confirmed that FedHome outperformed selected methods.

Table 3 Comparative summary of federated learning papers for generic health monitoring

References	Device used for data processing	Main challenges	Use cases	Prototype	Processing
[77]	Edge servers	Security and privacy	Sleep monitoring system	Yes	Training happens on the edge and the model is sent to the central server at a datacenter
[78]	Edge servers placed at hospital		Smart fitness	No	Training happens on the edge devices. Patient basic data and trained ML models are finally stored on the cloud
[79]	IoMT devices/ phones		Skin diseases detection	Yes	Training happens on the edge and the model is sent to the cloud
[80]	In-home edge servers	Safety of data, performance and latency	Remote health monitoring	Yes	Edge computing is used for training and cloud is used for aggregation
[81]	Processing is done at the fog	Data latency, quality of service (QoS)	Health monitoring/ electrocardiogram (ECG)	Yes	Fog computing is used for processing and finally patient data is sent to cloud for storage

(continued)

Table 3 (continued)

References	Device used for data processing	Main challenges	Use cases	Prototype	Processing
[72]	Edge servers placed near to devices	Resource-constrained devise	Contactless monitoring	No	Initially training is performed by the devices. If it fails due to resource constraints by devices, then training is performed at the edge. Finally, model aggregation is performed on the cloud
[73]	Edge servers and cloud	High memory usage and high computational costs	Person movement identification	Yes	Edge servers send the parameters (metadata) to the cloud for training ML model
[82]	Edge cloudlets	Latency and high-performance overhead	Obstructive sleep apnea (OSA) disease and diabetics	Yes	Edge cloudlets run ML models locally without sharing patients data
[74]	Edge servers	Recognition accuracy and user privacy	Human activity recognition	Yes	Training happens on the edge and the model is sent to the cloud. No mention of Federated learning

(continued)

Table 3 (continued)

References	Device used for data processing	Main challenges	Use cases	Prototype	Processing
[75]	Edge servers at the hospitals	Inference attacks	Monitor ECG signals to determine risk of illness	Yes	Training happens on the edge and the model is sent to the server located at the medical centre (no cloud)
[76]	Edge servers	Data privacy, heterogeneity and data skewness	Stress management	Yes	Training happens on the edge and the model is sent to the central server (no cloud)

Federated learning can result in high communication costs, due to frequent transmissions of data between the Federated learning server and clients [81]. Clients were grouped using neural networks, with a cluster leader assigned to each group. Unlike traditional models, this approach performed Federated Learning training exclusively among the cluster leaders, after which transfer learning was employed by the cluster leader to update its members with the trained Federated Learning model. Each member then fine-tuned the received model using its own data. Results indicated that the proposed model could reduce communication costs by up to 98.45%, with less than a 3% accuracy loss compared to conventional Federated Learning models.

Gupta et al. [72] developed an edge-based Federated Learning framework named FedCare, designed for resource-constrained devices, using a camera-based IoMT as an example. Traditional methods may exclude resource-constrained devices in Federated rounds, potentially delaying the process, despite these devices holding critical information. For instance, a person's vital data may contain abnormalities that should be included in the central dataset but might be excluded due to resource limitations. Unlike these traditional models, FedCare provided a method to include resource-constrained devices in the learning process. A new phase in the Federated Learning algorithm, called split learning, was introduced, allowing stragglers to collaborate with the nearest edge node. The model was tested using heterogeneous computing devices that extract vital signs from videos, resulting in a reduction of 3.6 h in training time compared to traditional approaches. The FedCare system achieved 87.29% and

60.26% test accuracy for non-constrained and constrained devices, respectively, with a global accuracy of 90.32% at the server.

A Federated Learning-based person movement identification (FL-PMI) model was proposed to address the high memory and computational costs associated with training unlabeled data from wearable devices [73]. The FL-PMI model utilized deep reinforcement learning (DRL) for auto-labeling data, which was then trained using Federated Learning, with only model parameters sent to edge servers instead of the entire sensor data. The bidirectional long short-term memory (BiLSTM) integrated into FL-PMI classified the data for various processes in the smart healthcare system. The proposed FL-PMI achieved 99.67% accuracy, reduced computation costs, memory usage, and data transmission. However, FL-PMI was not tested from security perspective and authors have proposed using blockchain technology to enhance the user's privacy as part of future work.

Medical devices can generate enormous amount of data, which are traditionally processed and stored on centralized servers [82]. This approach often results in significant delays, high performance overhead, and privacy concerns due to the transmission of patient's personal health data to centralized servers, exposing the system to security threats and compliance issues. To overcome these challenges, Gupta et al. [82] proposed a Federated Learning-based Anomaly Detection (AD) model that utilized edge cloudlet servers to run AD models locally without sharing patient data. Authors introduced a hierarchical Federated Learning approach that enabled data aggregation at multiple levels, supporting multi-party collaboration. A novel disease-based grouping mechanism was developed to group different AD models based on specific disease types.

A Federated Learning-based system called HARFLS was developed for wearable sensor-based human activity recognition to achieve high accuracy while protecting user privacy and sensitive information [74]. The Federated Averaging algorithm (FedAvg) employed by HARFLS allowed users to securely and collaboratively manage their activity recognition tasks. FedAvg is a communication-efficient algorithm for distributed training involving a large number of clients [74]. It exchanged updated weights rather than gradients, enabling local nodes to make multiple batch updates on their own data. However, recognition accuracy largely depended on the system's feature extraction capability. To capture sufficient features from human activity recognition (HAR) data, the authors designed a perceptive extraction network (PEN) as the feature extractor for each user. PEN comprised both a feature network and a relation network. Experimental results showed that PEN achieved the highest results among all HAR algorithms and Federated Learning feature extractors using four widely recognized datasets. Furthermore, HARFLS balanced recognition accuracy and system capacity using WISDM and PAMAP2 datasets.

Federated learning can solve challenges related to privacy leakage of patients encountered in AI models, however, these models are not resistant to inference attacks [75]. To counter this, Wang et al. [75] proposed proposed a privacy-preserving Federated Learning model to generate a global disease diagnosis model through multi-party collaboration. Authors used a Variational AutoEncoder (VAE) to reconstruct medical data, adding differential privacy noise to resist inference attacks. This data was then used to train local disease diagnosis models based on Federated Learning, thus preserving patient privacy. Additionally, authors designed an incentive mechanism to encourage patients to provide medical data. Experiments conducted on the MIT-BIH arrhythmia database demonstrated that the proposed mechanism ensured high accuracy for heart disease diagnosis and low success rates for adversarial inference attacks. The accuracy of the global disease diagnosis model was influenced by the process of local model aggregation in Federated Learning.

Jiang et al. [76], developed a Federated learning model for stress management which is considered a critical illness. In contrast to studies proposing a centralized cloud-based approach with potential data privacy concerns, the authors proposed decentralized Federated learning. However, existing Federated Learning algorithms must contend with data heterogeneity and skewness, as varying participant data can significantly degrade overall model performance. To address this, a tailored, low-overhead clustered Federated Learning technique for stress-level recognition was developed. This method outperformed two advanced baseline algorithms, improving accuracy by over 7 and 12%.

4.4 Federated Learning for Elderly Care

In-home health monitoring has garnered significant attention globally, particularly for the aging population. The vast amount of health data collected by Internet of Medical Things (IoMT) devices, combined with recent advancements in machine learning, has led to numerous success stories in smart healthcare. This section discusses the Federated learning use cases for elderly care, with a comparative summary shown in Table 4. Most of the papers have discussed about human activity recognition as the major area for elderly care use. Due to the sensitivity involved in elderly care most of the papers have recommended using local edge servers rather than the devices for processing, to avoid latency issues [83, 84] that can lead to catastrophic situations during accidents.

Time-sensitive health applications can significantly benefit from pre-processing real-time health data closer to the source, as this enhances performance and responsiveness for IoHT (Internet of Health Things) based applications [83]. In the "Non-invasive monitoring and health assessment of the elderly in a smart environment (RO-SmartAgeing)" project, authors integrated IoHT with edge computing to assess the benefits. The RO-SmartAgeing project implemented a fog-based architecture aimed at improving services for seniors. A demonstration was provided using a data analytics engine that could scan multiple data sources and collect information from smart devices, showing superior performance compared to cloud-based approaches. Moreover, it was demonstrated that the edge architecture offered greater security than cloud-based services.

Personalized machine learning models often fall short in fine-tuning parameters to accurately predict outcomes in health monitoring systems. To address this issue,

Table 4 Comparative summary of federated learning for elderly care

References	Device used for data processing	Main challenges	Use cases	Prototype	Processing
[83]	Edge servers	Latency	Generic RHM	No	Training happens on the edge and the model is sent to the cloud. No mention of Federated learning
[84]	Local edge devices	High communication costs and latency	Fall detection	Yes	Training happens on the edge devices whereas aggregated ML models are stored on the cloud
[85]	Edge server	Privacy and security attacks	Human activity recognition	No	Edge servers are used to run the ML models with human intervention when required
[86]	User devices	Limits global model architecture (i.e., heterogeneity)	Human activity recognition	No	Training performed on user devices and aggregation happens on the cloud
[87]	Local edge servers near the beacons	Object readability and privacy	Human activity recognition	No	Training happens on the edge devices whereas aggregated ML models are stored on the central server

Khoa et al. [84] introduced the Fed xData framework for contextual health monitoring in cloud-edge networks. The Fed xData framework featured a continuous data balancing structure enhanced by the Random Oversample method, addressing all data classes. The model resolved various challenges, including increasing personalization and handling non-independent and identically distributed (non-IID) data in health monitoring. When tested with datasets based on human actions, the model demonstrated high accuracy and reliability. Comparisons between different general centralized learning models, Federated Learning models, and the FedHome model revealed that Fed xData achieved the highest and most effective results.

To ensure secure monitoring and privacy preservation of patient data, Alawadi et al. [85] proposed a Federated Interactive Learning IoT-based Health Monitoring Platform. This platform incorporates multi-expert feedback as a 'Human-in-the-loop' strategy within an active learning framework to enhance the machine learning models used by clients. Experiments showed that the platform offered robust privacy-preserving features and could effectively counter adversarial attacks during incremental learning. Data leakage issues were addressed by allowing the model to share knowledge from local nodes without transferring the actual data.

Traditional federated learning requires identical models to ensure successful training on both local edge devices and global servers [86]. However, this uniformity can restrict the architecture of the global model due to the lack of heterogeneity in local models. To overcome this, Shaik et al. [86] proposed FedStack, a heterogeneous stacked Federated Learning framework. This framework decentralizes the machine learning process by enabling each device or client to train a machine learning algorithm locally on their private data. After evaluation, only the trained model predictions are sent to the global model, thus decentralizing the monitoring process and preserving privacy. The global model then aggregates the predictions from various devices or clients to update the central machine learning model using heterogeneous data. The proposed framework was tested by limiting one sensor input at a time to determine the optimal sensor placement on the human body for activity recognition. This approach successfully addressed the limitations of traditional Federated Learning architectures, which often face challenges due to differences in local model architectures, and it avoided complications in the global model's compilation.

Chen [87] discussed commercially available homecare systems that offer remote monitoring functionality. However, these systems typically require someone to be physically present to monitor the individual. To address this, the authors developed a system using beacon technology that automatically activated prepositioned cameras and sent alerts to family members or caregivers when potentially high-risk activities were detected. For instance, if a person remained in the bathroom beyond a certain time threshold, an alert would be sent to family members. The system's security was further enhanced with Federated Learning to isolate the user's personal data for recognition model training, thereby preventing the disclosure of sensitive information without additional verification certificates. The information about the elderly was stored on a local server, safeguarding home data and securing data transmission. Experimental results indicated that the proposed approach outperformed other object recognition systems.

5 Open Research Challenges

Federated Learning is widely regarded as one of the most promising technologies for addressing privacy and security challenges associated with centralized cloud architectures [58, 59, 62–65]. Although this technology shows great potential, the review suggests that Federated Learning is still in its early stages. There are several significant research gaps, as illustrated in Fig. 2, that require extensive investigation, particularly for RHM use cases.

5.1 Model Corruption

Though on one side Federated learning benefits from aggregating the training performed by local devices to improve the quality and performance of the models,

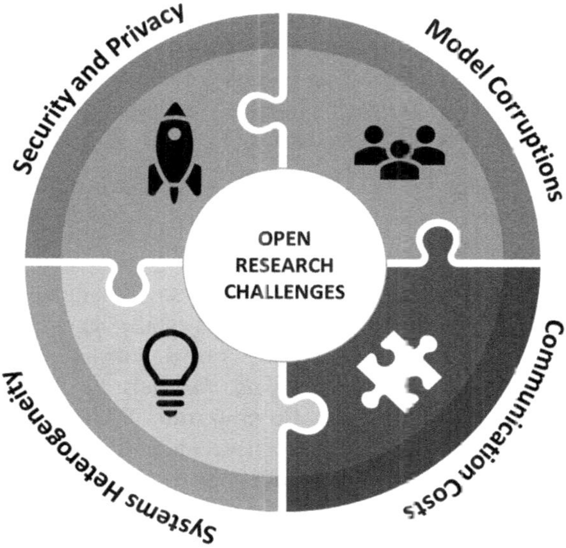

Fig. 2 Current research challenges

but on the other side it can still be susceptible to malicious attacks. Due to the nature of the technology where multiple devices participate in the learning process, even if few nodes in the network send poisonous model updates, or a malicious user contaminates the global model, the global model can be degraded. Due to the nature of the complex process involved in Federated learning where multiple devices send model updates, it can be difficult to trace which local device sent a poisonous update [88], with potential ramifications for the RHM use cases.

The primary threats to model integrity include [89]:

- **Data Poisoning Attacks**: Malicious participants can intentionally modify their local datasets or training processes to generate skewed model updates. When these updates are aggregated into the global model, they can introduce biases or vulnerabilities, effectively corrupting the model.
- **Model Poisoning Attacks**: Instead of manipulating the data, an adversary might directly alter the model updates they send to the server. This could be done in a way that the corrupted updates, once aggregated, significantly degrade the global model's performance, or introduce backdoors for exploitation.
- **Byzantine Faults**: These refer to situations where some nodes in the network behave arbitrarily, either due to software bugs, hardware failures, or external tampering, leading to erroneous updates being sent to the aggregator. Unlike malicious attacks, Byzantine faults are not necessarily intentional but can still corrupt the model [90].
- **Non-IID Data Distribution**: Even in the absence of malicious intent, the non-identically and independently distributed (non-IID) nature of data across devices can inadvertently lead to model corruption [91]. If the model updates from devices with skewed or unrepresentative data dominate the aggregation process, the global model may become biased or underperform on the overall data distribution.

- **Gradient Descent Vulnerabilities**: The process of aggregating gradients from multiple sources in FL can also be exploited to introduce corruption. An adversary could send significantly large gradients (gradient ascent rather than descent) to overpower the benign updates from other participants, steering the global model towards a suboptimal state.

To mitigate the risks of model corruption in FL, researchers and practitioners are developing and implementing several defensive strategies [89, 92, 93] which are still in early stages of implementation:

- **Robust Aggregation Algorithms**: Techniques such as Federated Averaging (FedAvg) can be enhanced with robust aggregation methods to detect and minimize the influence of outliers or significantly deviant updates, thereby reducing the impact of corrupted data or models.
- **Secure Multi-Party Computation (SMPC) and Homomorphic Encryption (HE)**: These cryptographic techniques can safeguard the aggregation process, ensuring that only valid updates contribute to the model while preventing individual updates from being disclosed to the aggregator or other participants.
- **Byzantine-Resilient Protocols**: These protocols are designed to be resilient against Byzantine faults by identifying and excluding updates from faulty or compromised nodes, ensuring that only trustworthy updates are aggregated.
- **Data and Model Auditing**: Implementing mechanisms to audit and verify the authenticity and integrity of the data and model updates before aggregation can prevent the inclusion of corrupted updates.
- **Differential Privacy**: Adding noise to the updates or the aggregation process can further obscure the contributions of individual nodes, making it harder for adversaries to target the model with corrupted updates successfully.

In summary, model corruption poses a significant challenge to the successful deployment of federated learning systems. Addressing this issue requires a multifaceted approach, combining advanced cryptographic techniques, robust aggregation methods, and continuous monitoring and auditing of model updates to ensure the integrity and performance of the global model.

5.2 Increased Cost of Communication

Federated learning mandates several devices to participate in the learning process and this can lead to massive communication overheads compared to a local centralized computation server. Communication within these networks can be far more costly than in conventional data center environments [44, 49–51]. To effectively fit a model to the data generated by devices in a federated network, it is crucial to develop communication-efficient methods that involve iteratively sending small messages or model updates during the training process, thereby optimizing the efficiency of model updates across the network.

Following are some of the key challenges:

- **Bandwidth Consumption**: Each training round in FL requires participating devices to upload their model updates and download the latest global model. Given the potentially large size of these updates, especially for deep learning models with millions of parameters, significant bandwidth is consumed. This is exacerbated in scenarios involving a large number of participants or when training complex models requiring frequent updates [44, 49].
- **Latency Issues**: The time taken to transmit updates from devices to the central server and back introduces latency [50]. This latency is influenced by the participants' network conditions, which can vary widely, especially in geographically distributed setups. High latency can slow down the training process, leading to longer times to convergence.
- **Synchronization Overhead**: Federated learning typically operates in synchronous rounds, where the server waits for updates from all or a subset of devices before proceeding to the next round. The need to synchronize these communications can introduce delays, particularly if some devices are slower due to poorer network conditions or computational capabilities [94].
- **Scalability Concerns**: As more devices participate in Federated Learning, the amount of communication increases, which can strain network resources and cause congestion, exacerbating bandwidth and latency issues.
- **Energy Consumption**: For battery-powered devices like smartphones or IoT devices, transmitting large updates over wireless networks can consume significant energy, posing a challenge.

To mitigate these increased communication costs, several strategies [50, 51, 95] are being explored within the industry, some which are listed below:

- **Model Compression**: Techniques such as quantization, pruning, and sparsification help reduce the size of model updates that need to be transmitted, conserving bandwidth and improving efficiency.
- **Communication Efficiency**: Methods like federated averaging (FedAvg) aim to decrease communication frequency by enabling more local computations before aggregation, balancing the trade-off between local computation and global communication.
- **Asynchronous Updates**: Allowing devices to send updates asynchronously can lower latency and synchronization overhead, making the learning process more resilient to varying network conditions.
- **Bandwidth Optimization**: Adaptive communication protocols that dynamically adjust the size and frequency of updates based on current network conditions can optimize bandwidth usage effectively.
- **Edge Computing**: Utilizing edge servers closer to data sources can reduce the distance that updates need to travel, decreasing latency and potentially improving bandwidth efficiency.

5.3 Systems Heterogeneity

Devices participating in federated learning differ in their storage, computational, and communication capabilities as they will have variations in hardware (central processing unit, random access memory), power (i.e., battery level) and network connectivity (e.g., some may use 5G while others may use Wi-Fi). Furthermore, the network size and system constraints typically mean that only a small fraction of devices are active at any given time. For instance, in a network of millions of devices, only a few hundred may be active simultaneously [96]. Some devices may also be unreliable, making it common for a device to drop out of model training during an iteration. These system-level characteristics introduce challenges, such as stragglers, which are more prevalent in Federated Learning scenarios than in traditional data center environments. Additionally, achieving reliable fault tolerance is challenging, as participating devices may disconnect before completing the training iteration [62, 63].

In other words, the variance in computational resources across devices introduces challenges in synchronizing and efficiently conducting the model training process. Devices with limited processing power or battery capacity may find it challenging to meet the computational demands of training complex models. This can lead to delays in model updates or, in some cases, cause these devices to drop out of the training process prematurely. Additionally, differences in network connectivity can result in varying data transmission rates, complicating the synchronization of model updates across the network.

Furthermore, the large scale of federated learning networks, which can involve millions of devices, often results in only a small fraction of these devices actively participating in model training at any given moment. For example, in a network of a million devices, it is common for only a few hundred to be actively contributing to model training during any specific iteration.

The phenomenon of stragglers—devices that lag in completing their assigned computations within the expected timeframe—is particularly evident in federated learning scenarios, especially those involving edge devices [97]. Stragglers can significantly slow down the overall training process, as the system may need to wait for these delayed updates before proceeding to the next training iteration. This issue is exacerbated in Federated learning contexts due to the wide variability in device capabilities and network conditions [97].

Achieving robust fault tolerance in such heterogeneous environments poses another significant challenge. Devices may unexpectedly drop out of the training process due to factors such as depleted battery life, network disconnections, or hardware failures. To maintain the effectiveness of the federated learning process despite device dropouts, it is crucial to implement sophisticated mechanisms that manage device participation and handle incomplete or missing updates without affecting model integrity or progress.

Addressing these challenges requires the development of advanced algorithms and protocols that can handle variations in device capabilities, optimize the use of computational and communication resources, and ensure the system's resilience to dropouts and other disruptions. This includes employing dynamic strategies for managing participants, using adaptive training methods that adjust to the capabilities of each device, and implementing robust fault tolerance measures to minimize the impact of slow or non-functional devices on the overall learning process.

5.4 Security and Privacy

Centralized architectures such as cloud are causing several concerns from privacy and security perspective and is one of the major reasons to see limited adoption of these technologies especially in healthcare [59, 62–67]. Due to the sensitivity of healthcare data, it is essential for decentralized healthcare systems to ensure the privacy and ownership of users' information. As a result, Federated Learning techniques are gaining popularity [79]. Several papers had discussed that Federated learning architectures are able to address privacy and security challenges by performing model training near to the devices and only sharing the model updates with others rather than the data itself. Although, there is merit in this architecture, however there is still a possibility that model parameters can be used to extract information about the local data. The trained models can be reverse engineered to understand the data that was used during training, compromising data privacy, and is termed as the poisoning attack. Thus, privacy preserving models needs to be created restricting the model manipulation to generate insights on private health data.

Three key technologies are essential for ensuring privacy in Federated Learning: Secure Multi-Party Computation (SMC), Differential Privacy (DP), and Homomorphic Encryption (HE). These technologies underpin the secure exchange of model parameters, preserving data privacy during FL.

Secure Multi-Party Computing (SMC):

SMC safeguards the input data of participants in cooperative tasks through encryption, ensuring data remains confidential throughout computation [98]. It enables privacy-preserving data processing, demonstrated by frameworks like Sharemind and applications in various fields including healthcare. In technical terms, SMC is defined as a cryptographic protocol that facilitates a group of parties, each holding private data, to collaboratively compute a specific function without disclosing their individual inputs to one another or to any external observer [98]. The essence of SMC lies in ensuring that, at the conclusion of the computation, no more information is disclosed to a party than what can be inferred from their own input and the output of the computation.

A Secure Multi-Party Computation (SMC) protocol is considered secure if it can replicate an ideal scenario in which a trusted third party (TTP) receives all inputs, performs the computation, and provides the results to the parties involved. The security of an SMC protocol is generally evaluated based on its ability to maintain privacy, correctness, independence of inputs, and fairness, even in the presence of adversaries. Semi-honest (or passive) adversaries adhere to the protocol but try to gain additional information, whereas malicious (or active) adversaries may deviate from the protocol to disrupt the computation or learn more about the inputs of honest participants.

SMC is widely applicable in areas that require privacy-preserving data analysis, such as secure voting systems, private set intersection, joint financial analysis, and collaborative machine learning. In these fields, parties aim to benefit from shared data without compromising the confidentiality of their individual datasets.

Despite its potential, SMC faces several challenges that can hinder its practical deployment, especially in federated learning (FL) and other collaborative computing environments [98, 99]:

- **Computational Overhead**—SMC protocols often involve complex cryptographic operations that can significantly increase computational overhead compared to non-private computations. This can lead to slower performance, especially for resource-constrained devices participating in federated learning.
- **Communication Cost**—The need for secure communication between parties in SMC can lead to increased communication costs. As parties exchange encrypted messages to perform computations, the amount of data transmitted can be substantially larger than in a non-secure computation, leading to higher bandwidth usage and potentially slower communication.
- **Scalability**: Scalability can be a challenge when extending Secure Multi-Party Computation (SMC) protocols to accommodate a large number of participants. As the number of parties increases, the complexity involved in managing secure communications and computations grows, potentially impacting performance and efficiency. This can make it difficult to apply SMC in scenarios with many participants, such as large-scale federated learning projects.
- **Integration with Existing Systems**—Integrating SMC into existing data processing and machine learning pipelines can be challenging due to the specialized cryptographic techniques SMC uses. Adapting SMC protocols to work efficiently with existing data formats, databases, and ML models requires additional effort and expertise.
- **Latency**: The computational and communication overhead introduced by SMC can result in increased latency in data processing and model training. This challenge is especially concerning for time-sensitive applications that demand real-time or near-real-time processing.
- **Usability and Accessibility**: Creating user-friendly interfaces and tools for SMC is crucial for widespread adoption. However, the complexity of SMC protocols can make them difficult for non-experts to understand and use effectively, which limits accessibility for many potential users.
- **Security vs. Efficiency Trade-off**: Balancing security and efficiency presents a significant challenge in SMC. High levels of security typically require complex cryptographic techniques, which can reduce system performance. Striking the right balance between security and performance is essential for practical applications.

- **Error Handling and Recovery**: In SMC, detecting and recovering from errors or failures in computation can be challenging due to the distributed nature of the computation and the encryption of data. Developing robust error handling and recovery mechanisms is essential for ensuring the reliability of SMC-based systems.

Differential Privacy (DP):

DP adds noise to data or computation results to prevent attackers from extracting individual data from aggregated information. It balances privacy with model performance, enabling FL systems to protect participant data against inference attacks [100].

From a technical perspective, DP is defined by the introduction of a privacy parameter, $\in$ (epsilon), which measures the level of privacy protection: a smaller $\in$ value corresponds to stronger privacy. A randomized algorithm A adheres to $\in$ -differential privacy if, for any two datasets D and D′ that differ by only one element, and for any possible outcome S in the output space of A, the following inequality is satisfied:

$$\mathbf{Pr}[\mathbf{A}(\mathbf{D}) \in \mathbf{S}] \leq \mathbf{e}\epsilon \times \mathbf{Pr}\left[\mathbf{A}\left(\mathbf{D}'\right) \in \mathbf{S}\right] \quad (5)$$

This condition ensures that the probability of a particular output does not significantly change with the modification of a single data point in the input dataset, thereby masking the contribution of individual participants.

Differential privacy achieves this protection by incorporating randomness into the data analysis process, often through mechanisms such as the addition of noise calibrated to the sensitivity of the query with respect to the input data. The choice of noise distribution (e.g., Laplace or Gaussian) and its parameters are critical in achieving the desired ϵ-DP level while balancing utility loss.

The (ϵ, δ)-differential privacy model builds on the basic differential privacy framework by permitting a small probability δ of greater deviations, which provides a more adaptable approach for practical applications where strict ϵ-differential privacy might be overly restrictive. Differential privacy has become essential in creating techniques for privacy-preserving data analysis, enabling a wide variety of applications, including statistical data release and federated learning. Its mathematical precision offers a reliable standard for ensuring privacy, making it a popular choice in both academic research and industry, where privacy is a critical concern.

Despite its strengths, differential privacy faces several challenges and gaps that need to be addressed:

- **Accuracy versus Privacy Trade-off**: A key challenge in differential privacy is balancing data accuracy and privacy. Adding noise to ensure privacy can reduce the accuracy of data or analysis outcomes. Finding the optimal balance where data remains useful while providing strong privacy protection is a critical issue.
- **Parameter Selection**: Choosing privacy parameters, such as the epsilon value in the epsilon-differential privacy model, is crucial and challenging. Choosing

inappropriate parameters can result in insufficient privacy protection or unnecessarily diminish data utility. Currently, there is a lack of standardized guidelines for setting these parameters across various contexts.

- **Complexity in Implementation**: Correctly implementing differential privacy requires a deep understanding of its theoretical foundations and careful attention to detail. Improper implementation may result in privacy breaches, which can hinder adoption, particularly for organizations that do not have specialized expertise in privacy-preserving technologies.
- **Data Heterogeneity**: The effectiveness of differential privacy can vary depending on the characteristics of the dataset. Ensuring consistent privacy protection while maintaining data utility across diverse data types and distributions in heterogeneous datasets is challenging.
- **Scalability**—Implementing DP in large-scale datasets or in Federated Learning environments, where data is distributed across numerous devices, can lead to scalability challenges. Efficiently managing the privacy budget in such settings, where multiple queries or analyses are performed, is non-trivial.
- **End-user Trust**—Building trust among end-users regarding the effectiveness and reliability of DP mechanisms is crucial. Users need assurances that their data is genuinely private, which requires transparent and understandable explanations of how DP works and its limitations.
- **Compliance and Regulation**—Aligning DP implementations with existing and emerging data protection regulations poses another challenge. To achieve widespread adoption, it is crucial that differential privacy methods comply with legal privacy standards while preserving data utility.
- **Long-term Privacy Guarantees**—Understanding the long-term implications of DP, especially in the context of advances in data analysis and machine learning techniques, is vital. There is a concern that future analytical methods could potentially undermine current DP protections.

Homomorphic Encryption (HE):

HE allows for operations on encrypted data, ensuring that computations on ciphertext yield the same results as if performed on plaintext. This encryption supports privacy-preserving computations in FL, facilitating secure model training and aggregation without exposing sensitive information [101].

Technically, HE is defined through the lens of algebraic structures. Let E be an encryption function that converts plaintext from space P to a ciphertext in space C, and D be a decryption function that converts ciphertext back to plaintext such that $D(E(p)) = p$ for all $p \in P$. For an operation $\oplus$ defined on P, Homomorphic Encryption (HE) permits a corresponding operation $\otimes$ on C such that for any p1, p2 $\in$ P:

$$\mathbf{D}(\mathbf{E}(\mathbf{p}1) \otimes \mathbf{E}(\mathbf{p}2)) = \mathbf{p}1 \oplus \mathbf{p}2 \tag{6}$$

An encryption scheme is considered homomorphic concerning the operation $\oplus$. If the scheme can handle an unlimited number of additions and multiplications, it is known as Fully Homomorphic Encryption (FHE). Conversely, if it supports only a limited number of operations or specific types of operations, it is referred to as Partially Homomorphic Encryption (PHE) or Somewhat Homomorphic Encryption (SHE).

The practical applications of HE are extensive, including secure cloud computing, privacy-preserving data analysis, and secure multi-party computation [98], among others. HE enables data to be encrypted and sent to a third party for processing, with the assurance that the third party cannot decipher the data but can still perform meaningful calculations on it.

The primary scientific challenge in HE is designing schemes that are both computationally efficient and capable of supporting a rich set of operations on ciphertexts. Advances in lattice-based cryptography and the development of optimized algorithms for arithmetic operations on encrypted data have significantly propelled the field, making HE a pivotal tool in the arsenal of privacy-preserving technologies.

Despite HE's potential to allow computations on encrypted data, several key challenges [101, 102] hinder its full-scale deployment:

- **Performance and Efficiency**—The computational intensity of HE operations significantly surpasses that of operations on plain data. This leads to efficiency issues, making HE impractical for real-time or resource-constrained environments without substantial optimization.
- **Complexity in Implementation**—Implementing HE schemes is inherently complex, requiring deep cryptographic knowledge. This complexity can hinder adoption, particularly for organizations that lack specialized expertise.
- **Scalability Concerns**—Scaling Homomorphic Encryption (HE) to handle an increasing number of participants or larger data volumes poses significant challenges.
- **Integration with Existing Systems**—Incorporating HE into established data processing and analytics pipelines demands considerable effort. The unique nature of HE operations necessitates adjustments in data handling practices, potentially disrupting existing workflows.
- **Balancing Security with Usability**—Achieving the highest levels of security through HE can sometimes compromise usability. The need for specialized knowledge to correctly implement and manage HE systems can limit its accessibility to a broader user base.
- **Latency Issues**—The additional computation required for HE operations introduces latency, which can be problematic for applications needing prompt data processing or real-time decision-making capabilities.
- **Adaptation to Diverse Use Cases**—Customizing HE solutions to fit various application scenarios requires an in-depth understanding of both the cryptographic aspects of HE and the specific domain needs, adding another layer of complexity to its adoption.
- **Error Handling and Data Recovery**—Detecting and correcting errors in computations on encrypted data is more difficult than in traditional computing environments. Creating effective strategies for error detection and correction is crucial to ensuring the reliability of systems based on HE.

6 Discussion

Federated learning is attracting significant interest in the healthcare sector due to its capability to protect data privacy. The ability of Federated learning to train a model across multiple decentralized edge devices, without exchanging data either with the central server, or other edge devices increases the efficiency of the model thereby making federated approach particularly suitable for sensitive domains, such as healthcare.

The Federated learning models have been widely used in the different healthcare use cases, and has demonstrated its importance even for human sensitive uses cases related to disease detection for critical illness, [58, 59, 62–65]. Several studies have explored the application of Federated Learning in electronic healthcare records [103, 104] medical imaging system [71, 103, 104] and drug discovery process [105] demonstrating advantages over traditional machine learning models. Numerous Federated Learning frameworks and models have been proposed, which can substantially reduce the computational load on IoMT devices and decrease the communication overhead between IoT devices and centralized servers. Additionally, these models address concerns related to security, privacy and latency [62, 70, 72, 73].

From a technical perspective, this study provides a detailed discussion of literature that extensively explores using Federated learning as a solution to overcome data privacy and security challenges for RHM use cases [59, 62–67, 74, 77–79, 85]. Other papers have discussed about solutions to address computation offloading over resource-limited edge servers [72, 84] and few others have addressed latency using federated learning models [80–84].

Overall, it seems that Federated learning is gaining a lot of traction, and there are promising solutions being developed to address the constraints related to security and privacy, latency and resource-limited edge servers across a magnitude of RHM use case. However, it has been observed that these solutions need further investigation and are not yet fully developed for adoption. On the other hand, as discussed in Sect. 4, there are still several areas that remains unexplored. Federated learning is prone to model corruption [89, 91] through malicious attacks such as data and model poisoning [89], Byzantine faults [90], and complications from non-IID data distribution IID, which can all undermine the performance and integrity of the global model. Additionally, Federated learning incurs high communication costs [44, 49, 81] due to the need for constant model updates among participating devices, leading to bandwidth consumption, latency [37] and scalability [38, 39] issues. The heterogeneity of systems, with variations in device capabilities and connectivity, further complicates synchronization and efficiency in model training. To mitigate these issues, strategies like robust aggregation algorithms, secure multi-party computation [98], homomorphic encryption, data and model auditing [82, 84], and differential privacy are being developed. Despite its potential to improve data privacy by localizing model training, Federated learning's susceptibility to security breaches [56] and the possibility of reverse engineering pose significant privacy concerns that need to be addressed with advanced protection technologies and innovative solutions.

Finally, the study has led us to the conclusion that Federated learning is still in the initial stages of implementation. The search for possible solutions and countermeasures to overcome these limitations is an open research challenge, which is fundamental to realize reliable, and accurate Federated learning based RHM applications.

7 Conclusion

Federated Learning represents a novel approach to training AI models by decentralizing the learning process. As an emerging distributed AI technique, it has garnered significant interest for its potential to enable privacy-enhancing and scalable RHM systems. This chapter examines the opportunities Federated Learning offers for RHM use cases through a comprehensive state-of-the-art review, providing an in-depth analysis of the various challenges linked to this technology. The motivation for this work stems from the absence of extensive reviews on the application of Federated Learning in RHM contexts.

This chapter reviews Federated Learning models for RHM, focusing on research in the RHM domain using this approach. It summarizes solutions that address challenges in security and privacy, latency, and resource-constrained devices through Federated Learning. However, several challenges remain unresolved, including poisoning attacks in Federated Learning, data reconstruction through reverse engineering of models, high communication costs, and system heterogeneity issues. These gaps can form the basis for future research directions.

References

1. Jawbone Inc: Jawbone fitness trackers. Jawbone Inc [Online]. Available: https://jawbone.com/up/trackers
2. Apple watch: Apple watch. Apple Inc [Online]. Available: https://www.apple.com/watch
3. Thakar, A.T., Pandya, S.: Survey of IoT enables healthcare devices. In: International Conference on Computing Methodologies and Communication (ICCMC), pp. 1087–1090 (2017)
4. Islam, S., Kwak, D., Kabir, M., Hossain, M., Kwak, K.: The internet of things for health care: a comprehensive survey. IEEE access **3**, 1–44 (2015)
5. Krishnan, B., Sai, S.S., Brata, M.S.: Real time internet application with distributed flow environment for medical IoT. In: International Conference on Green Computing and Internet of Things (ICGCIoT), pp. 832–837 (2015)
6. Sebastian, S., Jacob, N.R., Manmadhan, Y., Anand, V.R., Jayashree, M.J.: Remote patient monitoring system. Int. J. Distrib. Parallel Syst. (IJDPS) **3**(5) (2012)
7. Pantelopoulos, A., Bourbakis, N.G.: A survey on wearable sensor-based systems for health monitoring and prognosis. IEEE Trans. Syst. Man Cybern. Part C (Appl. Rev.) **40**(1), 1–12 (2010)
8. Paradiso, R., Loriga, G., Taccini, N.: A wearable health care system based on knitted integrated sensors. IEEE Trans. Inform. Technol. Biomed. **9**(3) (2005)

9. Navaz, A.N., et al.: Towards an efficient and Energy-Aware mobile big health data architecture. Comput. Methods Prog. Bio (2018)
10. Abdelmoneem, R.M., et al.: A cloud-fog based architecture for IoT applications dedicated to healthcare. IEEE ICC 1–6 (2019)
11. Giri, D., Obaidat, M.S., & Maitra, T.: Sechealth: Sechealth: An efficient fog based sender initiated secure data transmission of healthcare sensors for e-medical system. In: *GLOBECOM 2017–2017 IEEE Global Communications Conference*, pp. 1–6 (2017)
12. Min, M., Wan, X., Xiao, L., Chen, Y., Xia, M., Wu, D., & Dai, H.: Learning-based privacy-aware offloading for healthcare IoT with energy harvesting. IEEE Internet of Things J. **6**(3), 4307–4316 (2018)
13. Guo, H., Li, W., Nejad, M., Shen, C.C.: Access control for electronic health records with hybrid blockchain-edge architecture. In: IEEE International Conference on Blockchain (Blockchain) (2019)
14. Laplante, P.A., Laplante, N.: The internet of things in healthcare: potential applications and challenges. IT Prof. **18**(3), 2–4 (2016)
15. Alrubei, S.M., Ball, E.A., Rigelsford, J.M., Willis, C.A.: Latency and performance analyses of real-world wireless IoT-blockchain application. IEEE Sens. J. **20**(13), 7372–7383 (2020)
16. Mauro, C., Sandeep, K., Chhagan, L., Sushmita, R.: A survey on security and privacy issues of bitcoin. IEEE Commun. Surv. Tutor. **20**(4), 3416–3452 (2018)
17. Wadud, M.A.H., Bhuiyan, T.M.A.-U.-H., Uddin, M.A., Rahman, M.M.: A patient centric agent assisted private blockchain on hyperledger fabric for managing remote patient monitoring. In: International Conference on Electrical and Computer Engineering (ICECE), pp. 194–197 (2020)
18. Kazmi, H.S.Z., Nazeer, F., Mubarak, S., Hameed, S., Basharat, A., Javaid, N.: Trusted remote patient monitoring using blockchain-based smart contracts. In: The 14th International Conference on Broad-Band Wireless Computing, Communication and Applications (BWCCA) (2019)
19. Pongnumkul, S., Siripanpornchana, C., Thajchayapong, S.: Performance analysis of private blockchain platforms in varying workloads. International Conference on Computer Communication and Networks (ICCCN), pp. 1–6 (2017)
20. Qureshi, A., Dashti, W., Jahangeer, A., Zafar, A.: Security challenges over cloud environment from service provider prospective. Loud Comput. Data Sci. 12–20 (2020)
21. Jadeja, Y., Modi, K.: Cloud computing—concepts, architecture and challenges. In: International Conference on Computing, Electronics and Electrical Technologies, pp. 877–880 (2012)
22. Jolfaei, A.A., Aghili, S.F., Singelee, D.: A survey on blockchain-based IoMT systems: towards scalability. Inst. Electr. Electron. Eng. (IEEE) **9**, 148948–148975 (2021)
23. Singh, S., Jeong, Y.-S., Park, J.H.: A survey on cloud computing security: Issues, threats, and solutions. J. Netw. Comput. Appl. **77**, 200–222 (2016)
24. Shakil, K.A., Zareen, F.J., Alam, M., Jabin, S.: BAMHealthCloud: a biometric authentication and data management system for healthcare data in cloud. J. King Saud Univ. Comput. Inform. Sci. **33**(1), 57–64 (2020)
25. Moghadas, E., Rezazadeh, J., Farahbakhsh, R.: An IoT patient monitoring based on fog computing and data mining: cardiac arrhythmia usecase. Internet of Things **11** (2020)
26. Jenifer, A., Jeba, G., Paulraj, L., Kumar, N., Rozario, P., Amoli, R.: Edge-based heart disease prediction device using Internet of Things. In: International Conference on Applied Artificial Intelligence and Computing (ICAAIC), pp. 1500–1504 (2022)
27. Gia, T.N., Jiang, M., Rahmani, A.-M., Westerlund, T., Lilje, P.: Fog computing in healthcare internet of things: a case study on ECG feature extraction. In: IEEE International Conference on Computer and Information Technology; Ubiquitous Computing and Communications; Dependable, Autonomic and Secure Computing; Pervasive Intelligence and Computing, pp. 356–363 (2015)
28. Singh, M.B., Taunk, N., Mall, N.K., Pratap, A.: Criticality and utility-aware fog computing system for remote health monitoring. IEEE Trans. Serv. Comput. (2022)

29. Isa, I.S.B.M., El-Gorashi, T.E.H., Musa, M.O.I., Elmirghani, J.M.H.: Energy efficient fog-based healthcare monitoring infrastructure. IEEE Access **8**, 197828–197852 (2020)
30. Akrivopoulos, O., Chatzigiannakis, I., Tselios, C., Antoniou, A.: On the deployment of healthcare applications over fog computing infrastructure. In: IEEE 41st Annual Computer Software and Applications Conference (COMPSAC), pp. 288–293 (2017)
31. George, A., Dhanasekaran, H., Chittiappa, J., Challagundla, L., Nikkam, S., Abuzaghleh, O.: Internet of Things in health care using fog computing. In: IEEE Long Island Systems, Applications and Technology Conference (LISAT) (2018)
32. Dave, R., Seliya, N., Siddiqui, N.: The benefits of edge computing in healthcare, smart cities, and IoT. J. Comput. Sci. Appl. 23–34 (2021)
33. Dash, S., Biswas, S., Banerjee, D., Rahman, A.U.: Edge and fog computing in healthcare—a review. Scalable Comput. **20**(2) (2019)
34. Greco, L., Percannella, G., Ritrovato, P., Tortorella, F., Vento, M.: Trends in IoT based solutions for health care: moving AI to the edge. Pattern Recogn. Lett. **135**, 346–353 (2020)
35. Devarajan, M., Subramaniyaswamy, V., Vijayakumar, V., Ravi, L.: Fog-assisted personalized healthcare-support system for remote patients with diabetes. J. Ambient Intell. Human. Comput. 3747–3760 (2019)
36. Rajavel, R., Ravichandran, S.K., Harimoorthy, K., Nagappan, P., Gobichettipalayam, K.R.: IoT-based smart healthcare video surveillance system using edge computing. J. Ambient Intell. Human. Comput. **13**, 3195–3207 (2021)
37. Isa, I.S.M., Musa, M.O.I., El-Gorashi, T.E.H., Lawey, A.Q., Jaafar: Energy efficiency of fog computing health monitoring applications. In: 20th International Conference on Transparent Optical Networks (ICTON), pp. 1–5 (2018)
38. Hassen, H.B., Dghais, W., Hamdi, B.: An E-health system for monitoring elderly health based on Internet of Things and Fog computing. Health Inform. Sci. Syst. (2019)
39. Rahmani, A.M., Giac, T.N., Negash, B., Anzanpour, A., Azimi, I., Jiang, M., Liljebergc, P.: Exploiting smart e-Health gateways at the edge of healthcare Internet-of-Things: a fog computing approach. Futur. Gener. Comput. Syst. **78**, 641–658 (2018)
40. Debauche, O., Mahmoudi, S., Manneback, P., Assila, A.: Fog IoT for health: a new architecture for patients and elderly monitoring. Procedia Comput. Sci. **160**, 289–297 (2019)
41. Ning, Z., Dong, P., Wang, X., Hu, X., Guo, L., Hu, B., Guo, Y.: Mobile edge computing enabled 5G health monitoring for internet of medical things: a decentralized game theoretic approach. IEEE J. Sel. Areas Commun. **39**(2), 463–478 (2021)
42. Tuliab, S., Basumatarya, N., Gill, S.S., Kahania, M., Arya, R.C., Wander, G.S., Buyyaa, R.: HealthFog: an ensemble deep learning based smart healthcare system for automatic diagnosis of heart diseases in integrated IoT and fog computing environments. Future Gener. Comput. Syst. **104**, 187–200 (2020)
43. Wang, S., Tuor, T., Salonidis, T., Leung, K.K., Makaya, C.: Adaptive federated learning in resource constrained edge computing systems. IEEE J. Sel. Areas Commun. **37**(6), 1205–1221 (2019)
44. Bonawitz, K., Eichner, H., Grieskamp, W., Huba, D., Ingerman, A., Ivanov, V., Kiddon, C., Konečný, J., Mazzocchi, S., McMahan, H.B., Overveldt, T.V., Petrou, D., Ramage, D., Roselander, J.: Towards federated learning at scale: system design. Syst. Des. (2019)
45. Zhang, T., Gao, L., He, C., Zhang, M., Krishnamachari, B., Ave, A.S.: Federated learning for the internet of things: applications, challenges, and opportunities. IEEE Internet Things Mag. **5**(1) (2022)
46. Yahuza, M., Idris, M.Y.I.B., Wahab, A.W.B.A.: Systematic review on security and privacy requirements in edge computing: state of the art and future research opportunities. IEEE Access **8**, 76541–76567 (2020)
47. Mukherjee, M., Matam, R., Mavromoustakis, C.X., Jiang, H.: Intelligent edge computing: security and privacy challenges. IEEE Commun. Mag. **58**(9) (2020)
48. Ghimire, B., Rawat, D.B.: Recent advances on federated learning for cybersecurity and cybersecurity for federated learning for internet of things. IEEE Internet Things J. **9**(11), 8229–8249 (2022)

49. Konečný, J., McMahan, H.B., Yu, F.X., Richtárik, P., Suresh, A.T., Bacon, D.: Federated learning: strategies for improving communication efficiency. NIPS Workshop on Private Multi-Party Machine Learning (2017)
50. McMahan, H.B., Eider Moore, D.R., Hampson, S., Arcas, B.A.: Communication-efficient learning of deep networks from decentralized data. In: Proceedings of the 20th International Conference on Artificial Intelligence and Statistics (2017)
51. Yang, Q., Liu, Y., Chen, T., Tong, Y.: Federated machine learning: concept and applications. CM Trans. Intell. Syst. Technol. (2019)
52. Jiang, J., Hu, L., Hu, C., Liu, J., Wang, Z.: BACombo—bandwidth-aware decentralized federated learning. Electronics **9**, 440 (2020)
53. Wang, C., Liu, Z., Wei, H., Chen, L., Zhang, H.: Hybrid deep learning model for short-term wind speed forecasting based on time series decomposition and gated recurrent unit. Complex Syst. Model. Simul. **1**, 308–321 (2021)
54. Zhang, K., Song, X., Zhang, C.: Challenges and future directions of secure federated learning: a survey. Front. Comput. Sci. **16** (2022)
55. Feng, C., Liu, B., Yu, K., Goudos, S.K.: Blockchain-empowered decentralized horizontal federated learning for 5G-enabled UAVs. IEEE Trans. Indus. Inform. 3582–3592 (2022)
56. Dai, M., Xu, A., Huang, Q., Zhang, Z., Lin, X.: Vertical federated DNN training. Phys. Commun. **49** (2021)
57. Haghi, M., Thurow, K., Habil, D.I., Stoll, R., Habil, D.M.: Wearable devices in medical internet of things: scientific research and commercially available devices. Healthc. Inform. Res. (2017)
58. Baghersalimi, S., Teijeiro, T., Atienza, D., Aminifar, A.: Personalized real-time federated learning for epileptic seizure detection. IEEE J. Biomed. Health Inform. **26**(2) (2022)
59. Brophy, E., Vos, M.D., Boylan, G., Ward, T.: Estimation of continuous blood pressure from PPG via a federated learning approach. Sensors (2021)
60. Office for Civil Rights (OCR): Summary of the HIPAA privacy rule. Health Information Privacy (2022). [Online]. Available: https://www.hhs.gov/hipaa/for-professionals/privacy/laws-regulations/index.html
61. Naeem, A., Anees, T., Naqvi, R.A., Loh, W.-K.: A comprehensive analysis of recent deep and federated-learning-based methodologies for brain tumor diagnosis. J. Personalized Med. **12**(2), 275 (2022)
62. Yuan, B., Ge, S., Xing, W.: A federated learning framework for healthcare IoT devices. Distrib. Parallel Cluster Comput. (2020)
63. Sakib, S., Fouda, M.M., Fadlullah, Z.M., Abualsaud, K., Yaacoub, E.: Asynchronous federated learning-based ECG analysis for arrhythmia detection. In: IEEE International Mediterranean Conference on Communications and Networking (MeditCom), pp. 277–282 (2021)
64. Li, J., Meng, Y., Ma, L., Du, S., Zhu, H., Pei, Q.: A federated learning based privacy-preserving smart healthcare system. IEEE Trans. Indus. Inf. **18**(3), 2021–2031 (2022)
65. Wassan, S., Suhail, B., Mubeen, R., Raj, B., Agarwal, U., Khatri, E., Gopinathan, S., Dhiman, G.: Gradient boosting for health IoT federated learning. Inform. Technol. Healthc. Disaster Manage. (2022)
66. Vaid, K., Jaladanki, S.K., Xu, J., Teng, S., Kumar, A., Lee, S., Somani, S., Paranjpe, I., De Freitas, J.K., Wanyan, T., Johnson, K.W., Bicak, M., Klang, E., Kwon, Y.J.: Federated learning of electronic health records to improve mortality prediction in hospitalized patients with COVID-19: machine learning approach. JMIR Med. Inform. (2021)
67. Dou, Q., So, T.Y., Jiang, M., Liu, Q., Vardhanabhuti, V., Kaissis, G., Li, Z., Si, W., Lee, H.H.C., Yu, K., Feng, Z., Dong, L., Burian, E., Jungmann, F., Braren, R., Makowski, M., Kainz, B., Ruecker, D.: Federated deep learning for detecting COVID-19 lung abnormalities in CT: a privacy-preserving multinational validation study. NPJ Digit. Med. **4** (2021)
68. Nguyen, D.C., Ding, M., Pathirana, P.N., Seneviratne, A.: Federated learning for COVID-19 detection with generative adversarial networks in edge cloud computing. IEEE Internet Things J. **9**(12), 10257–10271 (2022)

69. Qayyum, A., Ahmad, K., Ahsan, M.A., Al-Fuqaha, A.: Collaborative federated learning for healthcare: multi-modal COVID-19 diagnosis at the edge. IEEE Open J. Comput. Soc. **3**, 172–184 (2022)
70. Fourati, L.C., Ayed, S.: Federated learning toward data preprocessing: COVID-19 context. In: IEEE International Conference on Communications Workshops (ICC Workshops), pp. 1–6 (2021)
71. Zhang, W., Zhou, T., Lu, Q., Wang, X., Zhu, C.: Dynamic-fusion-based federated learning for COVID-19 detection. IEEE Internet Things J. **8**(21), 15884–15891 (2021)
72. Gupta, A., Misra, S., Pathak, N., Das, D.: FedCare: federated learning for resource-constrained healthcare devices in IoMT system. EEE Trans. Comput. Soc. Syst. (2022)
73. Arikumar, K.S., Prathiba, S.B., Alazab, M., Gadekallu, T.R., Pandya, S., Khan, J.M., Moorthy, R.S.: FL-PMI: federated learning-based person movement identification through wearable devices in smart healthcare systems. In: Smart Healthcare Systems Based on the Internet of Things and Artificial Intelligence, p. 1377 (2022)
74. Xiao, Z., Xu, X., Xing, H., Song, F., Wang, X., Zhao, B.: A federated learning system with enhanced feature extraction for human activity recognition. Knowl.-Based Syst. **229** (2021)
75. Wang, X., Hu, J., Lin, H., Liu, W., Moon, H., Piran, M.J.: Federated learning-empowered disease diagnosis mechanism in the internet of medical things: from the privacy-preservation perspective. IEEE Trans. Indus. Inform. (2022)
76. Jiang, S., Firouzi, F., Chakrabarty, K.: Low-overhead clustered federated learning for personalized stress monitoring. TechRxiv (2022)
77. Fan, J., Wang, X., Guo, Y., Hu, X., Hu, B.: Federated learning driven secure internet of medical things. IEEE Wirel. Commun. **29**(2), 68–75 (2022)
78. Hakak, S., Ray, S., Khan, W.Z., Scheme, E.: A framework for edge-assisted healthcare data analytics using federated learning. In: IEEE International Conference on Big Data (Big Data), pp. 3423–3427 (2020)
79. Elayan, H., Aloqaily, M., Guizani, M.: Deep federated learning for IoT-based decentralized healthcare systems. In: International Wireless Communications and Mobile Computing (IWCMC), pp. 105–109 (2021)
80. Wu, Q., Chen, X., Zhou, Z., Zhang, J.: FedHome: cloud-edge based personalized federated learning for in-home health monitoring. IEEE Trans. Mob. Comput. (2020)
81. Chu, D., Jaafar, W., Yanikomeroglu, H.: On the design of communication-efficient federated learning for health monitoring. Distrib. Parallel Cluster Comput. (cs.DC) (2022)
82. Gupta, D., Kayode, O., Bhatt, S., Gupta, M., Tosun, A.S.: Hierarchical federated learning based anomaly detection using digital twins for smart healthcare. In: IEEE 7th International Conference on Collaboration and Internet Computing (CIC), pp. 16–25 (2021)
83. Ianculescu, M., Alexandru, A., Nicolau, N.-D., Neagu, G., Bica, O.: IoHT and edge computing, warrants of optimal responsiveness of monitoring applications for seniors. A case study. In: 22nd International Conference on Control Systems and Computer Science (CSCS), pp. 655–661 (2019)
84. Khoa, T.A., Nguyen, D.-V., Dao, M.-S., Zettsu, K.: Fed xData: a federated learning framework for enabling contextual health monitoring in a cloud-edge network. In: IEEE International Conference on Big Data (Big Data), pp. 4979–4988 (2021)
85. Alawadi, S., Kebande, V.R., Dong, Y., Bugeja, J., Persson, J.A., Olsson, C.M.: A federated interactive learning IoT-based health monitoring platform. Commun. Comput. Inform. Sci. **1450** (2021)
86. Shaik, T., Tao, X., Higgins, N., Gururajan, R., Li, Y., Zhou, X., Acharya, U.R.: FedStack: personalized activity monitoring using stacked federated learning. Knowl.-Based Syst. **257** (2022)
87. Chen, M.-Y.: Establishing a cybersecurity home monitoring system for the elderly. IEEE Trans. Indus. Inf. (2022)
88. Chan, P.P.K., He, Z.-M., Li, H., Hsu, C.-C.: Data sanitization against adversarial label contamination based on data complexity. Int. J. Mach. Learn Cybern. **6** (2020)

89. Yang, Y., Yang, H., Li, Y., Zhao, X.: A new clustering algorithm based on influence space. Expert Syst. Appl. **201**, 0957–4174 (2022)
90. Zhai, K., Ren, Q., Wang, J., Yan, C.: Byzantine-robust federated learning via credibility assessment on non-IID data. Math. Biosci. Eng. 1659–1676 (2022)
91. Wang, W., Zhang, M., Wu, Z., Chen, Q., Li, Y.: MDFD: study of distributed non-IID scenarios and Frechet distance-based evaluation. In: IEEE International Conference on Image Processing (ICIP), pp. 2300–2304 (2023)
92. Herath, C., Rahulamathavanv, Y.: Recursive Euclidean distance-based robust aggregation technique for federated learning. In: IEEE IAS Global Conference on Emerging Technologies (GlobConET) (2023)
93. Das, D.: Secure cloud computing algorithm using homomorphic encryption and multi-party computation. In: International Conference on Information Networking (ICOIN), pp. 391–396 (2021)
94. Yin, T., Li, L., Lin, W., Ma, D., Han, Z.: Grouped federated learning: a decentralized learning framework with low latency for heterogeneous devices. In: IEEE International Conference on Communications Workshops, pp. 55–60 (2022)
95. Imteaj, A., Thakker, U., Wang, S., Li, J., Amini, M.H.: A survey on federated learning for resource-constrained IoT devices. IEEE Internet Things J. **9**, 1–24 (2022)
96. Li, T., Sahu, A.K., Talwalkar, A., Smith, V.: Federated learning: challenges, methods, and future directions. IEEE Signal Process. Mag. **37**(3), 50–60 (2020)
97. Li, X., Qu, Z., Tang, B., Lu, Z.: Stragglers are not disasters: a hybrid federated learning framework with delayed gradients. In: IEEE International Conference on Machine Learning and Applications (ICMLA), pp. 727–732 (2022)
98. Hou, J., Hu, C., Yao, Z., Liu, S., Yuan, S., Guan, Z.: A verifiable secure multi-party computation scheme based on zero-knowledge proof for energy trading. In: International Conference on Intelligent Communication and Networking (ICN), pp. 145–149 (2023)
99. Sarhan, A.Y., Carr, S.: A highly-secure self-protection data scheme in clouds using active data bundles and agent-based secure multi-party computation. In: IEEE 4th International Conference on Cyber Security and Cloud Computing (CSCloud), pp. 228–236 (2017)
100. Han, L., Fan, D., Liu, J., Du, W.: Federated learning differential privacy preservation method based on differentiated noise addition. In: 8th International Conference on Cloud Computing and Big Data Analytics (ICCCBDA), pp. 285–289 (2023)
101. Mahmood, Z.H., Ibrahem, M.K.: New fully homomorphic encryption scheme based on multistage partial homomorphic encryption applied in cloud computing. In: 1st Annual International Conference on Information and Sciences (AiCIS), pp. 182–186 (2018)
102. Sendhil, R., Amuthan, A.: A descriptive study on homomorphic encryption schemes for enhancing security in fog computing. In: International Conference on Smart Electronics and Communication (ICOSEC), pp. 738–743 (2020)
103. Antunes, R.S., d. Costa, C.A., Arne Küderle, I.A.Y., Eskofier, B.: Federated learning for healthcare: systematic review and architecture proposal. ACM Trans. Intell. Syst. Technol. **13**(4), 1–23 (2022)
104. Kumar, Y., Singla, R.: Federated learning systems for healthcare: perspective and recent progress. Stud. Comput. Intell. **965**, 141–156 (2021)
105. Chen, S., Xue, D., Chuai, G., Yang, Q., Liu, Q.: FL-QSAR: a federated learning-based QSAR prototype for collaborative drug discovery. Bioinformatics **36**(22–23), 5492–5498 (2020)

Predicting Alzheimer's Disease: A Case Study Using Federated Learning Framework

Aman Sharma, Mohit Pal, Aditya Soni, and Rajni Mohana

Abstract Alzheimer's, as a neurodegenerative disease, has many challenges due to the complexity of this disease, with big data and the necessity of analyzing them fully to reach a precise prediction. The main challenges to conventional centralized approaches involve privacy and data silos, which impede collaboration and fully analyzing these very extensive, heterogeneous data sets. FL thus turns out to be a promising solution for model training on decentralized data while keeping the data privately owned. An FL framework has been developed expressly to predict AD progression with high accuracy, which is of prime importance in modern healthcare. This framework utilizes FL to facilitate collaboration among multiple healthcare institutions, each of which possesses local datasets of AD patients. Through a distributed learning paradigm, models are trained locally on the data of each institution without the need for data sharing. In its place, the model updates are centrally aggregated for privacy protection, yet benefiting from the wisdom accumulated across all participating institutions. The process of FL works by model training and aggregation through iterations for the constant improvement of the models without breach in the security of the data. This architecture fuses state-of-the-art federated learning with deep learning algorithms to learn complex patterns from the data. Experimental validation on real-world datasets shows how the proposed FL can make accurate and reliable predictions of AD progression. Moreover, all the analyses provided here show that privacy and confidentiality in the FL process are assured. Overall, FL

A. Sharma · M. Pal · A. Soni
Department of Computer Science Engineering, Jaypee University of Information Technology, Waknaghat, Himachal Pradesh, India
e-mail: aman.sharma@juitsolan.in

M. Pal
e-mail: 211484@juitsolan.in

A. Soni
e-mail: 211489@juitsolan.in

R. Mohana (✉)
Department of Computer Science, Amity School of Engineering and Technology, AUMP in Amity University Punjab, Mohali, Punjab, India
e-mail: rajnivimalpaul@gmail.com

R. Mohana et al. (eds.), *Federated Learning Applications in the Industrial Internet of Everything (IoE)*, Studies in Systems, Decision and Control 611,
https://doi.org/10.1007/978-3-031-99270-4_4

framework allows scalable, privacy-preserving AD prediction and, therefore, opens up ways toward collaborative research and personalized healthcare interventions against Alzheimer's Disease.

Keywords Federated learning · Alzheimer's · Data silos · Data integration · Privacy preservation · Local training · Neuroimaging

1 Introduction

Neurological disorders, especially Alzheimer's disease, present great challenges to healthcare structures worldwide. The emergence of mobile health technologies and wearable gadgets has enabled the gathering of various facts sorts, which include neuroimaging scans, genetic markers, and scientific critiques [1]. However, integrating those disparate information assets poses challenges because of privateness concerns and information heterogeneity. Additionally, medical datasets have a significant business value, making it less likely that they would be freely shared [2]. In order to implement ML models with multiple privacy-preserving methods, it is difficult to construct frameworks and infrastructural facilities [3]. In response, this chapter proposes an unconventional method that takes advantage of federated learning techniques for predicting neurodegenerative diseases, especially focusing towards Alzheimer's disease. By integrating information obtained from different sources but also preserving privacy of data through federated learning, the proposed approach aims at boosting prediction precision. This introduction paves way for exploring possibilities of the above stated innovative methods in transforming.

The current research indicates that the foundation is laid for examining the power of this innovative approach in transforming the expectation of neurological problems and improving the general outcomes for those people who experience outside effects. With a substantial number of global individuals, the occurrence of AD is expected to develop considerably because of an aging population, on one hand [4]. As a result, this is likely to create substantial challenges for both healthcare systems and the broader population. In addition, as a result of the increasing number of individuals who are developing the illness, it is critical to collaboratively recognize and correctly forecast it in a timely manner to enable practitioners to respond to those affected with the issues and administer the proper care that is tailored to their demands. On the other hand, innovative technologies are increasingly present in numerous areas of the economy, and, in recent times, systems and federations have emerged as conducive channels for anticipating and managing the AD.

Machine learning algorithms have demonstrated impressive abilities in examining intricate medical data and detecting patterns linked to the progression of Alzheimer's disease (AD). These algorithms can utilize a wide range of data sources, such as genetic information, brain imaging scans, cognitive assessments, and clinical records, to construct predictive models for assessing the risk of AD. However, traditional machine learning methods often find it difficult to ensure the confidentiality, integrity, and availability of data, especially when such as health information in medical centers.

Federated Learning (FL) [5], a method under privacy-preserving AI, allows for collaborative data utilization without sharing or revealing the data itself. It helps in overcoming these challenges by assisting in collaborative model training without bringing together private data.

Here, rather than consolidating all information into a centralized server, models are stored separately on single devices or servers where only model modifications (i.e., weights) are communicated to the central aggregator. While using the FL model, medical data/records remain with the individuals/patients and hospitals depending upon the type of application [6, 7] which ensures their privacy. This decentralized method upholds data privacy and confidentiality while enabling collaborative learning from various data origins. To predict Alzheimer's disease using federated learning techniques makes it possible for those involved in medical practice or academic research to take advantage of knowledge that comes from different sets but still adhering to regulations concerning confidentiality of client's information as well as morality.

1.1 Objectives of the Chapter

The objectives of the chapter are:

1. To develop a federated learning framework for predicting Alzheimer's disease;
2. To facilitate data privacy and confidentiality while using various data sources;
3. To integrate various information from different sources to improve the accuracy of our prediction;
4. And, to overcome the challenge of heterogeneity in medical data and privacy concerns.

1.2 Organization of the Chapter

The rest of the chapter is organized as: Sect. 2 considers the impact of Alzheimer's on patients. Section 3 delves into the use of federated learning approaches for predicting Alzheimer's disease. Section 4 explores practical applications in federated learning. Section 5 provides an in-depth analysis of advantages, implementation hurdles and healthcare uses of federated learning featuring subsections on benefits, challenges, and application. Existing federated learning methodologies for Alzheimer's prediction are recalled in Sect. 6. A case study will be presented in Sect. 7 with the dataset description, data visualization, preprocessing steps including model implementation such as training methods and evaluation procedures to be discussed under this chapter.

These include accuracy, precision, recall, F1-score and macro averages which are highlighted under Sect. 8 on evaluation metrics and their interpretation. Experimental results are analyzed under Sect. 9. And, finally, Sect. 10, concludes the chapter with future scope.

2 Alzheimer and Its Impact on Patients

Alzheimer's disease is a type of brain illness that is related to old age, and this illness comes with reduction in cognitive function, character change as well as memory loss. It is the most common cause of dementia making up more than 60% of cases [8, 9]. Alzheimer's disease is a form of brain damage usually seen in old people which leads to decreased intelligence, different personality and disorders in memories. In 2020, an estimated 5.8 million Americans aged 65 years or older had Alzheimer's disease and this number is projected to nearly triple to 14 million people by 2060 [10]. Alzheimer's disease mainly influences older individuals, normally going on after the age of sixty five, although early-onset cases can arise as young as the 30s or 40s.

In Alzheimer's disorder, the mind undergoes sluggish and irreversible adjustments, leading to the feature symptoms found in affected people. The key pathological hallmarks of the disease encompass the formation of plaques and tangles inside the brain. Plaques are clusters of a protein referred to as beta-amyloid, while tangles are shaped while any other protein referred to as tau accumulates and turns into tangled within neurons [11]. As those plaques and tangles gather, they disrupt the ordinary functioning of neurons, detrimental the communication among mind cells and lead to their eventual loss of life. The areas of the mind predominantly laid low by Alzheimer's disorder are the ones responsible for reminiscence, studying, and questioning.

While the exact causes of Alzheimer's disease are still not fully known, several risk factors have been identified. The greatest risk is the advanced age element in that the prevalence of Alzheimer's disease rises with advancing age [12]. Likewise, family history and genetics are significant factors that increase the risk of developing Alzheimer's disease. Genetic predispositions are critical, with mutations in genes such as APP (amyloid precursor protein), PSEN1 (presenilin 1), and PSEN2 (presenilin 2) being linked to early-onset familial Alzheimer's. Additionally, the APOE ε4 allele is a well-established genetic risk factor for late-onset Alzheimer's, influencing the metabolism and clearance of beta-amyloid in the brain. Beyond genetics, several modifiable risk factors are associated with Alzheimer's disease. Cardiovascular conditions, including hypertension and hypercholesterolemia, contribute to neurovascular damage and increased risk of cognitive decline. High blood pressure can lead to endothelial dysfunction and impaired cerebral blood flow, while elevated cholesterol levels can exacerbate amyloid plaque formation. Lifestyle factors such as smoking, obesity, and physical inactivity further compound the risk. Smoking induces oxidative stress and inflammation, while obesity and sedentary behavior are associated with insulin resistance and systemic inflammation, which are detrimental to brain health. Chronic illnesses like diabetes mellitus, particularly type 2 diabetes, are linked to increased amyloid deposition and vascular damage in the brain. Such things as head injuries from the past, including concussion and traumatic brain injury, can also raise the risk of Alzheimer's disease through promotion of neuroinflammation as well as worsening damage to neurons [13].

3 Use of Federated Learning Techniques in Prediction of Alzheimer Disease

Federated learning techniques in the prediction of Alzheimer's disease have been shown to allow for model training across decentralized data sources while maintaining privacy and data security [14]. This approach ensures that sensitive medical data remains localized and protected throughout the training process, addressing concerns about privacy and data sharing. Below is an in-depth explanation of how federated studying is applied in this context.

Federated gaining knowledge allows for version education across decentralized information resources, which includes cell gadgets or healthcare institutions. Each information supply keeps manipulation over its records, assuaging worries approximately privacy and information sharing. Initially, a fact is partitioned throughout specific resources, each containing records from numerous modalities inclusive of neuro-imaging scans, genetic markers, and scientific evaluations. This partitioning ensures that a sensitive medical fact stays localized. On every tool or data source, nearby fashions are skilled using the respective partition of facts. This training occurs independently, allowing every device to analyze from its neighborhood information even as maintaining privacy. After neighborhood schooling, the updated model parameters are aggregated to generate a worldwide model. This aggregation process normally involves techniques along with averaging or weighted aggregation to combine the knowledge learned from numerous sources effectively. The procedure of neighborhood education and model aggregation is iteratively repeated to refine the global version in addition. Each generation complements the model's predictive accuracy through leveraging insights from various facts assets. Federated learning guarantees privacy with the aid of keeping raw records localized and simplest sharing version updates [15]. Since statistics never leaves its source during education, sensitive clinical statistics stays blanketed all through the procedure. To validate the effectiveness of the federated mastering approach, cross-validation techniques are employed, in which subsets of information are held out for trying out functions. The worldwide model's performance is evaluated throughout those held-out datasets to make certain robustness and generalization. Federated getting to know lets in for continuous studying as new data turns into available. As extra statistics are accumulated from diverse sources, the global version may be updated iteratively to contain these new records, improving prediction accuracy over the years. By leveraging federated studying techniques, predictions of Alzheimer's ailment can enjoy the collective intelligence of numerous information sources whilst safeguarding character privacy, in the long run leading to extra accurate and reliable predictive fashions for early detection and intervention.

4 Practical Applications of Federated Learning Techniques

The improvement in the health sector has been a substantial one with Federated Learning, enabling different health institutions to operate together but still keeping private patient information in line with regulations such as the Health Insurance Portability and Accountability Act (HIPAA) and General Data Protection Regulation (GDPR) [16]. This technology has numerous functions in the field of medicine which include but not limited to; diagnosing diseases like cancer through deep learning algorithms developed by google (googleNet) also known as convolutional neural networks which have been applied successfully in several areas that require visual analysis of images or other sensory information by humans for instance diabetic retinopathy screening. The potential of federated learning to develop diagnostic models for illnesses from medical imaging data like X-rays, MRIs, and CT scans aggregated from different healthcare facilities is immense [17]. Distributed learning approaches would be used by drug manufacturers to speed up the process of developing medicines but also ensuring that specific chemical structures or any other research information do not leave their control as they train models using decentralized data sets. Figure 1 illustrates a general federated learning model.

Federated Learning has numerous programs throughout numerous domains. Here are some examples.

4.1 Healthcare

In healthcare, Federated Learning can be implemented to teach fashions for predicting and diagnosing diverse sicknesses, inclusive of neurological problems like Alzheimer's. By schooling models regionally on patient records, healthcare companies can ensure affected persons privateness even as they nevertheless cash in on a globally improved model that could be a useful resource in correct and early detection of sicknesses. By training locally on patient records, Federated Learning aids in developing models for disease prediction and diagnosis, including Alzheimer's disease, to ensure privacy while benefiting from global insights. It also utilizes aggregated medical imaging data (X-rays, MRIs, CT scans) to improve diagnostic capabilities without centralizing sensitive information. Figure 2 illustrates a hospital setting federated learning model.

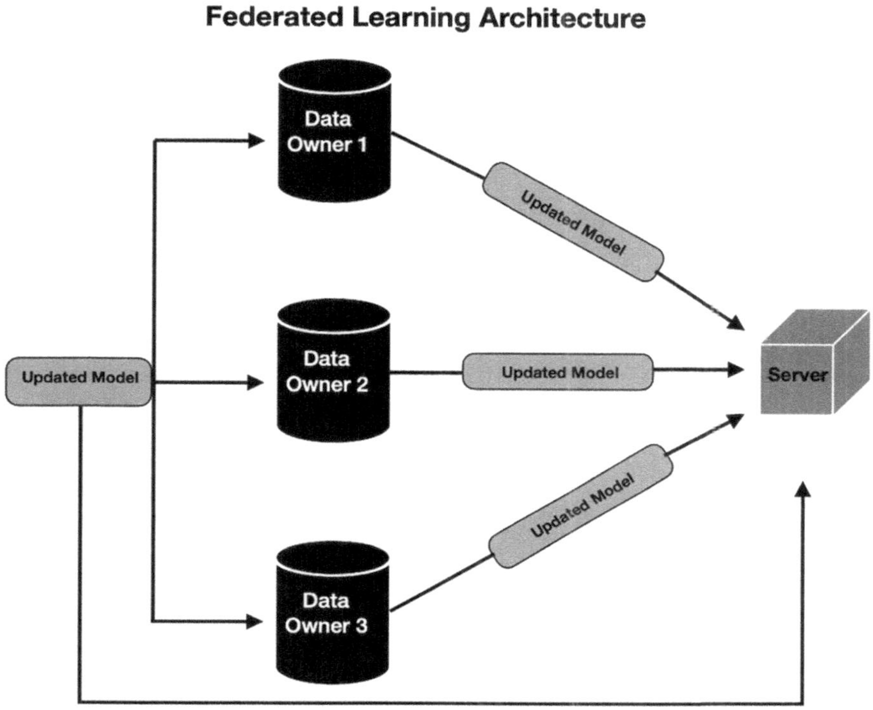

Fig. 1 General federated learning architecture

4.2 Financial Services

Federated learning is used in financial services to solve problems such as fraud detection, credit scoring and hazard estimation without sharing sensitive financial data. It enables multiple banks to train their models together for enhanced detection of fraud as well as making better decisions on credit. As such it allows banks to jointly train models without sharing sensitive financial data hence enhancing security and decision-making as well as improving fraud detection and credit scoring.

4.3 Internet of Things (IoT)

Federated learning allows learning models to be run on small quantities of data from appliances and sensors localized with the help of numerous IoTs. This can result in stepped forward forecasting, anomaly detection, and automation in numerous IoT applications, along with smart towns and industrial automation. Federated Learning is best suited for smart city infrastructure optimization and industrial automation

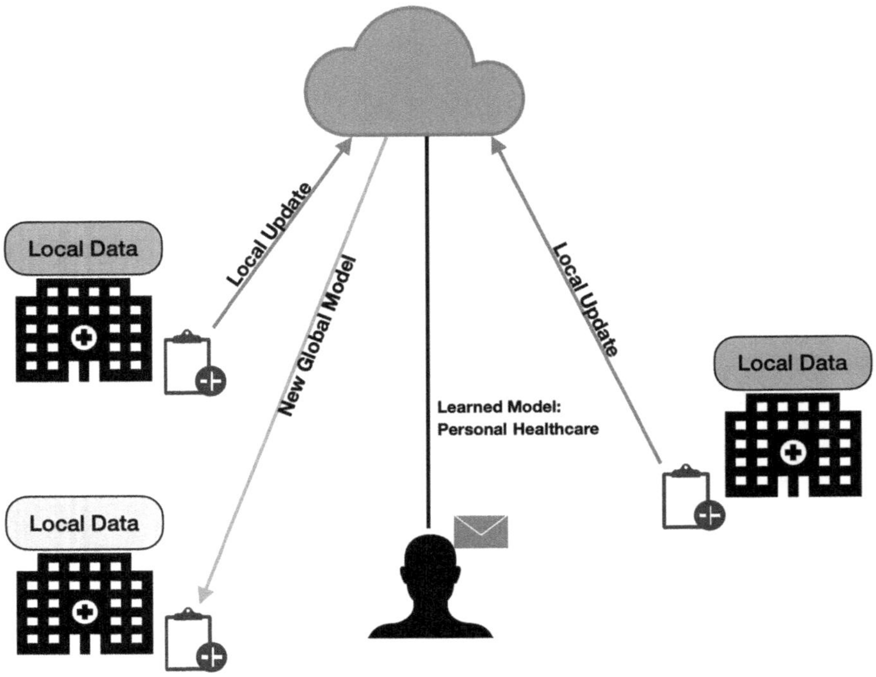

Fig. 2 Federated learning architecture at hospital setting

by training models on localized IoT data which improves forecasting and anomaly detection.

4.4 Natural Language Processing (NLP)

Federated Learning may be applied to teach models for herbal language processing duties, inclusive of speech reputation and language translation. By using facts dispersed across extraordinary gadgets and languages, fashions may be educated to perform with better accuracy and adaptability across more than one language and dialects. Through multiple devices and language sources it can leverage information that leads to greater accuracy and flexibility thereby improving speech recognition, language translation or any other NLP task.

4.5 Agriculture

Federated Learning is revolutionizing the agricultural sector by facilitating precision farming, forecasting crop yields, and detecting pests while safeguarding farmers' privacy and proprietary farm data. Its applications in agriculture are diverse. Farmers can optimize the efficiency of their inputs, including water, fertilizers, and pesticides, by leveraging real-time data from sensors, drones, and satellite imaging, thus maximizing crop yields while minimizing environmental impact. Integrating sensors' data with that from drones and climate sources supports precision farming as well as early crop disease and pest detection besides weather forecasting in agriculture thus ensuring maximum yields while minimizing environmental impact.

4.6 Latest Developments

The conventional Federated Learning (FL) seeks to safeguard private information by disclosing solely the model parameters away from the raw data. Nevertheless, recent investigations have uncovered loopholes which enable detractors to draw sensitive information from these parameters. Consequently, more resilient encryption and obfuscation approaches for reinforcing security in FL systems were explored. Among the achievements are the updated model improvement, data privacy measures have been enhanced, and the blockchain has been utilized for safe, participative training. Another important challenge is fairness among participants. In a federated environment, there might be different data quality and computing abilities among the participants leading to differences in model performance. Some of the recent works have aimed at creating fairer aggregation methods and used approaches such as meta-learning and multi-task approach to solve these problems.

5 Detailed Analysis of Federated Learning Approach Towards Alzheimer: Advantages, Implementation Challenges, and Healthcare Applications

Federated Learning (FL) offers potential advantages in the field of Alzheimer's disease research and healthcare, providing new solutions to long-existing issues of data privacy, security, and scalability [18]. Since the frequency of Alzheimer's disease has been on the rise, there are increased concerns to formulate efficient diagnostic and therapeutic means. This analysis describes Federated Learning in research on Alzheimer's disease, it examines the advantages as well as the subtle implementation challenges and diverse health care applications. The aim of this analysis is to provide insights that could help in future research undertakings and inflation of clinical practice concerning Alzheimer's disease management using FL.

5.1 Advantages of Federated Learning in Alzheimer's Disease Research

FL is a good new solution that healthcare data management can use to address key issues in Alzheimer's disease studies. It does this by providing a decentralized way of training models that ensures privacy for the patient through making sure such important medical information stays where it belongs either on someone's personal gadgets or company's computers. This localization not only minimizes the risk of data breaches and unauthorized access to data storages but also ensures compliance with strict regulatory standards like HIPAA and GDPR. Furthermore, FL's cooperative structure allows for smooth model training across various academic centres and research centres, thus improving expandability and enabling the use of bigger and more varied datasets. Using this capacity, FL helps to create tailored diagnostic and prognostic models that are specific to individual patient attributes and progression patterns of the disease, thus moving forward the field of personalized medicine in managing Alzheimer's disease. FL has many benefits. For instance, it may increase in size or expand itself in order to accommodate a larger number of machines or devices thus enhancing the robustness and generalization of the model [19]. Additionally, in the areas such as gaming and robotics alongside deep reinforcement learning, federated learning has the potential to tackle scalability issues and privacy concerns associated with distributed data [20].

5.2 Implementation Challenges of Federated Learning in Alzheimer's Research

Federated Learning (FL) in healthcare settings is challenging especially for Alzheimer's research. One of the main issues is data heterogeneity, where various healthcare providers and research institutes have to do a lot of work to put together different sources appropriately. Variation in the form of data, quality, and patients can impact training and generalization of models hence there is a need for strong data preprocessing methods. Federated learning requires a central server to communicate often with client devices to come together with model updates. In cases where network infrastructure is poor or uncertain, this can use up too much data. The model parameters are updated iteratively by FL. At every iteration, there are two main steps: To update model parameters with site-specific data locally and to transmit these local model parameter updates to a central server for aggregation into a new set of model parameters [21] that requires constant communication between the master station and the participating devices, which highly adds to the communication overhead especially when the resources are limited.

To tackle these problems, one must focus on optimizing protocols facilitating communication between different devices as well as efficiently utilizing available bandwidth. Moreover, due to the way in which data is distributed globally and various devices provide updates at different time intervals, the integration of these models presents some difficulties. For precise and dependable model results, it is very necessary for convergence algorithms and coordination strategies to be adjusted accordingly. In federated learning, ensuring robustness and fault tolerance is a considerable hurdle. Various device malfunctions, interrupted connections, security attacks, as well as malicious actions can lead to undesirable training processes and final model quality [22]. Device breakdowns, interrupted connections or malicious actions may negatively affect the training process and quality of the final model. Additionally, ensuring regulatory compliance on Federated Learning approaches are essential in order that the model can comply meticulously with the data privacy and security regulations while still maintaining performance and utility. It is important to overcome these obstacles before we can realize total application of FL technique in transforming Alzheimer's research and health provision.

5.3 Healthcare Applications of Federated Learning

FL models can accurately identify persons with increased chances to be infected with Alzheimer's disease based on data streams integration [23]. To speed up discovering new medications and developing treatments datasets from pharma developers and research institutes are employed. Utilizing different datasets from pharmaceutical companies and research institutions makes searching for new drugs as well as creating therapies. Collaborative model training enables identification of possible drug targets and evaluation of treatment efficacy on different patient's parts, hence advancing Alzheimer's treatment frontier for good. In healthcare, FL models help in the diagnosing of Alzheimer's disease and planning individualized patient care which is essential for healthcare. This is because FL models help clinicians by providing access to patient data during decision-making, thereby facilitating informed treatment decisions hence, improving patients' treatment outcome [24]. FL's primary goal is to facilitate participant collaboration so they can create a better model than they could alone [25]. Moreover, FL supports conducting a variety of studies at the level of the population with regards to looking after the frequency of Alzheimer's disease, its growth and treatment results that will eventually help inform policies and strategies for population care effectively. FL has multiple healthcare applications that will most likely bring about major progress in Alzheimer's disease research and patient care also signaling a fresh age in personalized data-based healthcare distribution.

6 Federated Learning Existing Techniques/Methodology for Alzheimer Disease Prediction

In this section, we have reviewed several researches and observed the techniques used by the researchers using similar dataset to get an accurate federated learning model for neurological disorder prediction.

Lei et al. [26] introduced a novel hybrid federated learning framework for training neural networks with unlabeled data, achieving accuracy levels between 72.8 and 91.2% on datasets obtained from various hospitals, including ADNI-1, ADNI-2, ADNI-3, AIBL, and MCAD. Ouyang et al. [27] conducted a study on the implementation of multi-modal federated learning systems for Alzheimer's monitoring, utilizing data from ADNI-1, ADNI-2, and ADNI-3. By combining hybrid federated learning and structural MRI imaging, their approach has achieved accuracy rates ranging from 87.5 to 95.6%. Qian et al. [28] proposed a federated deep learning model specifically tailored for brain MRI scanning, utilizing sMRI imaging and hybrid federated learning techniques. The proposed method demonstrated accuracy levels ranging from 72.4 to 91.6% on ADNI datasets. Mandawkar et al. [29] introduced a federated learning environment for Alzheimer's detection using MRI images and a flamingo-based deep CNN classifier. The approach showed accuracy rates ranging from 76.5 to 91.0% on datasets from ADNI-1, ADNI-2, and ADNI-3. Stripelis et al. [30] proposed federated deep learning for the detection of Alzheimer's disease in multi-cohort brain MRI data, achieving impressive accuracy levels between 96 and 98% on ADNI-1, ADNI-2/GO, and ADNI-3 datasets using a 3D-CNN within a federated learning setting. Mitrovska et al. [31] focused on secure federated learning for Alzheimer's disease detection using T1-weighted MPRAGE scans with 3 T scanners. By employing FedAvg and SecAgg algorithms, authors achieved accuracy rates ranging from 76 to 91% on ADNI datasets. Castro et al. [32] proposed a federated learning system incorporating biometric medical image authentication for Alzheimer's diagnosis using datasets from OASIS, ADNI-1, ADNI-2, and ADNI-3. Their CNN-based approach demonstrated recall, F1-score, and accuracy levels ranging from 86.32 to 91.94%. Yao et al. [33] conducted research on utilizing artificial intelligence for diagnosing Alzheimer's disease through the analysis of brain MRI images obtained from PET and T1-weighted MRI scans. The CNN-based method successfully achieved accurate diagnoses, however, detailed performance metrics were not disclosed (Table 1).

7 Case Study/Implementation

This chapter provides essential details about the dataset utilized in the research, enhancing comprehension of the study's findings. The selection and analysis of the dataset were conducted meticulously. Additionally, the chapter introduces the

Table 1 Comparison of existing federated learning techniques

S. no	Author(s)	Goal	Dataset	Performance parameter	Results
1	Lei et al. [26]	Hybrid FL framework to train networks via unlabeled data	ADNI-1, ADNI-2, ADNI-3, AIBL & MCAD (from 4 different hospitals)	BANet, HFL, sMRI images scanning	72.8–91.2%
2	Ouyang et al. [27]	Deployment of multi-modal federated learning systems for Alzheimer's monitoring	ADNI-1, ADNI-2 & ADNI-3	HFL, sMRI	87.5–95.6%
3	Qian et al. [28]	Federated deep learning model for brain MRI scanning	ADNI-1, ADNI-2, ADNI-3	Structural Magnetic Resonance Images (sMRI), HFL	72.4–91.6
4	Mandawkar et al. [29]	Federated learning environment for Alzheimer's detection	ADNI-1, ADNI-2, ADNI-3	MRI Images, flamingo based deep CNN classifier	76.5–91.0%
5	Stripelis et al. [30]	Federated deep learning for detecting AD in multi-cohort brain MRI	ADNI 1, ADNI 2/GO, ADNI 3	T-1 weighted brain MRI Data, 3D-CNN in federated learning environment	96–98%
6	Mitrovska et al. [31]	Secure federated learning for Alzheimer's disease detection	ADNI-1, ADNI-2, ADNI-3	T1-weighted MPRAGE scans with 3 T scanners using FedAvg & SecAgg	76–91%
7	Castro et al. [32]	A federated learning system with biometric medical image authentication for Alzheimer's diagnosis	OASIS, ADNI-1, ADNI-2, ADNI-3	RGB MRI image, Recall, F1-score, T1-weighted MRI scans using CNN	86.32–91.94%
8	Yao et al. [33]	Artificial intelligence based diagnosis of Alzheimer's disease with brain MRI images	ADNI 1, ADNI 2, ADNI 3	PET, T1-weighted MRI scanned images using CNN	Not specified

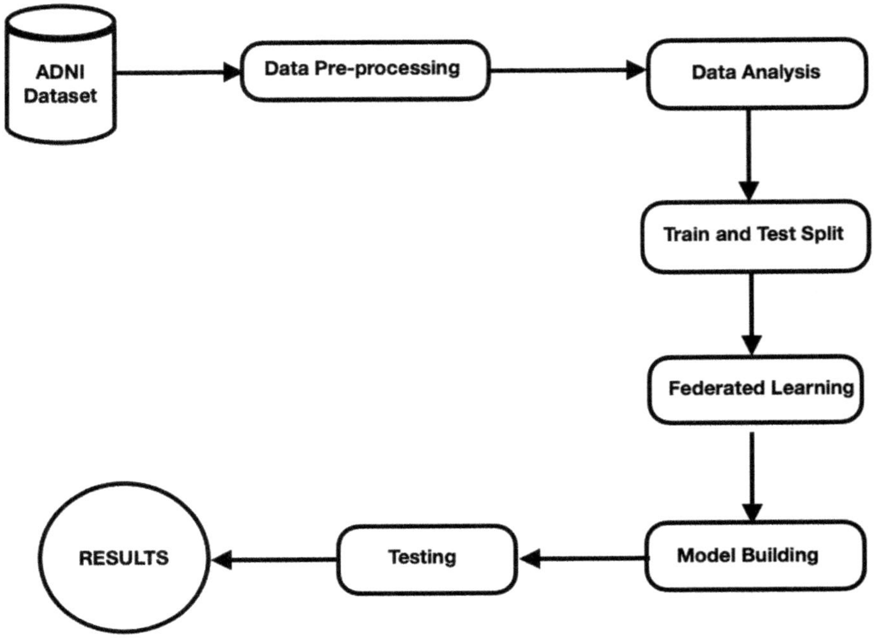

Fig. 3 The architecture of proposed model

proposed model, which is a product of thorough research and analysis. Furthermore, Figure 3 illustrates the architecture of the proposed federated learning model.

7.1 Dataset Description

The Alzheimer's Dataset used in this research has been taken from [34] which is collected from Alzheimer's Disease Neuroimaging Initiative (ADNI) database. The project utilizes a dataset consisting of 36,984 MRI images that are divided into four distinct categories: Alzheimer's Disease (AD), Cognitively Normal (CN), Early Mild Cognitive Impairment (EMCI), and Late Mild Cognitive Impairment (LMCI). Table 2 provides a summary of the number of images of each class in the dataset. This diverse collection of images provides a comprehensive overview of different cognitive conditions, which can be beneficial for the creation and assessment of machine learning models designed to diagnose and comprehend neurodegenerative diseases such as Alzheimer's disease. Figure 4 provides a bar plot for visualization of the number of images of each class in the dataset.

Table 2 Number of samples in the ADNI dataset

Class	No. of samples
AD	8960
CN	6464
EMCI	9600
LMCI	8960

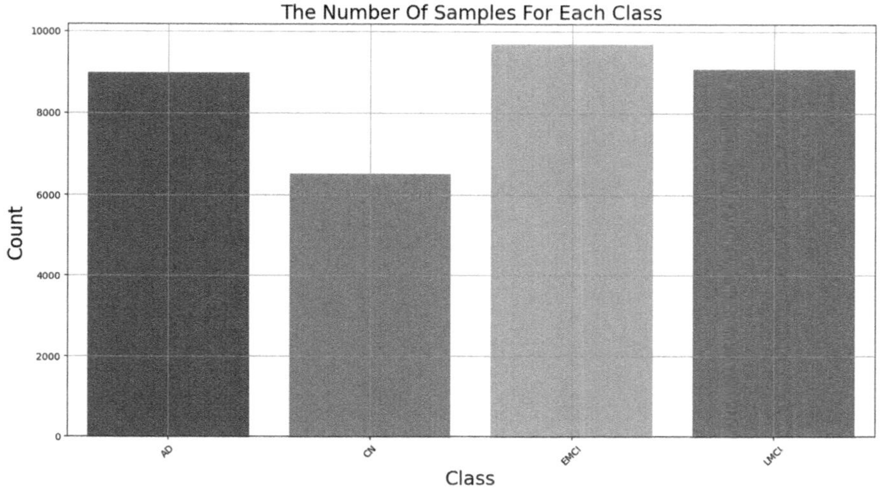

Fig. 4 Number of samples of each class in dataset

7.2 Data Visualisation

A crucial step in building a model is visualising the data, which is what we've done in this section. By showing trends and outliers and transforming data into a more understandable format, data visualisation makes storytelling easier. The purpose of data visualisation is to transform raw data into a comprehensible format, enabling the communication of trends and anomalies. By eliminating unnecessary data noise and highlighting the most relevant information, visualisation effectively narrates a story. Figure 5 depicts the histogram showcasing the distribution of the sample image dataset, illustrating the frequency of occurrence of various image sizes measured in kilobytes (KBs). Figure 6 depicts how pixel intensities, ranging from 0 to 255, are distributed across different images, providing insight into the color characteristics of the sample dataset.

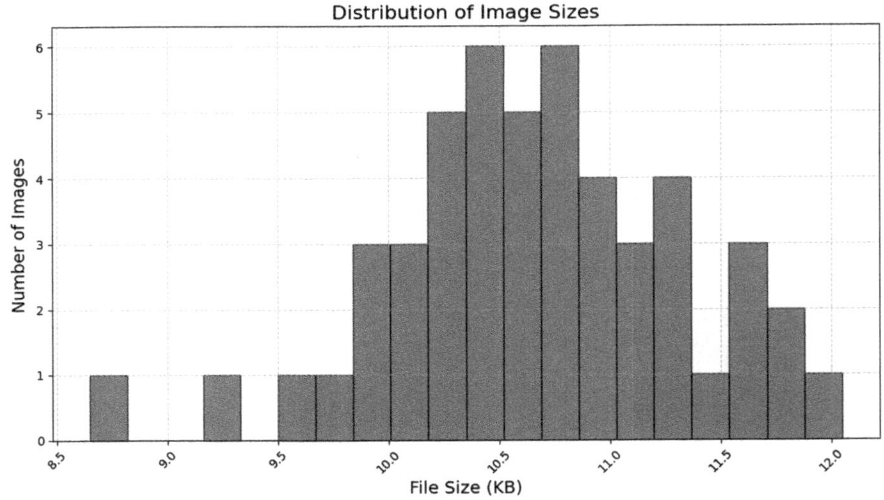

Fig. 5 Distribution of images with respect to sizes

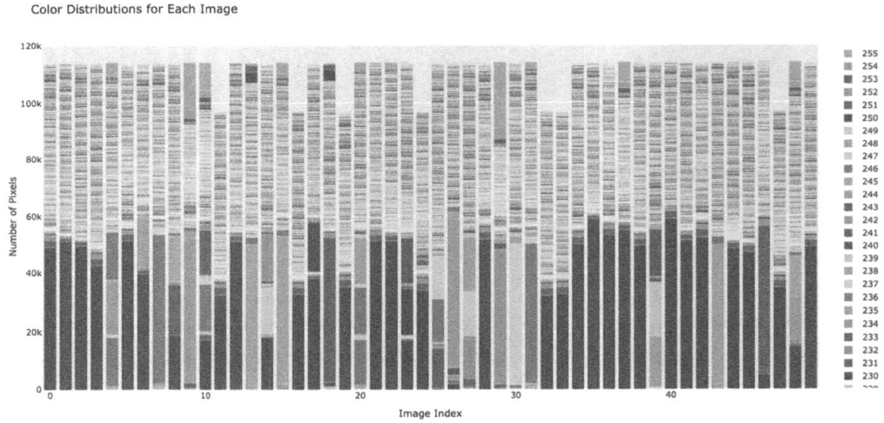

Fig. 6 Distribution of color for each image (x-axis: Number of Pixels; y-axis: Image Index)

7.3 Data Pre-processing

Data pre-processing involves a number of computerized techniques used to put together input data and ensure that it is ready for target learning models. To ensure that the final images contains only relevant information for the task at hand, data preprocessing is important for cleaning noise out of your data, detecting and correcting missing values as well as incorporating it [35]. Preprocessing of image datasets often includes resizing to make them uniform, scaling pixel values for consistency,

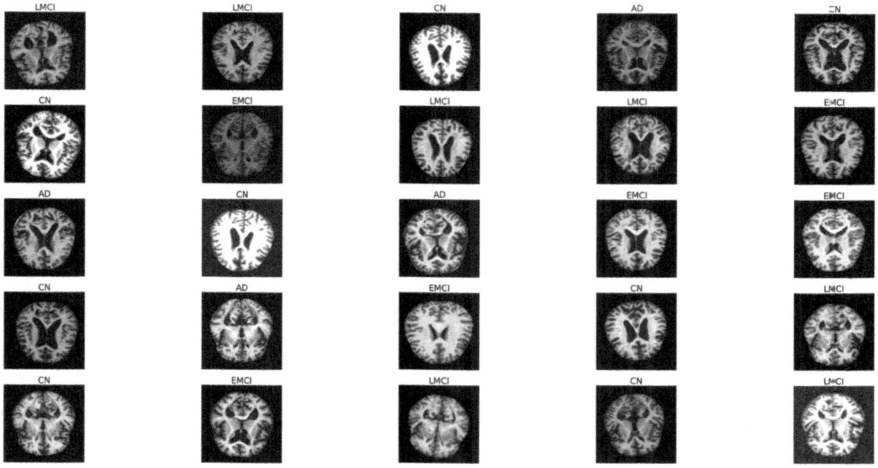

Fig. 7 Resized sample images of dataset

appending rotations, flips or crops for more examples if necessary among others. In addition to this, techniques like segmentation may be employed to isolate and extract specific regions of interest within the images. Segmentation helps in partitioning the images into meaningful segments, allowing the model to focus on relevant areas and ignore irrelevant background information.

7.3.1 Image Resizing

The dataset includes original images of different dimensions, especially those gathered from the real world, hence resizing MRI images to fit a target size is considered in terms of model implementation and computation. All images are converted into sizes of $32 \times 32 \times 3$, indicating height, width, and number of channels used, respectively. Figure 7 represents the resized sample images of the dataset.

7.3.2 Label Encoding

Label encoding is an essential preprocessing step that must be performed prior to obtaining a solution from machine learning tasks, particularly when dealing with categorical data like class labels encountered in image classification. In our dataset, we have images of four classes for which each of the corresponding labels represents the category that it falls under but due to the same, it's necessary that we have to change them into numerical representations as this is what machine learning algorithms mostly understand. Label encoding is used to accomplish this task where each unique label in the dataset is assigned a different integer value and hence converting

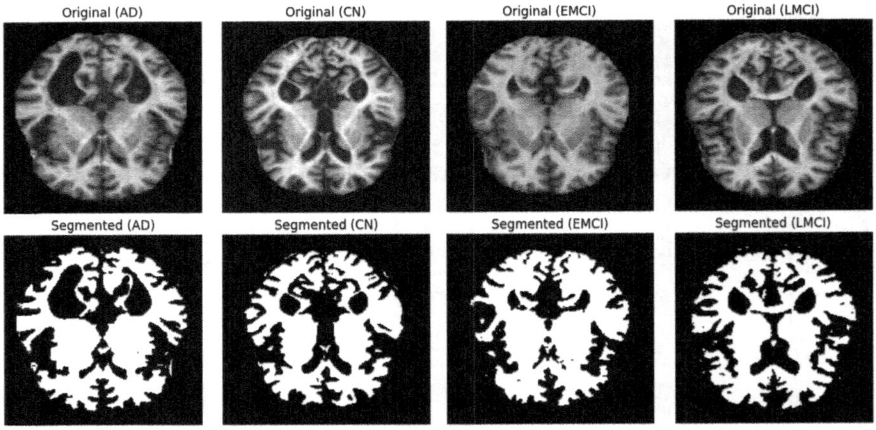

Fig. 8 Original image compared with pre-processed image

them into numerical representation that can be easily processed by machine learning models.

7.3.3 Image Segmentation

Segmentation techniques play a vital role in various applications, including medical imaging, autonomous driving, satellite imagery analysis, and more. It involves partitioning an image into multiple segments or regions to simplify its representation and facilitate further analysis. Unlike object detection, which identifies objects within an image as bounding boxes, image segmentation provides pixel-level granularity by assigning a label to each pixel indicating its region or object category. Figure 8 shows segmented images compared to original image data.

7.3.4 Data Splitting

The dataset was divided into a training set and a testing set to train the models and assess their performance. The training set was used to fit the models, while the testing set was used to generate predictions, which were evaluated against the original labels to measure the effectiveness of the approach. The data splitting was performed at the subject level to define the training, validation, and test sets. To train and test the proposed model, the entire dataset was split into 80% for training data and 20% for testing and validation data. After preprocessing the image dataset, a data split was conducted, where 80% of the data was assigned to training, 10% to testing, and 10% to validation.

7.4 Implementation of Models

In the proposed system, the client initializes machine learning models distributed across the different decentralized edges devices or servers, each with local data samples. Instead of raw data transfer, model updates as gradients or weights are exchanged and combinations to improve the biggest model. Figure 9 describes the implemented model.

7.4.1 Model Framework

Using TensorFlow Keras API, the proposed model uses a simple convolution neural network (CNN).

Model consist of:

- Input layer starting with convolution layer featuring 32 filters size (3, 3) ReLU activated.
- Next max-pooling layer with size (2, 2).
- Flattening layer for flatten previous output.
- Dense layer with 128 units ReLU activation as a major component!
- Output with 10 units softmax activations for classification tasks ten classes.

7.4.2 Training Method

- Training is orchestrated by the 'fit' function within the 'FlowerClient' class!
- Initially model setup and compiled with Adams optimizer and sparse categorical cross entropy loss.

```
Model: "sequential_14"
_________________________________________________________________
 Layer (type)                Output Shape              Param #
=================================================================
 conv2d_13 (Conv2D)          (None, 30, 30, 32)        896

 max_pooling2d_13 (MaxPooli  (None, 15, 15, 32)        0
 ng2D)

 flatten_14 (Flatten)        (None, 7200)              0

 dense_29 (Dense)            (None, 128)               921728

 dense_30 (Dense)            (None, 10)                1290

=================================================================
Total params: 923914 (3.52 MB)
Trainable params: 923914 (3.52 MB)
Non-trainable params: 0 (0.00 Byte)
```

Fig. 9 Description of implemented model parameters

- Image data loaded from specific dataset folder 'dataset_folder' using 'load_images_and_labels'.
- Each image resized by (32, 32) with 'resize_image'.
- Model trains for one epoch on a dataset with 'model.fit'.
- Updated model weights are returned finally.

7.4.3 Evaluation Procedure

- In the 'FlowerClient' class, 'evaluate' carries evaluate procedures.
- Model initializes provided parameters compiled.
- Testing image data, subset of training data shows purpose, loaded and resized.
- Model undergo evaluation with this test datasets 'model.evaluate'.
- Evaluation result reports with loss, sample count and accuracy percentage.

7.4.4 Client Generation

- 'client_fn' constructs an instance for the 'FlowerClient' class creating a Flower client specifically.

7.4.5 ClientApp Legacy Mode

- 'ClientApp' class managed and executed Flower client.
- Legacy mode client initiated 'start_client', connecting specified server address commencing Flower client operations!

8 Evaluation Metrics and Interpretation

The evaluation of the model's performance employs accuracy, precision, recall, and F1 score, which are fundamental metrics in assessing classification models. These metrics are essential for understanding the model's ability to correctly classify instances of Alzheimer's Disease and its various stages.

8.1 Accuracy

One of the most common measures used for evaluating the effectiveness of a classifier is the measure of correctly classified instances. This metric quantifies the proportion of accurately identified results out of the total number of instances. Specifically, it is calculated as a percentage of correctly classified samples.

$$Accuracy\ (in\ Percentage) = \frac{TP + TN}{TP + TN + FP + FN} \times 100 \quad (1)$$

8.2 Precision

The Precision value is calculated by dividing the number of correctly predicted positive rates by the total predicted positive rates. A Precision value of 1 indicates a good classifier. In this particular scenario, it represents the proportion of predicted AD cases that are actually AD patients.

$$Precision\ (in\ Percentage) = \frac{TP}{TP + FP} * 100 \quad (2)$$

8.3 Recall or True Positive Rate (TPR)

It describes the possibility of a classifier accurately forecasting a positive outcome in the presence of an illness. A recall value of 1 signifies an effective classifier. Within the medical field, recall holds greater significance as a fundamental metric compared to precision.

$$Recall\ (in\ Percentage) = \frac{TP}{TP + FN} * 100 \quad (3)$$

8.4 F1-Score

The F1 score is a metric that represents the weighted average of precision and recall, also known as the harmonic mean. A perfect F1 score of 1 indicates the best performance, while a score of 0 signifies the worst. In F-measures, the true negatives (TNs) are not factored into the calculation. The F1 score can be computed by combining precision and recall in a balanced manner, as shown below:

$$F1\ Score\ (in\ Percentage) = 2 * \frac{Recall * Precision}{Recall + Precision} * 100 \quad (4)$$

8.5 *Macro Averages*

To assess the model's overall performance across all classes, macro averages of these metrics are calculated. The macro average is the unweighted mean of the metric calculated for each class. The formulas for macro average precision, recall, and F1 score are given below:

$$Macro\ Average\ Precision = \frac{1}{N}\sum_{i=1}^{N} Precision_i \tag{5}$$

$$Macro\ Average\ Recall = \frac{1}{N}\sum_{i=1}^{N} Recall_i \tag{6}$$

$$Macro\ Average\ F1 - Score = \frac{1}{N}\sum_{i=1}^{N} F1\ Score_i \tag{7}$$

where N is the number of classes.

9 Experimental Results Analysis

In this section, the performance of proposed model trained on the Alzheimer's Disease Neuroimaging Initiative (ADNI) dataset is analysed. The evaluation of the model's performance relies on several key metrics: accuracy, precision, recall, and F1 score. These metrics are essential for understanding the model's ability to correctly classify instances of Alzheimer's Disease and its various stages. Figures 10 and 11 displays the training loss and accuracy of client 1 and client 2 model over 10 epochs, highlighting significant fluctuations. The training loss starts high, around 25, and although it decreases overall, it exhibits substantial variability across epochs. By the 10th epoch, the loss approaches zero, indicating better fitment to the training data. Similarly, the training accuracy shows erratic behavior, with sharp increases and decreases, eventually reaching around 0.9.

The model's macro average metrics are:

- Macro Average Precision: 0.8305
- Macro Average Recall: 0.8300
- Macro Average F1 Score: 0.8301

These values indicate that, on average, the model performs well across all classes, without being biased towards any particular class. The balanced and high F1 scores demonstrate the model's ability to accurately and consistently classify instances of Alzheimer's disease and its various stages. Table 3 provides a summary model's performance across the four classes. This balanced performance is crucial for medical

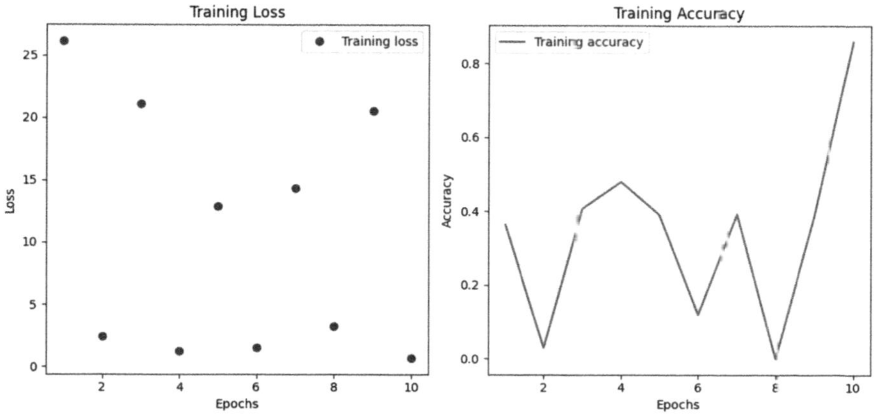

Fig. 10 Loss and accuracy curve of client 1

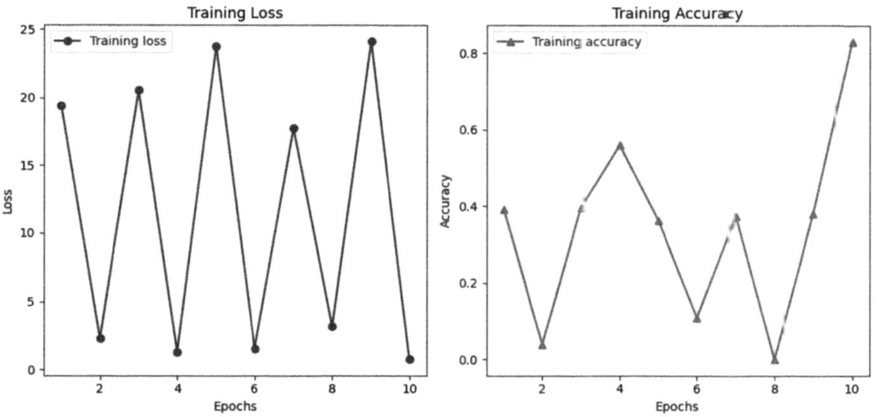

Fig. 11 Loss and accuracy curve of client 2

diagnosis tasks, where both false positives and false negatives can have serious implications. The high precision ensures that most of the predicted positive cases are true positives, while the high recall ensures that most actual positive cases are identified. Figure 12 shows the barplot depicting the precision, recall and F1-score of each individual class.

Table 3 Model's performance across the four classes

Class	Precision	Recall	F1-score
AD	0.85	0.83	0.84
CN	0.80	0.82	0.81
EMCI	0.82	0.83	0.82
LMCI	0.84	0.84	0.84

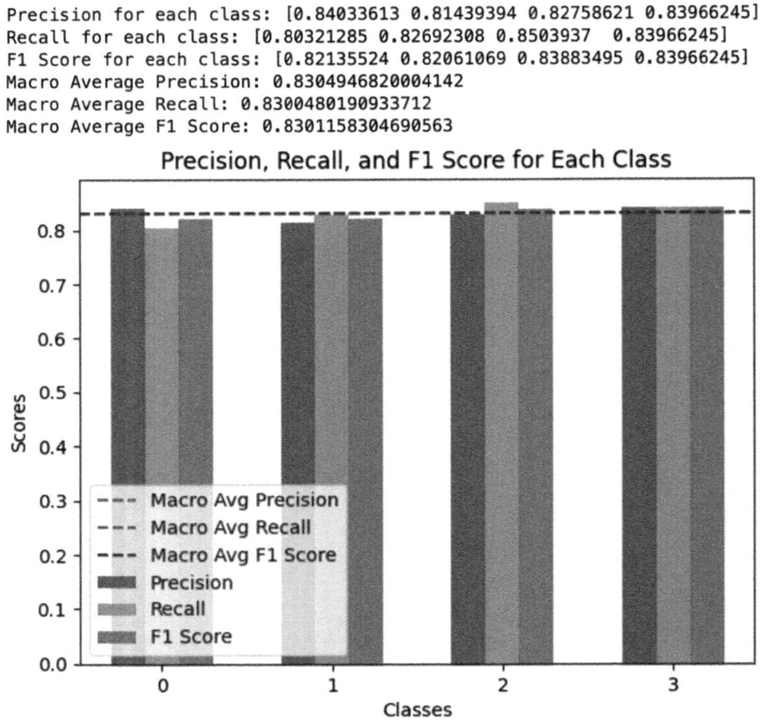

Fig. 12 Precision, Recall and F1 score of each class

10 Conclusion and Future Scope

This study has shown important results regarding how effective federated learning is in healthcare information systems. We could do more research about it in different ways though. The results might be confirmed or complemented through further empirical studies across a wider spectrum of health care domains and involving larger groups of institutions. Examining the capabilities of Federated Learning frameworks to handle more extensive and more varied datasets might enhance their practicality in healthcare settings. Furthermore, further investigation is needed to improve the methods for maximizing model convergence and reducing communication overhead in federated learning configurations. There exists significant potential for progress in the diagnosis and treatment of Alzheimer's disease. To enhance its applicability and reliability, the approach will undergo testing on larger datasets. Additionally, integration of multimodal inputs—neuroimaging, genetic markers, and clinical variables—provides broad knowledge in AD development in a single framework for better accuracy and robustness of prediction. By combining these modalities, predictive models can take advantage of the complementary information to improve early detection and differential diagnosis of Alzheimer's disease. Furthermore, to enhance robustness and fault tolerance, it is important to develop advanced algorithms along with strong aggregation techniques to manage device failures, interrupted connections and malicious actions.

Continuous research and development efforts are crucial in overcoming these challenges and improving how federated learning is rolled out in healthcare settings. The study highlights the significance of ongoing collaboration among academics, healthcare professionals and regulatory entities to enhance and broaden the use of federated learning. This partnership should develop standardized procedures, address confidentiality matters and ensure conformity with the rules at hand to create a strong and morally excellent foundation for the operation of Federated Learning among other things. In Conclusion, this study has shown great potential for FL within healthcare, most notably in Alzheimer's disease. Yet there remains a lot more work that needs to be done. What needs to be done therefore includes furthering studies across various parameters; as well as incorporating more than one type of data set while enhancing on technicalities involved in frameworks supporting FL. If such processes are followed up, there is no doubt that FL will become dependable and popularly used in hospitals that may lead to improved ways of diagnosing patients and better administering medicine.

References

1. Alzheimer's Disease International and McGill University: World Alzheimer Report 2021: Journey Through the Diagnosis of Dementia. ADI - World Alzheimer Report 2021, Sept. 21, 2021. https://www.alzint.org/resource/world-alzheimer-report-2021
2. Kaissis, G., Makowski, M., Rückert, D., Braren, R.: Secure, privacy-preserving and federated machine learning in medical imaging. Nat.e Mach. Intell. (2020). https://doi.org/10.1038/s42256-020-0186-1
3. Li, Q., Wen, Z., Wu, Z., Hu, S., Wang, N., Li, Y., Liu, X., He, B.: A survey on federated learning systems: vision, hype and reality for data privacy and protection. IEEE Trans. Knowl. Data Eng. (2023). https://doi.org/10.1109/tkde.2021.3124599
4. Association, A.: Alzheimer's disease facts and figures. Alzheimers Dement. **17**(3), 327–406 (2021)
5. McMahan, H., Moore, E., Ramage, D., Hampson, S., Arcas, B., Li, Q., Wen, Z., Wu, Z., Hu, S., Wang, N., Li, Y., Liu, X., He, B.: Communication-efficient learning of deep networks from decentralized data (2023). https://doi.org/10.48550/arxiv.1602.05629
6. Nguyen, D., Pham, Q.-V., Pathirana, P., Ding, M., Seneviratne, A., Lin, Z., Dobre, O., Hwang, W.-J., McMahan, H., Moore, E., Ramage, D., Hampson, S., Arcas, B., Li, Q., Wen, Z., Wu,

Z., Hu, S., Wang, N., Li, Y., Liu, X., He, B.: Federated learning for smart healthcare: a survey (2023). https://doi.org/10.48550/arxiv.2111.08834
7. Kumar, Y., Singla, R., Nguyen, D., Pham, Q.-V., Pathirana, P., Ding, M., Seneviratne, A., Lin, Z., Dobre, O., Hwang, W.-J., McMahan, H., Moore, E., Ramage, D., Hampson, S., Arcas, B., Li, Q., Wen, Z., Wu, Z., Hu, S., Wang, N., Li, Y., Liu, X., He, B.: Federated learning systems for healthcare: perspective and recent progress (2023). https://doi.org/10.1007/978-3-030-70604-3_6
8. Burns, A., Iliffe, S.: Alzheimer's disease. BMJ (2009). https://doi.org/10.1136/bmj.b158. Erratum in: BMJ (2009) Apr 1. https://doi.org/10.1136/bmj.b1349. PMID: 19196745
9. Lam, B., Masellis, M., Freedman, M., Stuss, D., Black, S.: Clinical, imaging, and pathological heterogeneity of the Alzheimer's disease syndrome. Alzheimers Res. Ther. (2013). https://doi.org/10.1186/alzrt155. PMID: 23302773; PMCID: PMC3580331
10. Matthews, K., Xu, W., Gaglioti, A., Holt, J., Croft, J., Mack, D., McGuire, L.: Racial and ethnic estimates of Alzheimer's disease and related dementias in the United States (2015–2060) in adults aged $\geq$65 years. Alzheimer's Dement. (2018). https://doi.org/10.1016/j.jalz.2018.06.3063
11. Sevigny, J., Chiao, P., Bussière, T., Weinreb, P., Williams, L., Maier, M., Dunstan, R., Salloway, S., Chen, T., Ling, Y., O'Gorman, J., Qian, F., Arastu, M., Li, M., Chollate, S., Brennan, M., Quintero-Monzon, O., Scannevin, R., Arnold, H., Engber, T., Rhodes, K., Ferrero, J., Hang, Y., Mikulskis, A., Grimm, J., Hock, C., Nitsch, R., Sandrock, A.: The antibody aducanumab reduces Aβ plaques in Alzheimer's disease. Nature (2016). https://doi.org/10.1038/nature19323
12. Livingston, G., Huntley, J., Sommerlad, A., Ames, D., Ballard, C., Banerjee, S., Brayne, C., Burns, A., Cohen-Mansfield, J., Cooper, C., Costafreda, S., Dias, A., Fox, N., Gitlin, L., Howard, R., Kales, H., Kivimäki, M., Larson, E., Ogunniyi, A., Orgeta, V., Ritchie, K., Rockwood, K., Sampson, E., Samus, Q., Schneider, L., Selbæk, G., Teri, L., Mukadam, N.: Dementia prevention, intervention, and care: 2020 report of the lancet commission. Lancet (2020). https://doi.org/10.1016/s0140-6736(20)30367-6
13. Norton, S., Matthews, F., Barnes, D., Yaffe, K., Brayne, C.: Potential for primary prevention of Alzheimer's disease: an analysis of population-based data. Lancet Neurol. (2014). https://doi.org/10.1016/s1474-4422(14)70136-x
14. Kairouz, P.: Advances and Open Problems in Federated Learning (2019). https://doi.org/10.48550/arXiv.1912.04977
15. Danek, B., Makarious, M., Dadu, A., Vitale, D., Lee, P., Singleton, A., Nalls, M., Sun, J., Faghri, F.: Federated learning for multi-omics: a performance evaluation in Parkinson's disease. Patterns (2024). https://doi.org/10.1016/j.patter.2024.100945
16. Reddy, K., Gadekallu, T.: A comprehensive survey on federated learning techniques for healthcare informatics. Comput. Intell. Neurosci. (2023). https://doi.org/10.1155/2023/8393990
17. Willemink, M., Koszek, W., Hardell, C., Wu, J., Fleischmann, D., Harvey, H., Folio, L., Summers, R., Rubin, D., Lungren, M.: Preparing medical imaging data for machine learning. Radiology (2020). https://doi.org/10.1148/radiol.2020192224
18. Li, T., Sahu, A., Talwalkar, A., Smith, V.: Federated learning: challenges, methods, and future directions. IEEE Signal Process. Mag. (2020). https://doi.org/10.1109/msp.2020.2975749
19. Khalil, K., Mamun, M., Sherif, A., Elsersy, M., Imam, A.-A., Mahmoud, M., Alsabaan, M.: A federated learning model based on hardware acceleration for the early detection of Alzheimer's disease. Sensors (2023). https://doi.org/10.3390/s23198272
20. Nagy, B., Hegedűs, I., Sándor, N., Egedi, B., Mehmood, H., Saravanan, K., Lóki, G., Kiss, Á.: Privacy-preserving Federated Learning and its application to natural language processing. Knowl.-Based Syst. (2023). https://doi.org/10.1016/j.knosys.2023.110475
21. Pan, W., Xu, Z., Rajendran, S., Wang, F.: An adaptive federated learning framework for clinical risk prediction with electronic health records from multiple hospitals. Patterns (2024). https://doi.org/10.1016/j.patter.2023.100898

22. Choi, G., Cha, W., Lee, S., Shin, S.: Survey of medical applications of federated learning. Healthc. Inform. Res. (2024). https://rb.gy/t9qvv6
23. Weiner, M., Aisen, P., Jack, C., Jagust, W., Trojanowski, J., Shaw, L., Saykin, A., Morris, J., Cairns, N., Beckett, L., Toga, A., Green, R., Walter, S., Soares, H., Snyder, P., Siemers, E., Potter, W., Cole, P., Schmidt, M.: The Alzheimer's disease neuroimaging initiative: progress report and future plans. Alzheimer's Dement. (2010). https://doi.org/10.1016/j.jalz.2010.03.007
24. Rieke, N., Hancox, J., Li, W., Milletari, F., Roth, H., Albarqouni, S., Bakas, S., Galtier, M.N., Landman, B.A., Maier-Hein, K.H., Ourselin, S., Sheller, M.J., Summers, R.M., Trask, A., Xu, D., Baust, M., Cardoso, M.J.: The future of digital health with federated learning. NPJ Digit. Med. (2020). https://doi.org/10.1038/s41746-020-00323-1
25. Gupta, M., Sharma, P., Kalra, R.: Federated Learning and Artificial Intelligence in E-Healthcare, pp. 104–118 (2023)
26. Lei, B., Liang, Y., Xie, J., Wu, Y., Liang, E., Liu, Y., Yang, P., Wang, T., Liu, C., Du, J., Xiao, X., Wang, S.: Hybrid federated learning with brain-region attention network for multi-center Alzheimer's disease detection. Pattern Recogn. (2024). https://doi.org/10.1016/j.patcog.2024.110423
27. Ouyang, X.: Design and Deployment of Multi-Modal Federated Learning Systems for Alzheimer's Disease Monitoring (2023).https://doi.org/10.1145/3581791.3597505
28. Qian, C., Xiong, H., Li, J.: FEDEFO: a personalized federated deep forest framework for Alzheimer's disease diagnosis (2023). https://doi.org/10.18293/seke2023-013
29. Mandawkar, U., Diwan, T.: Alzheimer disease classification using tawny flamingo based deep convolutional neural networks via federated learning. Imaging Sci. J. (2023). https://doi.org/10.1080/13682199.2023.2172524
30. Stripelis, D., Dhinagar, N., Romero, R., Thomopoulos, S., Thompson, P., Ambite, J.: Federated deep learning for detecting Alzheimer's disease in multi-cohort brain MRI. Alzheimer's Dement. (2023). https://doi.org/10.1002/alz.065998
31. Mitrovska, A., Safari, P., Ritter, K., Shariati, B., Fischer, J.: Secure federated learning for Alzheimer's disease detection (2024). https://doi.org/10.3389/fnagi.2024.1324032
32. Castro, F., Impedovo, D., Pirlo, G.: A federated learning system with biometric medical image authentication for Alzheimer's diagnosis. In: International Conference on Pattern Recognition Applications and Methods (2024). https://doi.org/10.5220/0012550200003654
33. Yao, Z., Wang, H., Yan, W., Wang, Z., Zhang, W., Wang, Z., Zhang, G.: Artificial intelligence-based diagnosis of Alzheimer's disease with brain MRI images. Eur. J. Radiol. (2023). https://doi.org/10.1016/j.ejrad.2023.110934
34. Ahmed, A.: FDATA ADNI Dataset. Kaggle (2024). https://www.kaggle.com/datasets/ahmedashrafahmed/fdata-adni-dataset
35. García, S., Ramírez-Gallego, S., Luengo, J., Benítez, J., Herrera, F.: Big data preprocessing: methods and prospects. Big Data Anal. (2016). https://doi.org/10.1186/s41044-016-0014-0

Improving Healthcare Privacy and Efficiency with Federated Learning

Dheeraj Sonkhla and Amit Chauhan

Abstract Medical image processing is crucial in modern healthcare by aiding diagnosis, treatment planning, and disease monitoring. However, practical analysis of medical images often necessitates large, diverse datasets, which pose significant privacy and data-sharing challenges. Federated learning (FL), a decentralized machine learning approach, enables the collaborative training of the model across multiple institutions without sharing sensitive patient data. This chapter comprehensively analyzes federated learning algorithms, techniques, and architectures specifically designed for healthcare applications. It examines the technological and infrastructural prerequisites and regulatory and ethical factors and presents various case examples to illustrate the practical implementation of these approaches. The chapter aims to correlate the application of federated learning and techniques to medical image processing, focusing on improving privacy and collaboration in healthcare settings. It also addresses typical constraints and obstacles associated with federated learning, including communication overhead, data heterogeneity, and algorithm convergence. The FL paradigm offers a solution to these problems by enabling algorithms to learn from diverse medical imaging datasets across several locations through collaborations facilitated by a central server. The findings indicate that federated learning holds considerable promise in transforming the healthcare sector by utilizing extensive medical data while upholding patient information confidentiality. Ongoing research and development are needed to harness its potential fully, ultimately resulting in enhanced patient care, medical research advancements, and increased healthcare system efficiency.

D. Sonkhla (✉)
Department of Computer Science, Himachal Pradesh University Regional Centre, Dharamshala, Himachal Pradesh, India
e-mail: dheeraj.sonkhla15@gmail.com

A. Chauhan
School of Computer Science Engineering and Technology, Bennett University, Greater Noida, Uttar Pradesh, India
e-mail: chauhanamit37@gmail.com

R. Mohana et al. (eds.), *Federated Learning Applications in the Industrial Internet of Everything (IoE)*, Studies in Systems, Decision and Control 611,
https://doi.org/10.1007/978-3-031-99270-4_5

Keywords Federated learning · Image processing · Federated averaging · Vertical federated learning · Horizontal federated learning

1 Introduction

1.1 Overview of Medical Image Processing in Healthcare

Medical image processing has significantly advanced healthcare by offering critical insights into the structure and function of the human body. It involves various methods for acquiring, analyzing, and interpreting images from medical imaging technologies such as X-rays, MRI, CT scans, ultrasound, and PET scans. These images are essential in clinical decision-making, aiding in the diagnosis, treatment planning, and monitoring of various medical conditions [1].

Recent technological advancements have resulted in the production of vast amounts of medical image data. This data contains valuable information that can enhance patient outcomes and drive medical research forward. However, effectively analyzing and interpreting medical images often necessitates access to large, diverse datasets representing a broad spectrum of patient populations and medical conditions.

1.2 Challenges Associated with Data Sharing and Privacy in Medical Imaging

While access to diverse medical image datasets is crucial for advancing healthcare research and improving patient care, sharing such data poses significant privacy and security challenges. Medical images contain sensitive information about patients' health, making them subject to stringent regulations and privacy laws, such as the Health Insurance Portability and Accountability Act (HIPAA) in the United States [2].

Traditional data-sharing methods in medical imaging, such as centralized repositories or cloud-based platforms, raise concerns about data privacy and security. Centralized storage of medical images increases the risk of data breaches and unauthorized access, potentially compromising patient confidentiality [1].

Additionally, data sharing across institutions faces obstacles due to regulatory barriers, technical challenges, and issues related to data ownership and governance. Healthcare organizations are understandably cautious about sharing sensitive patient data due to legal and ethical considerations, which often results in limited access to diverse datasets for research and development.

1.3 Introduction to Federated Learning as a Solution

Federated learning (FL) presents a promising solution to the challenges of data sharing and privacy in medical imaging. FL is a decentralized machine learning approach that facilitates collaborative model training across multiple institutions without exchanging raw data. Instead of centralizing data, federated learning allows institutions to train models locally on their own datasets while sharing only model updates or gradients with a central server [3, 4].

This method preserves data privacy and confidentiality, addressing concerns associated with data sharing in medical imaging. Institutions can collaborate and leverage each other's data without compromising patient privacy or breaching regulatory requirements [5]. Federated learning enables healthcare institutions to gain insights from diverse datasets while safeguarding sensitive patient information.

In the realm of medical image processing, federated learning provides a scalable and privacy-preserving approach for training machine learning models using distributed data sources. This allows institutions to enhance diagnostic algorithms, develop personalized treatment plans, and advance medical research without needing to share sensitive patient data.

1.4 Objectives of the Chapter

The objectives of the chapter are:

1. To provide a comprehenive overview of federated learning and its potential applications in healthcare;
2. To compare federated learning with traditional centralized machine learning approaches in terms of privacy, efficiency, and accuracy;
3. To explore the benefits of federated learning in addressing privacy concerns in medical image processing;
4. To examine the challenges and limitations of federated learning in a healthcare context;
5. And, to propose potential future research directions for advancing federated learning in medical image processing.

1.5 Organization of the Chapter

The rest of the chapter is organized as: Sect. 2 discusses foundations of federated learning. Section 3 enlightens federated learning algorithms and techniques. Section 4 highlights technological and infrastructural prerequisites. Section 5 illustrates regulatory and ethical considerations, while Sect. 6 focuses on practical implementation and case examples followed by challenges and solutions in Sect. 7. And, finally, Sect. 8 concludes the chapter with future scope.

2 Foundations of Federated Learning

2.1 *Explanation of Federated Learning Principles and Concepts*

Federated learning (FL) is a decentralized machine learning approach that enables model training across multiple edge devices or institutions while keeping the training data decentralized. In traditional centralized machine learning, data is collected and aggregated in a central server or data center, where a global model is trained using the combined dataset. However, in federated learning, the training process occurs locally on each device or institution's data, and only model updates are shared with a central server for aggregation [4–6].

The key principles and concepts of federated learning include:

- **Decentralization**: Federated learning distributes the process of model training across multiple devices or institutions, allowing each participant to train the model locally on their data.
- **Privacy Preservation**: Federated learning ensures data privacy by keeping sensitive data localized and only sharing model updates with a central server. This decentralized approach minimizes the risk of exposing raw data to unauthorized access or breaches.
- **Collaborative Learning**: Federated learning enables collaboration among multiple participants without the need to share raw data. Institutions can collectively improve the performance of machine learning models by sharing insights and model updates while preserving data privacy.
- **Model Aggregation**: In federated learning, model weights/updates or gradients computed locally on each device or institution are aggregated at a central device to update the global model. This process allows the global model to benefit from insights learned from diverse datasets without centralizing data.

2.2 *Architecture of Federated Learning*

This section discusses the structure of federated learning and detailing each type of architecture separately.

2.2.1 Horizontal Federated Learning

Figure 1 illustrates a typical structure for a horizontally federated learning system. Multiple organizations possess diverse clients, collectively referred to as K respondents, which hold identical data formats. These participants collaborate to train a machine learning (ML) model with the support of a cloud server or parameters through this framework [5]. It is presumed that the edge devices are precise and the

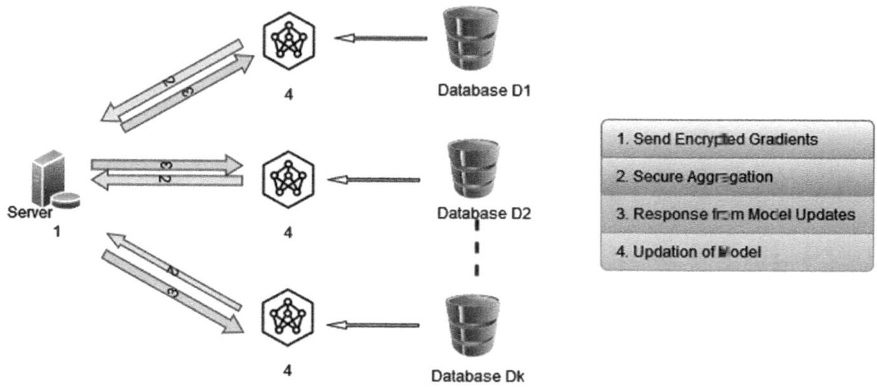

Fig. 1 Horizontal federated learning architecture

server is dependable but interested in learning. Therefore, no data loss is permitted from any sample respondents [7, 8]. Typically, the preparation process for such a program includes the following four phases:

- **Local Computation**: Edge devices perform local computations of the training gradient, utilizing techniques such as encryption [7] and confidential sharing [9], and subsequently transmit masked outcomes to the cloud server.
- **Secure Aggregation**: An extra secure and encrypted form of aggregation is implemented on the server without the need to access any raw data from the edge devices.
- **Server Feedback**: After the server obtains the combined outcomes, it returns the aggregated results to the edge devices [5].
- **Model Update**: Edge devices subsequently adjust their learning by integrating the decrypted gradients.

These four steps are continuously iterated to minimize the loss function as the training progresses. The horizontal architecture operates independently of any specific machine learning algorithm, and all edge devices utilize the final model parameters [5].

2.2.2 Vertical Federated Learning

Vertical federated learning is a collaborative approach to machine learning where multiple parties, typically organizations or entities, jointly train a model using their respective datasets without directly sharing sensitive information. Unlike horizontal federated learning, where entities have the same type of data, vertical federated learning involves parties with different types of data, often with some overlapping attributes [7].

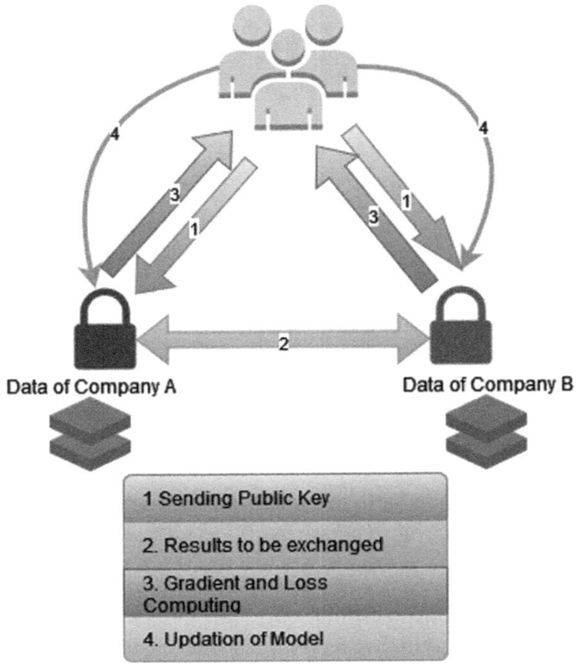

Fig. 2 Vertical federated learning architecture [8]

Figure 2 illustrates a standard framework for a vertical federated learning network. In vertical federated learning, data remains decentralized and encrypted throughout training to ensure privacy and security. A trusted third party or secure computation techniques are often employed to facilitate communication and coordination between the parties without compromising data confidentiality [10, 11].

This approach is particularly useful in situations where organizations need to collaborate on predictive modeling tasks while adhering to strict data privacy regulations or confidentiality agreements. By enabling collaboration without data sharing, vertical federated learning allows entities to leverage each other's datasets while preserving control over their proprietary information.

2.3 Comparison with Traditional Centralized Machine Learning Approaches

Traditional centralized machine learning approaches aggregates data from multiple sources in a central location for model training. While this method can produce high-quality models due to access to large, centralized datasets, it presents several challenges, especially in healthcare settings.

2.3.1 Data Privacy Concerns

In centralized data storage, there are significant vulnerabilities for healthcare data:

- **Data Breach Risk**: Centralized repositories are prime targets for cyberattacks, which can expose sensitive patient information.
- **Unauthorized Access**: The risk of unauthorized access is higher in centralized systems, increasing the potential for data misuse and privacy violations.
- **Data Retention Challenges**: Centralized systems often struggle with effective data retention policies, which leads to the unnecessary storage of sensitive information.

2.3.2 Regulatory Compliance

Centralized data management is subject to stringent regulations like HIPAA [2].

- **Compliance Burden**: Complying with centralized data storage regulatory requirements can be complicated and require a lot of resources.
- **Legal Risks**: Healthcare organizations that fail to comply with regulations may face significant fines and legal responsibilities.
- **Audit Challenges**: Regular audits are essential to ensure compliance with data privacy regulations, leading to increased operational costs.

2.3.3 Data Quality and Heterogeneity

Centralized data collection can introduce challenges related to data quality and heterogeneity.

- **Data Quality Issues**: Data that is kept in one central place often has mistaken and missing information, and this can make models work less well.
- **Data Heterogeneity**: Information from different places might look and work differently, which can make it hard to put together and analyze.
- **Data Silos**: Centralization can lead to data silos, preventing a comprehensive view of patient information and hindering research.

2.3.4 Computational Efficiency and Scalability

Centralized machine learning models can face limitations in terms of computational efficiency and scalability.

- **Computational Costs**: Training big models with data all in one place can be very expensive because it needs really powerful computers.
- **Scalability Issues**: Dealing with more and more data can be hard and cause things to slow down.

- **Model Training Time**: It takes a long time to train complicated models using centralized data, which can delay when the models are put into use and when they are updated.

2.3.5 Model Interpretability and Explainability

Centralized models often lack transparency and interpretability.

- **Black Box Models**: Complex models can be hard to grasp, so it is tough to explain how they make decisions.
- **Trust Issues**: Not being clear about how AI makes medical decisions can make people not trust it.
- **Regulatory Requirements**: Some rules say AI systems need to be explainable, which can be hard with centralized models.

Comparing these points with the advantages of federated learning can help show the benefits of this decentralized approach for analysing healthcare data.

2.4 Advantages of Federated Learning in Healthcare Settings

- **Privacy Preservation**: Federated learning allows healthcare institutions to work together to train models without sharing raw data. This helps keep sensitive patient information private. They can improve model performance while following the rules and doing the right thing [5].
- **Data Sovereignty**: Federated learning enables healthcare organizations to maintain control and ownership of their data, ensuring compliance with data governance policies and regulations. They can take part in collaborative research initiatives without giving up control over their datasets.
- **Scalability and Efficiency**: Federated learning enables distributed model training across multiple devices or institutions, allowing efficient use of computational resources. Institutions can utilize their existing infrastructure for model training without the need for centralized data storage or processing [6].
- **Diverse Data Sources**: Federated learning helps healthcare organizations use different types of data from places like hospitals, clinics, research centers, and wearable devices. This kind of learning can make models better at working with a variety of patients and medical issues, so they can perform more reliably.
- **Collaborative Research**: Federated learning helps healthcare institutions work together to speed up medical research and innovation. They can create and check machine learning models for things like diagnosing diseases, planning treatments, and customizing medicine.

In healthcare, federated learning helps keep patient data private while allowing different institutions to work together to train models using their combined knowledge.

3 Federated Learning Algorithms and Techniques

3.1 *Overview of Federated Learning Algorithms Tailored for Medical Imaging*

Federated learning algorithms for medical imaging use decentralized model training and are designed to work with healthcare data from different places. They focus on privacy, efficiency, and reliability. Some important federated learning algorithms for medical imaging are:

- **Federated Averaging (FedAvg)**: FedAvg is a basic algorithm in federated learning. In this process, each contributing device or institution trains a local model on its data and sends the model updates to a central server [12]. The central server then combines the model updates using weighted averaging to update the global model. FedAvg can be adjusted for medical imaging tasks by using techniques to handle data differences and privacy concerns.
- **Secure Aggregation**: Secure aggregation methods are used to keep data private when combining models. These methods use special techniques like homomorphic encryption and differential privacy to combine model updates from different people securely, without showing the original data to the main server. Secure aggregation is very important in healthcare, where keeping data private is crucial.
- **Transfer Learning**: Transfer learning methods can be used in federated learning to utilize pre-trained models and share knowledge across organizations. Instead of training models from the beginning on local datasets, participating organizations can adjust pre-trained models using their data, while sharing only the updated model parameters with the central server. This process reduces the computational load and speeds up model convergence in federated learning scenarios [8].

3.2 *Techniques for Efficient Model Aggregation and Updating*

Efficiently combining and updating models is really important for federated learning to work well in healthcare. This is especially true when large medical imaging datasets are spread out across different institutions. There are a number of techniques that can make combining and updating models more efficient in federated learning:

- **Gradient Compression**: Gradient compression techniques reduce the communication overhead associated with transmitting model updates from local devices to the central server. These techniques compress the gradients computed during local model training before transmission, reducing the size of the transmitted data while preserving model accuracy. Gradient compression is essential for federated learning in resource-constrained environments with limited bandwidth or communication bandwidth [13–15].
- **Asynchronous Updates**: Asynchronous federated learning lets devices or institutions update the main model independently and at their own pace. Instead of syncing updates at set times, asynchronous federated learning allows devices to update the main model as soon as their local training is done. This approach can improve how fast the model converges and how scalable the federated learning algorithms are, especially in changing and diverse environments.
- **Adaptive Learning Rates**: Adaptive learning rate techniques change the learning rate based on how quickly each device or institution is learning. These techniques adjust the learning rate for each participant to help them learn faster and improve model performance. Adaptive learning rates work well in federated learning with different datasets and resources, where using the same learning rate for everyone may not give the best results [15].

3.3 *Addressing Data Heterogeneity and Privacy Concerns*

Data differences and privacy worries are big issues in federated learning, especially in healthcare. Medical imaging datasets vary in the ways images are taken and the types of patients. There are ways to deal with these issues in federated learning for medical imaging.

- **Data Preprocessing and Standardization**: Data preprocessing methods, including normalization, resampling, and feature extraction, can mitigate the impact of data heterogeneity across institutions. Standardizing and preprocessing medical imaging datasets before model training can ensure consistency and compatibility across distributed datasets for participating institutions.
- **Privacy-Preserving Algorithms**: Privacy-protecting federated learning methods use strategies like differential privacy, federated averaging with secure aggregation, and federated learning with encrypted model updates to keep sensitive patient information safe during model training and aggregation. These methods ensure that patient privacy is maintained throughout the federated learning process, even when working with multiple institutions [10].
- **Data Augmentation**: Data augmentation methods create new artificial samples and variations from current medical image datasets to increase diversity and strength. By adding artificial samples to the training data, organizations can manage data differences and enhance model performance without exposing real patient data. Additionally, these methods can address overfitting problems and boost the performance of collaborative learning models in medical imaging tasks.

4 Technological and Infrastructural Prerequisites

4.1 Infrastructure Requirements for Implementing Federated Learning in Healthcare Institutions

To introduce federated learning in healthcare institutions, a sturdy infrastructure is needed to facilitate distributed model training and collaboration across multiple sites. Several essential infrastructure prerequisites for federated learning in healthcare encompass:

- **High-Performance Computing (HPC) Resources**: Healthcare facilities necessitate access to high-performance computing resources such as multi-core processors, GPUs, and cloud computing infrastructure to facilitate computationally intensive model training tasks. High-performance computing (HPC) resources allow for efficient parallelization of model training and aggregation across distributed datasets, thereby expediting convergence and scalability in federated learning [3].
- **Network Connectivity**: Dependable network connectivity is vital for enabling communication and data exchange between participating institutions and a central server in federated learning systems. High-speed internet connections, secure Virtual Private Networks (VPNs), and dedicated communication channels guarantee smooth transmission of model updates and data exchanges while minimizing delays and communication overhead [4–6].
- **Data Storage and Management**: Healthcare facilities need reliable systems to store and manage large medical imaging datasets securely. Different types of storage solutions and data management tools make it easier to store, access, and share medical images across multiple locations while keeping the data safe and in line with regulations.
- **Secure Computing Environments**: Federated learning systems should be deployed in secure computing environments protected by firewalls, intrusion detection systems, and access control mechanisms to safeguard against unauthorized access, data breaches, and cyber threats. Secure computing environments isolate federated learning workloads from external threats and ensure the integrity and confidentiality of patient data throughout the model training process.

4.2 Considerations for Data Preprocessing and Standardization

Data preparation and standardization are important things to think about in federated learning to make sure the data is consistent, compatible, and of good quality across different datasets. In healthcare, some important things to consider for data preparation and standardization are:

- **Data Cleaning and Quality Assurance**: Healthcare information often has mistaken and missing data, which can make it hard to rely on. To make sure the data is good and stays the same across all the different pieces, it's important to use methods like finding unusual numbers, filling in missing data, and fixing errors.
- **Normalization and Standardization**: Medical pictures can be different because they use different machines, methods, and resolutions. To make sure that our computer programs can understand and use these pictures, we need to make them similar. We do this by making their brightness and size the same and by matching them up with each other.
- **Feature Extraction and Selection**: Methods like breaking down images, finding important parts, and simplifying data are helpful for getting the right information from medical images and making it easier for computers to process the information without getting too specific. There are also ways to pick out the most useful parts and get rid of any extra stuff in models used for cooperative learning.
- **Data Expansion**: Data expansion techniques such as geometric transformations, image rotations, and random cropping generate synthetic variations of medical imaging data to enhance diversity and robustness. Data augmentation improves model generalization and reduces overfitting by exposing federated learning models to a broader range of data distributions and variations.

4.3 Security Measures to Protect Sensitive Patient Data

Security protocols are crucial for safeguarding sensitive patient information and maintaining adherence to regulatory standards such as HIPAA in federated learning systems. Some key security measures to protect patient data in federated learning include:

- **Data Encryption**: Federated learning systems should encrypt sensitive patient data during transmission and storage using encryption techniques such as SSL/TLS for data in transit and AES encryption for data at rest. Encryption ensures privacy and veracity of patient data and protects against unauthorized access and data breaches.
- **Access Control and Authentication**: Federated learning systems need to use controls and authentication methods to limit access to patient data based on user roles, privileges, and credentials. Role-based access control (RBAC), multi-factor authentication (MFA), and secure authentication protocols like OAuth and OpenID Connect are important to make sure that only allowed users can access patient data and join in model training activities [17].
- **Data Anonymization and De-identification**: When using federated learning systems, it's crucial to take out any personal patient information before training and analyzing models. Techniques like hashing, tokenization, and anonymization algorithms help keep patient data private and follow data protection rules [5].

- **Audit Trails and Logging**: Federated learning systems need to keep detailed records and logs to track who accesses patient data, model training activities, and system events. Audit logs help with investigating, reporting compliance, and responding to security incidents or data breaches [17].

These factors ensure that federated learning systems in healthcare facilities are strong, secure, and compliant. They also allow for collaborative model training and analysis while keeping patient information private and confidential.

5 Regulatory and Ethical Considerations

5.1 Compliance with Healthcare Regulations Such as HIPAA

Following the rules of healthcare regulations, like the Health Insurance Portability and Accountability Act (HIPAA), is very important when using federated learning in healthcare settings. HIPAA sets national standards for protecting people's health information and applies to certain entities like healthcare providers, health plans, and healthcare clearinghouses. Several important things to think about for following HIPAA rules in federated learning include:

Data Security and Encryption: HIPAA requires strong security measures to protect patients' electronic health information when it is sent or stored. Federated learning setups must use encryption methods like SSL or TLS to encrypt the data while it's being sent and while it's stored, which ensures privacy and security.

Access Control and Authorization: HIPAA rules require that organizations limit access to ePHI using access controls and authorization systems based on the principle of least privilege. For federated learning systems, it's important to enforce role-based access control (RBAC) and detailed access policies to make sure that only authorized individuals can access patient data and take part in model training [16].

- **Audit Trails and Logging**: HIPAA mandates the establishment of audit trails and logging mechanisms to track access to ePHI and monitor system activity for unauthorized access or breaches. Federated learning implementations should maintain comprehensive audit logs of data access, model training activities, and system events to facilitate forensic analysis and compliance reporting.
- **Data Minimization and Retention**: HIPAA mandates that organizations limit how much electronic protected health information (ePHI) they gather, use, and keep to what is absolutely needed for the intended purpose [2]. Federated learning systems should follow data minimization rules by only collecting and keeping the data needed for training, analysis, and validation of models.

5.2 Ethical Guidelines for Handling Patient Data in Federated Learning

Ethical rules are important for guiding the proper use of patient information in federated learning. Ethical considerations in federated learning include principles such as keeping information private, being open about the process, treating everyone fairly, and taking responsibility [17]. Some important ethical rules for handling patient data in federated learning include:

- **Agreement**: People involved in federated learning projects need to give their okay before sharing their data for model training and analysis. The agreement should clearly explain why the data is being shared, what the risks and benefits are, and what measures are in place to protect the data during model training in federated learning.
- **Privacy**: Privacy preservation is important in federated learning systems. Techniques such as secure aggregation, differential privacy, and encrypted model updates help protect patient data during the learning process.
- **Transparency and Accountability**: In federated learning programs, it's important to be open and accountable by clearly explaining how data is used and the results of the models to everyone involved. Open communication helps to build trust and confidence in federated learning systems and allows everyone to understand the impact of sharing data and training the models.
- **Ensuring Fairness and Addressing Bias**: When creating and assessing federated learning models, it's important to consider fairness and address biases related to patient demographics, socioeconomic factors, and healthcare disparities. Using techniques such as fairness constraints, bias mitigation algorithms, and model interpretability methods can help tackle fairness and bias issues in federated learning models.

5.3 Legal Frameworks and Data Ownership Issues

When it comes to federated learning, it's important to consider the laws and rules about who owns and controls patient data that is shared across different organizations. Some important legal frameworks and data ownership issues in federated learning include:

- **Data Ownership and Governance**: Federated learning projects need to clearly decide who owns the data and how it should be used and shared. Agreements, policies, and frameworks should cover issues like data ownership, intellectual property rights, and data management [2, 5].
- **Interoperability and Interconnectivity**: Federated learning systems should make sure that different institutions can easily share data, train models, and work together. This means setting up standards, formats, and frameworks to ensure that diverse healthcare systems can work together smoothly.

- **Cross-border Data Transfer**: Federated learning projects that involve international collaboration or data sharing across different regions must comply with regulations for transferring data across borders and international laws that protect data. To ensure that the transfer of data across borders is both legal and secure, it is important to use methods like standard contractual clauses (SCCs), binding corporate rules (BCRs), and adequacy decisions.
- **Liability and Accountability**: Federated learning systems should set up ways to handle legal responsibilities and liabilities connected to data breaches, privacy violations, and model errors. This includes creating agreements, clauses, and insurance policies to reduce legal risks and make sure everyone involved is accountable.

6 Practical Implementation and Case Examples

6.1 Case Studies Demonstrating Successful Implementations of Federated Learning in Medical Imaging

Case studies show how federated learning has been successfully used in medical imaging, highlighting its practical benefits and real-world applications.

- **Covid-19 Screening using X-ray images**: Feki et al. [18] presents a joint framework that allowed multiple medical facilities to use deep learning to screen for COVID-19 without sharing patient information. Authors emphasized the use of chest X-ray images as a potential tool for diagnosing COVID-19, given the virus's impact on the respiratory system [19–22]. It tackles the challenges posed by medical data privacy regulations such as HIPAA and GDPR [23], which prohibit the centralization of patient data. The suggested federated learning model demonstrates comparable results to traditional models trained with centralized data sharing, even with non-independent and non-identically distributed (non-IID) and unbalanced data distributions.
- **Federated Learning for MRI Reconstruction**: A team of hospitals collaborated to create a federated learning system for MRI reconstruction [24]. Each hospital shared MRI data from different patient groups and imaging methods. This collaboration led to the development of strong and widely applicable reconstruction algorithms. The federated learning system enhanced MRI reconstruction accuracy and speed while maintaining patient privacy by keeping sensitive imaging data localized.
- **Multi-Center Collaboration for Cancer Diagnosis**: Oncology centers in multiple countries worked together to develop federated learning models for diagnosing cancer e.g. breast cancer, using histopathology images [25, 26]. By combining different datasets from various patient groups and healthcare systems, the federated learning models achieved excellent performance in detecting and categorizing cancer [27–32]. The collaboration between multiple centers made it

possible to create strong diagnostic models without centralizing patient data or compromising data privacy.

- **Remote Monitoring Using Wearable Devices**: Federated learning was utilized to monitor cardiovascular health remotely using wearable devices and mobile apps. Patients used wearable sensors to continuously monitor vital signs such as heart rate, blood pressure, and oxygen levels [33]. Federated learning models were trained on data from different patients, which combined insights from the wearable sensors while ensuring patient privacy. This remote monitoring system allowed for early detection of cardiovascular issues and facilitated timely interventions for patients at risk [34].

6.2 *Real-World Examples of Improved Efficiency and Privacy Protection*

Real-world examples demonstrate how federated learning improves efficiency and privacy protection in medical imaging applications:

- **Enhanced Model Performance**: Federated learning allows collective model training across multiple institutions, leveraging diverse datasets to improve model performance and generalization. By aggregating insights from distributed data sources, federated learning models achieve higher accuracy and robustness compared to models trained on single-site datasets.
- **Privacy-Preserving Model Training**: Federated learning preserves patient privacy by keeping sensitive medical imaging data localized and only sharing model updates with a central server. This decentralized method ensures that patient data remains protected throughout the model training process, reducing the risk of data breaches and unauthorized access.
- **Secure Data Sharing and Collaboration**: Federated learning facilitates secure data sharing and collaboration among healthcare institutions, enabling the development of predictive models, treatment plans, and diagnostic algorithms without centralizing patient data. By maintaining data sovereignty and control, federated learning fosters trust and collaboration among healthcare providers while safeguarding compliance with regulatory requirements and ethical guidelines [16].
- **Efficient Resource Utilization**: Federated learning optimizes resource utilization by distributing model training across multiple devices or institutions, reducing the computational burden on individual healthcare systems. By leveraging existing infrastructure and computational resources, federated learning maximizes efficiency and scalability in medical imaging applications while minimizing costs and resource constraints [35].

These case studies and real-world examples illustrate the tangible benefits and applications of federated learning in medical imaging, including improved efficiency, privacy protection, and collaboration among healthcare institutions. By leveraging federated learning techniques, healthcare providers can enhance patient care, accelerate medical research, and drive innovation in healthcare delivery

7 Challenges and Solutions

7.1 Addressing Communication Overhead and Scalability Challenges

Communication overhead and scalability challenges are important factors to consider in federated learning, especially in large-scale healthcare applications with distributed datasets and different environments. There are several methods that can help tackle these challenges:

- **Communication Compression**: Communication compression methods reduce the size of transmitted data during model aggregation and updating, cutting down on communication overhead. Methods like quantization, sparsification, and differential compression compress model updates before transmission, decreasing bandwidth needs and enhancing communication efficiency [6].
- **Distributed Computing**: Federated learning makes use of distributed computing systems like Apache Spark and TensorFlow Extended (TFX) to run model training and aggregation across many nodes or clusters. Distributed computing helps use computational resources effectively and enables scalable model training on large datasets without needing centralized infrastructure.
- **Edge Computing**: Edge computing moves computer processing closer to where data is created, allowing local model training and inference on devices like smartphones, wearable sensors, and medical devices. Using federated learning models on edge devices reduces communication overhead and improves real-time data processing. This enhances scalability and responsiveness in healthcare applications [5, 6].

7.2 Dealing with Data Heterogeneity and Algorithm Convergence Issues

Data differences and algorithm convergence problems create difficulties in federated learning, especially when working with different datasets and computational environments. Several methods can solve these challenges:

- **Federated Transfer Learning**: Federated transfer learning techniques leverage pre-trained models and knowledge transfer across institutions to bootstrap model training and accelerate convergence. By fine-tuning pre-trained models using local datasets, participating institutions can adapt transfer learning techniques to address data heterogeneity and improve model generalization [8].
- **Adaptive Aggregation Strategies**: Adaptive combining methods change combining weights and update frequencies depending on how participating devices or institutions come together [36]. Approaches such as adjusting learning speeds, momentum, and federated optimization algorithms adapt to optimize the speed of coming together and model performance in different environments.
- **Data Sampling and Weighting**: Data sampling and weighting techniques help in balancing the contribution of data sources based on their relevance and representativeness during model training. By sampling and weighting data sources according to their contribution to the global objective function, federated learning algorithms can reduce the effects of data heterogeneity and improve convergence stability across distributed datasets [9].

7.3 Proposed Solutions and Ongoing Research Directions

Suggestions have been proposed and current research areas in federated learning are concentrating on handling communication overhead, scalability challenges, data differences, and algorithm convergence issues.

- **Communication-Efficient Algorithms**: Current studies are working on making federated learning algorithms that use less communication, reducing the need for exchanging data while still maintaining model accuracy and convergence speed. Methods like decentralized optimization, asynchronous updates, and adaptive communication protocols are being looked into to improve how efficiently communication works in federated learning.
- **Robustness and Generalization**: Our research is focused on making federated learning models strong and adaptable in different situations. We are looking into methods like meta-learning, using multiple models together, and improving optimization algorithms to make the models better at handling different data and computer setups.
- **Privacy-Preserving Techniques**: Ongoing research in federated learning is looking into creating ways to keep data private while training models across different datasets. This includes using secure aggregation, differential privacy, and homomorphic encryption to enhance data confidentiality and integrity.
- **Standardization and Benchmarking**: Efforts are currently in progress to create standard evaluation measures, data sets, and procedures for federated learning research. Standardized benchmarks and evaluation frameworks will help ensure comparisons and reproducibility of federated learning algorithms. This will advance the field and encourage collaboration among researchers and practitioners.

8 Conclusion and Future Scope

The use of federated learning in medical image processing is proving to be very effective. It protects data privacy while allowing collaboration, enabling the development of strong and accurate models using distributed datasets. Federated learning has great potential in personalized medicine, early disease detection, and collaborative research. Though there are challenges, the benefits in terms of data security, efficiency, and model performance strongly support integrating federated learning into future healthcare systems. As technology advances, this approach is set to redefine the landscape of medical image analysis and drive innovation in healthcare delivery. Federated learning is on track to revolutionize healthcare by making the most of decentralized data. Through collaborative model development, this technology can significantly advance personalized medicine by enabling tailored treatment plans based on individual patient profiles. Real-time health monitoring, supported by federated learning on edge devices, will enable proactive disease management and early intervention. Additionally, by enabling collaborative disease surveillance and the integration of multi-modal imaging data, federated learning can improve public health outcomes and diagnostic accuracy. As the healthcare landscape evolves, continuous model improvement through federated learning will ensure that medical solutions remain adaptable and effective in addressing emerging challenges.

References

1. Guan, H., Yap, P.T., Bozoki, A., Liu, M.: Federated learning for medical image analysis: a survey. Pattern Recogn. 110424 (2024)
2. US Department of Health and Human Services, "HIPAA," (2020). https://www.hhs.gov/hipaa/index.html
3. McMahan, B., Moore, E., Ramage, D., Hampson, S., Arcas, B.A.: Communication-efficient learning of deep networks from decentralized data. In: Artificial Intelligence and Statistics, pp. 1273–1282. PMLR (2017)
4. Konečný, J., McMahan, H.B., Ramage, D., Richtárik, P.: Federated optimization: Distributed machine learning for on-device intelligence (2016). arXiv preprint arXiv:1610.02527
5. KhoKhar, F.A., Shah, J.H., Khan, M.A., Sharif, M., Tariq, U., Kadry, S.: A review on federated learning towards image processing. Comput. Electr. Eng. **99**, 107818 (2022)
6. Li, L., Fan, Y., Tse, M., Lin, K.Y.: A review of applications in federated learning. Comput. Ind. Eng. **149**, 106854 (2020)
7. Aono, Y., Hayashi, T., Wang, L., Moriai, S.: Privacy-preserving deep learning via additively homomorphic encryption. IEEE Trans. Inf. Forensics Secur. **13**(5), 1333–1345 (2017)
8. Yang, Q., Liu, Y., Chen, T., Tong, Y.: Federated machine learning: concept and applications. ACM Trans. Intell. Syst. Technol. (TIST) **10**(2), 1–19 (2019)
9. Bonawitz, K., Ivanov, V., Kreuter, B., Marcedone, A., McMahan, H.B., Patel, S., Ramage, D., Seth, K.: Practical secure aggregation for privacy-preserving machine learning. In: Proceedings of the 2017 ACM SIGSAC Conference on Computer and Communications Security, pp. 1175–1191 (2017)
10. Scannapieco, M., Figotin, I., Bertino, E., Elmagarmid, A.K.: Privacy preserving schema and data matching. In: Proceedings of the 2007 ACM SIGMOD International Conference on Management of Data, pp. 653–664 (2007)

11. Liang, G., Chawathe, S.S.: Privacy-preserving inter-database operations. In: International Conference on Intelligence and Security Informatics, pp. 66–82. Springer Berlin Heidelberg, Berlin, Heidelberg (2004)
12. Chen, M., Shlezinger, N., Poor, H.V., Eldar, Y.C., Cui, S.: Communication-efficient federated learning. Proc. Natl. Acad. Sci. **118**(17), e2024789118 (2021)
13. Wicaksana, J., Yan, Z., Yang, X., Liu, Y., Fan, L., Cheng, K.T.: Customized federated learning for multi-source decentralized medical image classification. IEEE J. Biomed. Health Inform. **26**(11), 5596–5607 (2022)
14. Dinh, C.T., Tran, N., Nguyen, J.: Personalized federated learning with Moreau envelopes. Adv. Neural Inform. Process. Syst. **33**, 21394–21405 (2020)
15. Tan, A.Z., Yu, H., Cui, L., Yang, Q.: Towards personalized federated learning. IEEE Trans. Neural Netw. Learn. Syst. **34**(12), 9587–9603 (2022)
16. Li, X., Gu, Y., Dvornek, N., Staib, L.H., Ventola, P., Duncan, J.S.: Multi-site fMRI analysis using privacy-preserving federated learning and domain adaptation: ABIDE results. Med. Image Anal. **65**, 101765 (2020)
17. Chiruvella, V., Guddati, A.K.: Ethical issues in patient data ownership. Interact. J. Med. Res. **10**(2), e22269 (2021)
18. Feki, I., Ammar, S., Kessentini, Y., Muhammad, K.: Federated learning for COVID-19 screening from chest X-ray images. Appl. Soft Comput. **106**, 107330 (2021)
19. Malik, H., Naeem, A., Naqvi, R.A., Loh, W.K.: DMFL_Net: a federated learning-based framework for the classification of COVID-19 from multiple chest diseases using X-rays. Sensors **23**(2), 743 (2023)
20. Akter, M., Moustafa, N., Lynar, T., Razzak, I.: Edge intelligence: federated learning-based privacy protection framework for smart healthcare systems. IEEE J. Biomed. Health Inform. **26**(12), 5805–5816 (2022)
21. Kandati, D.R., Gadekallu, T.R.: Genetic clustered federated learning for COVID-19 detection. Electronics **11**(17), 2714 (2022)
22. Park, S., Ye, J.C.: Multi-task distributed learning using vision transformer with random patch permutation. IEEE Trans. Med. Imaging **42**(7), 2091–2105 (2022)
23. General Data Protection Regulation, "GDPR," (2019). https://gdpr-info.eu/
24. Feng, C.M., Yan, Y., Wang, S., Xu, Y., Shao, L., Fu, H.: Specificity-preserving federated learning for MR image reconstruction. IEEE Trans. Med. Imaging **42**(7), 2010–2021 (2022)
25. Chowdhury, A., Kassem, H., Padoy, N., Umeton, R., Karargyris, A.: A review of medical federated learning: applications in oncology and cancer research. In: International MICCAI Brainlesion Workshop, pp. 3–24. Springer International Publishing, Cham (2021)
26. Naeem, A., Anees, T., Naqvi, R.A., Loh, W.K.: A comprehensive analysis of recent deep and federated-learning-based methodologies for brain tumor diagnosis. J. Personalized Med. **12**(2), 275 (2022)
27. Wang, Q., Zhou, Y.: FedSPL: federated self-paced learning for privacy-preserving disease diagnosis. Brief. Bioinform. **23**(1), bbab498 (2022)
28. Kalapaaking, A.P., Stephanie, V., Khalil, I., Atiquzzaman, M., Yi, X., Almashor, M.: Smpc-based federated learning for 6g-enabled internet of medical things. IEEE Netw. **36**(4), 182–189 (2022)
29. Aminifar, A., Shokri, M., Rabbi, F., Pun, V.K.I., Lamo, Y.: Extremely randomized trees with privacy preservation for distributed structured health data. IEEE Access **10**, 6010–6027 (2022)
30. Guo, Y., Liu, F., Cai, Z., Chen, L., Xiao, N.: FEEL: A federated edge learning system for efficient and privacy-preserving mobile healthcare. In: Proceedings of the 49th International Conference on Parallel Processing, pp. 1–11 (2020)
31. Zhang, D.K., Toni, F., Williams, M.: A federated cox model with non-proportional hazards. In: Multimodal AI in Healthcare: A Paradigm Shift in Health Intelligence, pp. 171–185. Springer International Publishing, Cham (2022)
32. Repetto, M., La Torre, D.: Federated learning through goal programming: a computational study in cancer detection. In: 2022 5th International Conference on Signal Processing and Information Security (ICSPIS), pp. 80–85. IEEE (2022)

33. Wu, Q., Chen, X., Zhou, Z., Zhang, J.: Fedhome: cloud-edge based personalized federated learning for in-home health monitoring. IEEE Trans. Mob. Comput. **21**(8), 2818–2832 (2020)
34. Nguyen, D.C., Pham, Q.V., Pathirana, P.N., Ding, M., Seneviratne, A., Lin, Z., Hwang, W.J.: Federated learning for smart healthcare: a survey. ACM Comput. Surv. (Csur) **55**(3), 1–37 (2022)
35. Yuan, B., Ge, S., Xing, W.: A federated learning framework for healthcare IoT devices (2020). arXiv preprint arXiv:2005.05083
36. Zhang, F., Kreuter, D., Chen, Y., Dittmer, S., Tull, S., Shadbahr, T., Schut, M., Asselbergs, F., Kar, S., Sivapalaratnam, S., Williams, S., Schönlieb, C.B.: Recent methodological advances in federated learning for healthcare. Patterns **5**(6) (2024)

Securing Collaborative Model Training: Navigating Privacy Challenges in Federated Learning for Industrial IoT

C. V. Suresh Babu and G. Suruthi

Abstract Federated Learning (FL) has emerged as a promising paradigm with the emergence of the Industrial Internet of Everything (IoE), where connected devices work together to optimize industrial processes. FL enables machine learning models to be trained across decentralized and distributed edge devices without centralizing sensitive data. FL presents a set of particular security and privacy problems that require careful study, notwithstanding its benefits in terms of data privacy and less transmission overhead. In-depth analysis of the security and privacy issues raised by the use of federated learning in the context of industrial IoE is provided in this chapter. The chapter includes a thorough analysis of the body of research, stressing the weaknesses and dangers particular to the decentralized character of FL in industrial settings. The chapter also explores the regulatory environment and compliance standards controlling data privacy in industrial sectors, offering insights into how Federated Learning systems can comply with current standards and laws. This work aims to facilitate the development and deployment of secure and privacy-preserving Federated Learning systems in industrial environments, fostering the adoption of cutting-edge machine learning techniques while safeguarding sensitive information. It does this by identifying potential threats and proposing mitigation strategies.

Keywords Centralizing sensitive data · Decentralized · Federated learning · Industrial internet of everything · Machine learning · Privacy preserving · Security concerns

C. V. Suresh Babu (✉) · G. Suruthi
Department of Information Technology, Hindustan Institute of Technology and Science, Padur, Chennai, India
e-mail: pt.cvsuresh@hindustanuniv.ac.in

G. Suruthi
e-mail: gsuruthi10@gmail.com

R. Mohana et al. (eds.), *Federated Learning Applications in the Industrial Internet of Everything (IoE)*, Studies in Systems, Decision and Control 611,
https://doi.org/10.1007/978-3-031-99270-4_6

1 Introduction

With the advent of the Industrial Internet of Everything (IoE), Federated Learning (FL) has become a promising concept where interconnected devices collaborate to optimize industrial processes [34]. Unlike traditional machine learning models, which require central data collection, FL allows machine learning models to be trained across decentralized and distributed edge devices framework [1]. Thus, it preserves data privacy while reducing transmission overheads. Nonetheless, FL is associated with unique security and privacy concerns majorly in industrial IoE cases. This implies that there are several vulnerabilities and risks related to this decentralized nature of FL that need a great deal of focus on those issues [3, 36]. This chapter examines the security and privacy issues surrounding the use of FL in an industrial IoE setting; it goes through the previous studies pointing out particular threats posed by decentralised architecture of FL in industry settings as well as examining how they operate within regulated environments. Besides, the research discusses regulatory environment and compliance standards for data protection governing industrial areas so that it offers insights on how FL systems can be made comply with present regulations or legislations existing at any given time.

1.1 Objectives of the Chapter

The objectives of the chapter are:

1. To identify and analyze potential security and privacy threats in FL applications within industrial IoE;
2. To review and synthesize existing research on FL security and privacy in industrial contexts;
3. To examine regulatory and compliance standards relevant to data privacy in industrial sectors;
4. To propose strategies for mitigating identified risks and ensuring compliance with regulatory standards;
5. And, to facilitate the secure and privacy-preserving deployment of FL systems in industrial environments.

1.2 Organization of the Chapter

The rest of the chapter is organized as: Sect. 2 enlightens Federated Learning in Industrial IoT: Complexity and Integration. Section 3 discusses Security and Privacy issues in Federated Learning. Section 4 discusses Mitigation Strategies. Section 5 highlights regulatory compliance. Section 6 focuses on roles and collaboration for secure ecosystem followed by continuous improvement and vigilance in Sect. 7. And, finally, Sect. 8, concludes the chapter.

2 Federated Learning in Industrial IoT: Complexity and Integration

2.1 Overview of Federated Learning

Federated Learning (FL) is a technique of machine learning which addresses privacy, security and data ownership issues. It does this by enabling joint training of models on distributed edge devices such as smartphones and IoT sensors. The approach is most applicable to industries like healthcare, banking and manufacturing where enterprises can harness data from different points without risking sensitive information. For instance, privacy is ensured through decentralised model training in FL which minimizes chances of data breach or unauthorized access.

Privacy is one of the biggest concerns when it comes to adopting FL because it reduces the risk; not only does this apply to the exposure but also fear about privacy [6–8]. It also empowers users with control over their own data hence they can decide when and how to use them for training models. Besides cyber-attacks prevention, keeping information on local gadgets helps tighten up security measures against potential breaches [7, 14, 19] (Fig. 1).

Federated learning (FL) has become increasingly popular in different industries because of its ability to train machine learning models without sharing sensitive data, which is important for sectors that are bounded by strict privacy and regulatory requirements contexts [4, 20, 25]. An excellent case in point is healthcare where FL follows rules such as the Health Insurance Portability and Accountability Act (HIPAA) to improve predictive algorithms and patient outcomes. Nevertheless, it is essential to note that implementing FL in such delicate and intricate environments has some challenges [9] hence necessitating secure aggregation, strong communication procedures, and specialized algorithms specifically designed for federated learning [2, 15].

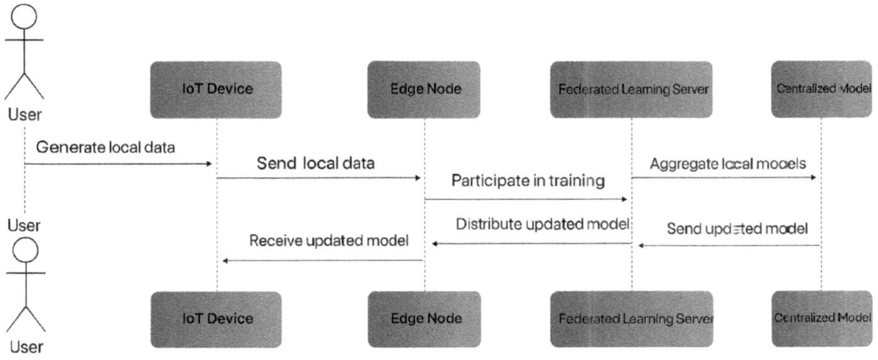

Fig. 1 Use case diagram for privacy and security model

2.2 The Importance of Federated Learning in Privacy-Sensitive Industries

- **Healthcare**: In health care, patient information is highly personal and subject to tight regulations such as HIPAA in America. Through this approach, healthcare providers can use patient records from various sites without necessarily having to get access to their individual data [5]. This partnership enhances the accuracy as well as generalizability of models for tasks like say diagnosing diseases or predicting treatments while still preserving confidentiality of patients. As an illustration, FL can be used to improve fraud detection algorithms by considering transaction patterns without revealing individual transaction records thus preserving patient confidentiality [6, 37].
- **Financial Services**: Similarly, as in healthcare, the financial industry deals with highly-sensitive data such as transaction records, personal finance information, fraud detection data. In Europe for example, regulations such as the General Data Protection Regulation (GDPR) restrict data sharing and privacy [29]. Financial institutions are able to come up with strong fraud detection models through using FL to merge datasets from different banks and other financial institutions but still not disclosing customers' individual transactions [26]. Consequently, this improves detecting fraudulent activities while ensuring adherence to regulatory standards.

2.3 Challenges in Implementing Federated Learning in Industrial IoT

- **Different Kinds of Data**: Diverse sort of data created by various industrial processes.
- **Privacy and Security**: Guaranteeing data privacy and security in a decentralized system.
- **Restricted Network**: Low bandwidth and network limitations in industrial contexts.
- **Different Types of Devices**: Devices are varied, as well as various sensors, with differing capabilities.
- **Inequitable Distribution of Information**: Data availability is unevenly spread across devices.
- **Integration of Models**—Combining model updates from disparate sources is hard to do.
- **Interoperability**: Making different devices fit seamlessly into federated learning frameworks.
- **Flexible Settings**—Adaptation of federated learning approaches for diverse industrial contexts.
- **Observing Guidelines**: Keeping up with sector-specific rules and standards.

- **Managing the Influx of Data Streams**: More IoT devices will connect, resulting in more data streams that require being managed somehow.
- **Fault Tolerance**: Making sure the system remains reliable even if there are device failures or connection losses [15, 32].

2.4 Integration of Federated Learning into Industrial Environments

Centralized machine learning has been revolutionized by Federated Learning, but it still faces many challenges in industrial settings. The limitations of industrial settings need to be recognized as well as the advantages of centralized machine learning. Traditional systems use a centralized data centre for their operations which is not able to work properly due to low bandwidth and latency issues. Therefore, with Federated Learning every model can be customized according to the capabilities of each device so that they can all participate in training even if some have limited resources such as power or memory [16]. Another thing about this method is that it allows for sharing pre-trained models through federated transfer learning and then local fine-tuning them where necessary. In industrial settings data imbalance is often experienced and this may negatively affect model performance unless adjustments are made by modifying federated learning algorithms to handle such imbalances better while maintaining performance levels across models trained from different subsets of data having various amounts per class label. Real time adaptation and learning are important in dynamic environments, however, federated learning does not readily accept these conditions without frequent manual inputs [31].

3 Security and Privacy Issues in Federated Learning

3.1 Data Privacy Preservation

- **Onsite information handling**: Security is attained in the federated learning through local data processing, eliminating central data collection; this does computations on the raw data locally, thereby ensuring its safety.
- **Addition of Noise**: Another strategy for preserving privacy is by making use of random noise or perturbations that are superimposed on private training records to avoid private about those records from being exposed during the model update process.
- **Encryption Approaches**: There are various ways in which model updates can be secured such as homomorphic encryption, secure multi-party computation (SMPC), and protective aggregation. These techniques make sure that even when the updates are sent, no one will have access to how the model was trained [6].

- **Weighted Averaging**: With this method, global changes and weighted averaging methods ensure user's contributions become more hidden by covering individual entries. Thus, one can still consolidate a secure model using such tactics without jeopardizing personal privacy.
- **Client-Side Anonymization**: Preprocessing client data before it's used for training helps protect against any unintentional exposure of sensitive information.
- **Regulatory Compliance**: It is important for Federated Learning implementations to comply with regulations such as GDPR. This includes being open, accountable and responsible in relation to the data privacy rights of individuals and giving its participants authority over their data.
- **Continuous Monitoring**: Regular identification and mitigation of privacy risks are necessary. This requires continuous monitoring and auditing procedures to ensure compliance with legal requirements such as GDPR [26].

3.2 Inadvertent Data Leakage

- **Differential Privacy**: Adding noise to data through differential privacy techniques can help prevent accidental leakage of information which could enable re-identification of specific persons.
- **Secure Aggregation Methods**: The risk of unintended disclosure is reduced by these methods, ensuring that no individual contribution is revealed during aggregation.
- **Participant Diversity**: By having a diverse set of participants, privacy can be protected from being inferred from aggregated results in an effort to safeguard against possible breaches when it comes to this policy.
- **Regular Audits**: Conducting routine audits and monitoring activities aids in catching gaps or malpractices that result into deliberate or inadvertent data leakages
- **Adversarial Defense Techniques**: The resilience of the system against attacks can be increased using adversarial training techniques such that it becomes difficult for adversaries to take out confidential information.
- **Transparency and Documentation**: Being transparent and documenting everything about data handling is a way towards building trust among participators in terms of safety of the system.
- **Data Minimization**: Reducing data usage to only what is needed decreases the risk of sensitive information being exposed, thereby enhancing individual privacy protection [21].

3.3 Model Compromise Attempts (Model Poisoning)

- **Skewed or False Updates**: Adversaries can introduce skewed or false updates into the federated learning process, which undermines both security and accuracy of aggregated global model.
- **Motivations for Attacks**: Manipulation, sabotage or extraction of data may lead to model poisoning each with diverse impacts such as loss of credibility on system security, lower model precision, or exposure of private information.
- **Outlier Detection**: Applying outlier detection mechanisms may be useful in identifying and removing malicious contributions so as to keep up with dependability within a model.
- **Reputation Systems**: Developing reputational systems of participants could be used to evaluate the trustworthiness of the contributions, promote deserving contributors and penalize those who are trying to breach it [38].

3.4 Secure Access and Communication Channels

- **Confidentiality and Integrity**: Maintaining the confidentiality and integrity of data during transmission is crucial in Federated Learning. This involves implementing secure communication channels to protect data from interception or alteration.
- **Access Controls**: Strict access controls and authentication processes are essential
- **Encryption Techniques**: Employing encryption mechanisms like Virtual Private Networks (VPNs) or Transport Layer Security (TLS) protocols guarantees that information remains safe while moving between the central server and individual participants.
- **Participant Trust**: Encrypted communications combined with secure communications channel contribute to the participant's trust which subsequently leads to a more robust collaboration in terms of data sharing [1].

4 Mitigation Strategies: A Holistic Solution Framework

4.1 The Secure Federated Learning Framework

The complex security and privacy issues that come with federated learning are addressed by the complete and reliable Secure Federated Learning Framework (Fig. 2). This approach combines strong defense mechanisms like model robustness training and outlier detection with a variety of privacy-preserving strategies like differential privacy and secure aggregation. To guarantee that participants have control over their data, it places a strong emphasis on responsibility, openness,

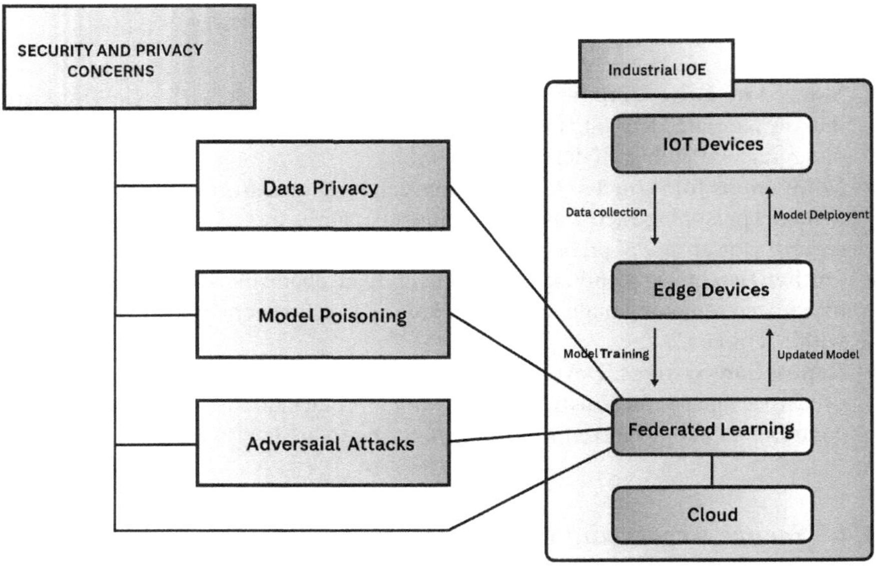

Fig. 2 Potential threats and mitigation strategies

and adherence to data protection laws. Trust and security are improved by participant reputation systems and secure communication channels. A safer and more considerate environment for sharing data and training models is promoted by the Secure Federated Learning Framework, which offers a comprehensive approach to collaborative machine learning [7].

4.2 Leveraging Differential Privacy Techniques

The differential privacy feature of federated learning is vital for securing the model aggregation part. It does this by injecting some controlled noise into database query responses such that it becomes hard for attackers to extract exact information about the individual data points. Differential privacy underpins federated learning due to several reasons. One, it preserves privacy where providers should be private, especially when all participants have their own local data storage in a distributed fashion as in FL (Federated Learning). Two, no sensitive information can be exposed during the combining process because of it. Inference attacks prevention is also done by differential privacy which refers to those attacks where an adversary infers private facts from examining outputs of models trained on them; this is achieved through ensuring that specific values are not disclosed by the produced model while countering such threats as they make extraction of valuable details hard from these kinds of outputs for malicious actors [39].

4.3 Advantages and Challenges of Using Differential Privacy

Differential privacy is a good method for the federated learning system since it provides strong mathematical guarantees, strikes a balance between data privacy and model utility, prevents insider attacks and complies with laws like GDPR. It gives assurances on paper that no one's data involved will be compromised mathematically but allows us to find the right trade-off between private data about individuals or groups on one hand and useful general models applicable across populations or organizations on another. However there lie some challenges such as accuracy versus privacy tradeoff; noise calibration; user experience. It can be difficult to achieve this trade-off because if too much noise is added then the global model may lose its quality. Noise addition in differential privacy should be calibrated correctly which requires understanding of the use case and data at a deeper level. Furthermore, in applications where model predictions are critical, excessive noise could lead to suboptimal user experience [10].

4.4 Implementing Secure Communication Protocols

In Federated Learning, secure communication protocols are very important for protecting data during transmission and preventing unauthorized access or tampering. These protocols involve choosing secure transport protocols like TLS or SSL, enabling end-to-end encryption to secure data from participants to the central server and vice versa, and using VPNs for distributed participants.

Secure authentication and access control is also important where strong mechanisms should be used such as username-password combinations or multi-factor authentication so as to restrict access to data and model updates. Encrypting data at rest is equally important because it ensures that even if physical access is gained then the information will still be safe.

Communication between participants and the central server needs to be done through secure APIs which have authentication, authorization, and rate limiting features in order prevent misuse as well as attacks against them. When dealing with web-based applications it's advisable to always use HTTPS as your secure data transfer protocol. Regular software updates along with patches are necessary in order to prevent vulnerabilities from being exploited by attackers.

Network monitoring alongside intrusion detection systems can identify suspicious activities or unauthorized access in real-time thereby providing early warnings together with enabling rapid response towards security incidents. Regular security audits coupled with penetration testing can help identify vulnerabilities within a federated learning system [17].

Training participants, administrators, and developers on security best practices can enhance overall security posture. Compliance with data protection regulations, such as GDPR in Europe, is essential for protecting participant privacy. By integrating secure communication protocols into Federated Learning, you can safeguard data and model updates, maintain participant privacy, and reduce the risk of unauthorized access or tampering, ultimately creating a more secure and trustworthy environment for collaborative machine learning [27].

4.5 Data Anonymization and Making Approaches

Federated Learning is a machine learning model training system that protects privacy of individual data sources by using data encryption while allowing them to participate in collaborative model training. Data encryption refers to changing the form of data so as to hide sensitive information but still make it usable for machine learning purposes. Differential anonymity is considered as the best method for encrypting data since it involves adding noise to models or inputs such that no attacker can tell which person did what contribution. Another technique in this domain is called Federated Averaging with Weighted Contributions where updates from different users are combined together into one global model thus reducing the chances of exposing any user's dataset. What local transformations mean is that before sending their updated models back to the central server, participants have the opportunity apply various operations on them like dimensionality reduction, feature selection or even aggregation across datasets etc. Feature engineering may involve creating new features from original ones, removing personally identifiable information (PII) or de-identifying risk factors associated with reidentification attacks against released healthcare records etc. (World Health Organization). Secure multi-party computation (MPC) algorithms can be used for computing with encrypted data to ensure privacy in collaborative model training. Homomorphic encryption enables operations on encrypted data without decrypting it, thereby preserving privacy. Data shuffling reshuffles the arrangement of data points within a dataset to make it harder to track back specific data points to individual participants. Secure multi-party computation and homomorphic encryption are among the cryptographic techniques that also safeguard data privacy by allowing secure processing without exposing raw information [30].

The choice of which anonymization technique to go for depends on the particular use case, privacy needs, and nature of data. Understanding the balance between utility of information and its confidentiality is critical in Federated Learning success with privacy preservation [12].

5 Ensuring Regulatory Compliance

5.1 Regulatory Landscape in Industrial IoT

The Industrial Internet of Things (IIoT) regulatory environment is a complex and dynamic terrain that crosses multiple areas. It worries about data security and privacy as well as compliance with local and EU data protection rules and regulations such as the GDPR. IIoT must also comply with industry-specific cybersecurity guidelines and standards, as well as industry-wide interoperability standards from groups like the OPC Foundation and IIC. IIoT systems are impacted by safety and environmental regulations, which guarantee adherence to environmental requirements in resource-intensive businesses and safety standards like those set by OSHA. The legal framework for IIoT is also shaped by rules pertaining to intellectual property, import/export restrictions, radio frequency constraints, and government incentives. To ensure success in navigating this terrain, a thorough strategy combining technical know-how and legal knowledge is needed [23].

5.2 Navigating Legal and Regulatory Requirements

For companies operating in the Industrial Internet of Things (IIoT) sector, navigating the intricate legal and regulatory requirements is a critical task. Companies that want to flourish need to deal with a variety of issues, such as data security and privacy laws that differ depending on the location and demand that personal data be handled carefully. To protect systems from threats, industry-specific norms and cybersecurity requirements must be adhered to. In order to preserve operational integrity and prevent potential liabilities, compliance with import/export, environmental, and safety rules is crucial. Radio frequency restrictions should be followed by IIoT devices that have wireless communication capabilities. While utilizing government incentives might provide strategic advantages, intellectual property and patent considerations require careful attention. A multidisciplinary strategy comprising legal experts, compliance officials, and technical professionals is necessary for success in IIoT in order to successfully traverse the complex regulatory landscape [32].

5.3 Aligning Federated Learning Practices with Regulations

A significant opportunity and difficulty in the Industrial Internet of Things (IIoT) is to align Federated Learning approaches with regulations. There are special compliance issues with federated learning, a decentralized machine learning technique where models are trained locally on edge devices and only aggregated insights are shared. First and foremost, handling personal data carefully is required under data privacy laws like the GDPR in the European Union. Federated Learning fits very nicely with these rules since it permits local processing without disclosing raw data. Organizations must however take precautions to make sure that aggregated insights cannot be used to re-identify specific individuals, which may require using additional privacy-preserving measures [35].

Furthermore, cybersecurity laws and guidelines—like the US's NIST Cybersecurity Framework—are crucial to the IIoT, and federated learning can lessen security concerns by minimizing the need to transfer sensitive data across networks. Ensuring the resilience of the federated learning infrastructure is crucial, particularly when combining model updates from edge devices. Adherence to safety standards is crucial when it comes to quality and safety laws, particularly in sectors like manufacturing. It is necessary to make sure that models follow pertinent safety standards such as IEC 61508 and do not jeopardize safety-critical systems when they are deployed on edge devices in order to integrate Federated Learning with safety practices [24].

Since resource utilization and energy consumption are major IIoT concerns, environmental restrictions also apply. When used effectively, federated learning can minimize the need for centralized data processing and transmission, which can assist improve energy consumption. Organizations must respect patents and other people's intellectual property rights while safeguarding their Federated Learning algorithms and models. While addressing IP regulations, cooperative efforts and open-source initiatives may provide a means of advancing Federated Learning jointly. Finally, since federated learning can boost competitiveness and innovation while resolving regulatory issues, government programs, subsidies, and incentives may promote its use in the IIoT [18].

6 Roles and Collaboration for a Secure Ecosystem

6.1 Role of Stakeholders: Data Owners, IoT Manufactures, and Experts

To create a secure Internet of Things (IoT) ecosystem, collaboration and education are important. In this process, there are three main players; data owners, IoT manufacturers and experts. Data owners should establish strict rules for privacy and access to data which should be implemented as well. System and device development is taken by IoT manufacturers who should ensure that security features are incorporated into their products right from the design stage. They must also share information about vulnerabilities and updates for safeguarding the environment while responding rapidly to new threats that come along their way. Additionally, professionals such as researchers, academia cyber security specialists among others can offer assistance in terms of coming up with new security measures as well as identifying weak points or areas that need improvement based on best practices. Therefore, what is needed most is cooperative strategy because it will help deal with dynamic nature of threats against systems plus ensuring sustainability through long term safety within the internet things realm. Accordingly, education acts as an enabler towards achieving these goals by enlightening stakeholders about emerging risks and what should be done hence leading into more reliable IoTs which are also safe from unauthorized access [33].

6.2 *Importance of Collaboration in Mitigating Security Risks*

Security concerns in the IoT ecosystem need collaboration. Since IoT systems are interconnected, they have a wide attack surface. Thus, sharing knowledge and generating industry standards and best practices become mandatory. This kind of cooperation makes the combined wisdom of different players available to the entire system. Therefore, overall security is improved while individual gadgets and networks get protected. Another benefit is that it enhances incident response capabilities. When people work together in dealing with security breaches; their speed in thinking on their feet increases as well as how fast they act efficiently towards them. This can involve planning joint approaches for reacting to incidents among stakeholders or even exchanging information about threats which may help reduce attacks impact thus enabling enterprises fortify their defenses [13]. Furthermore, it promotes development of educational programmes too. Sometimes there are many challenges faced by single stakeholder such as keeping up with current trends in security risk management brought about by complex nature and variety within the Internet-of-Things environment. In light of this fact let training manuals be created through collaboration and workshops should be organized so that participants can share knowledge concerning these issues besides creating educational programs which equip people with skills necessary for identifying and mitigating against various types of security threats related to the internet of things [20].

7 Continuous Improvement and Vigilance

7.1 *Dynamic Nature of Security Challenges*

The security challenges of the Industrial Internet of Things (IIoT) are influenced by a rapidly changing technological landscape and interconnectedness of industrial systems. Major stumbling blocks include increasing attack surfaces, ever more sophisticated cyber threats, and continually evolving regulatory frameworks. The advent of technology in IIoT is rapid such that old systems become complex requiring assessment and update of legacy system security on a regular basis. Secure new technology might be integrated with outdated infrastructure resulting in weaknesses and difficulties in terms of safety. Supply chain risks are also dynamic; hence continuous monitoring as well as security updates should be done because they come from different countries through software or components. To safeguard critical industrial systems and data, it is important to have a proactive strategy that emphasizes ongoing risk assessment, system updating as well as awareness about cybersecurity. Governments continuously enhance their regulations while trade associations revise standards so that they can match up with changing security landscapes all over again. Resilience at large will require frequent evaluation plus updating legacy system security too. All these points to one thing: there should always be development coupled with attention to detail owing to the never static nature surrounding threats for IIoT securities in conclusion [15].

7.2 *Incorporating Feedback and Enhancing Security Measures*

Maintaining a strong and resilient ecosystem in the context of the Industrial Internet of Things (IIoT) requires a constant and iterative process that includes incorporating input and improving security measures. Feedback systems are essential for spotting weaknesses, evaluating threats, and drawing conclusions from security events. Numerous sources, like as security audits, penetration tests, incident reports, and threat intelligence sharing, may provide this input. Above all, they have the ability to patch and update IoT systems and devices on a regular basis to fix known vulnerabilities and new threats. In addition, organizations should also spend money on staff awareness and training initiatives to make sure that security best practices are broadly accepted and adhered to [25].

7.3 *The Role of Regular Security Audits and Monitoring*

Maintaining the integrity and resilience of Industrial Internet of Things (IIoT) systems requires regular security audits and monitoring. These procedures are crucial in the dynamic IIoT environment, where networks and devices are linked and the threat landscape is ever-changing, for a number of reasons. To begin with, security audits offer an organized and methodical approach to assess the security of IIoT systems. Organizations can find vulnerabilities, misconfigurations, and other dangers that might otherwise go undetected by conducting periodic evaluations. These audits assist in evaluating the efficacy of current security measures and making sure they are compliant with industry best practices and emerging threats [11].

Secondly, for real-time threat detection and response, ongoing monitoring is essential. Because IIoT systems are always in use, attackers find them to be appealing targets. Monitoring tools are able to spot anomalies, suspicious activity, and possible security breaches as they happen. By taking a proactive stance, companies can minimize the effects of security issues by responding quickly. Regular security audits and monitoring also assist companies in staying in compliance with industry rules and guidelines. Regular security assessments and monitoring are mandated by numerous regulatory frameworks, including those that oversee data protection. To sum up, frequent security audits and monitoring play a crucial role in IIoT maintenance of a strong and flexible security framework. In addition to spotting threats and vulnerabilities, these procedures also help with compliance and promote a security-aware culture [18].

7.4 Regulatory Environment, Compliance and Promoting a Security-Aware Culture

The regulatory background in industrial sectors concerning data privacy adds more complexity to FL systems' deployment. To prevent legal consequences and use data ethically, existing standards and laws must be complied with. Integrating compliance measures into FL systems demands understanding the regulatory landscape and aligning with industry standards. To make sure that FL systems can be effectively deployed in industrial settings, it is very important promote security aware culture. Regular security audits and continuous monitoring help in identifying vulnerabilities as well as ensuring compliance with security standards. These practices also cultivate a sense of security mindedness, which stimulates all actors to have strong regard for privacy and data safeguarding efforts [4].

8 Conclusion

Federated learning in Industrial Internet of Things (IIoT) requires comprehensive strategies such as privacy-enhancing technologies, secure communication channels, access management systems, surveillance activities, openness and continuous evaluation to facilitate collaborative model training and data security [14]. Such an approach enables IIoT devices to collectively improve AI models without aggregating their private data thereby requiring strong encryption techniques and data anonymization methods. Preprocessing PII removal and encrypting sensitive information on IIoT devices prevent breaches and unauthorized access. Secure communication tunnels and authentication mechanisms are crucial in thwarting man-in-the-middle attacks and eavesdropping; while role-based permissions/access control measures limit participation only by authenticated machines/persons. Learning systems from emerging cyber threats and maintain data confidentiality. Early identification of security events and risk mitigation require continuous monitoring and anomaly detection. The focus of future research must be on development of improved security mechanisms, sophisticated encryption techniques, secure communication protocols, effective aggregation methods as well as privacy-preserving algorithms to protect

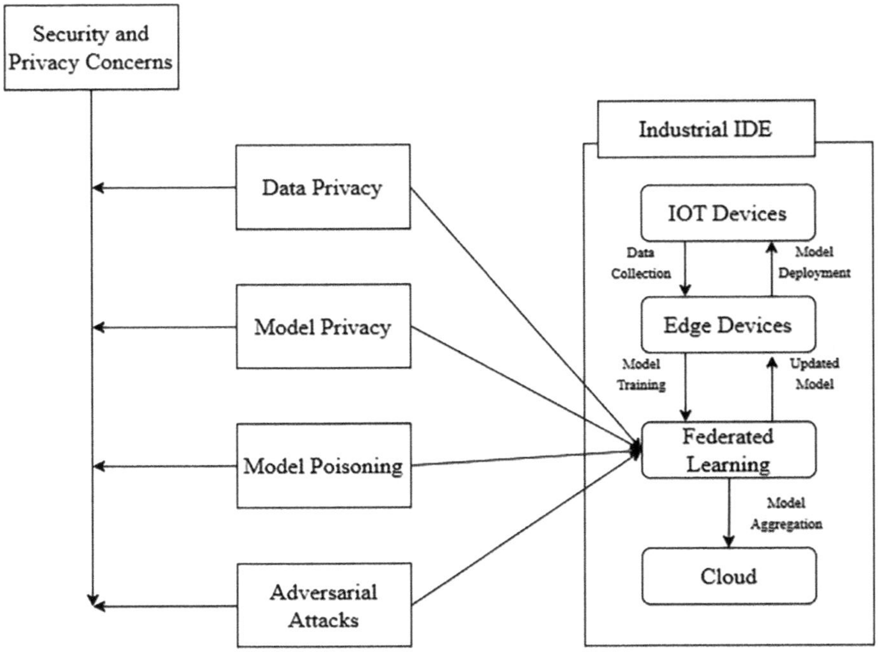

Fig. 3 Futuristic model of Industrial of everything

federated learning systems against emerging cyber threats and preserve data confidentiality. To make sure that federated learning models are robust enough, efficient defense mechanisms against adversarial attacks are needed. The issue of scalability in large-scale industrial applications calls for methodologies optimizing communication efficiency, model convergence speed and overall resource utilization along with dynamic security policies adapting to changing threat environments [23]. Improved interpretability and transparency in federated learning models are important for generating stakeholder trust in industrial applications [19] (Fig. 3).

One of the problems facing federated learning systems which are Industry IoT-specific is how to address specific objectives and restrictions, including diverse security and privacy concerns in different sectors of the Industrial IoE. Consequently, this multidimensional research agenda would impact greatly on the development of secure, privacy-preserving, and effective federated learning systems for Industrial IoE [15]. This is a joint responsibility in IIoT where federated learning holds great possibilities for collaborative machine learning while maintaining a safe and respectful environment. Creating an atmosphere of trust and respect requires applying stringent security measures that are complemented by sophisticated encryption methods as well as strict data anonymization and transparency standards [28]. The realization of federated learning's promise in IIoT will be achieved through collaboration efforts that preserve data protection, trust and respect [22].

References

1. Abadi, M., Bonawitz, K., Talwar, K.: Deep learning with differential privacy. In: Proceedings of the 2016 ACM SIGSAC Conference on Computer and Communications Security (2016)
2. Bonawitz, K., Eichner, H., Grieskamp, W., Huba, D., Ingerman, A., Ivanov, V., Kiddon, C., Konečný, J., Mazzocchi, S., McMahan, B., Van Overveldt, T., Sun, B.: Towards federated learning at scale: system design (2019). arXiv preprint. https://arxiv.org/abs/1902.01046
3. Calmon, F.P., Wei, D., Vinzamuri, B., Varia, M.: LEAF: A platform for secure federated learning. In: Proceedings of the 2019 ACM SIGSAC Conference on Computer and Communications Security (2019)
4. Cui, H., Radosavljevic, V., Chou, F.C., Lin, T.H., Nguyen, T., Huang, T.K., Schneider, J., Djuric, N.: Multimodal trajectory predictions for autonomous driving using deep convolutional networks. In: Proceedings of the 2019 International Conference on Robotics and Automation (ICRA), pp. 2090–2096 (2019)
5. Chen, Y., Qin, X., Wang, J., Yu, C., Gao, W.: FedHealth: a federated transfer learning framework for wearable healthcare. IEEE Intell. Syst. **35**(4), 83–93 (2020). https://doi.org/10.1109/MIS.2020.2975363
6. Chen, Z., Lv, N., Liu, P., Fang, Y., Chen, K., Pan, W.: Intrusion detection for wireless edge networks based on federated learning. IEEE Access **8**, 217,463–217,472 (2020). https://doi.org/10.1109/ACCESS.2020.3047125
7. Dwork, C.: Differential privacy. In: Proceedings of the 33rd International Conference on Automata, Languages and Programming (2006)
8. Feldstein, S.: The road to digital unfreedom: how artificial intelligence is reshaping repression. J. Democr. **30**(3), 40–52 (2019)
9. Geyer, R.C., Klein, T., Nabi, M.: Differentially private federated learning: A client level perspective (2017). arXiv preprint. https://arxiv.org/abs/1712.07557
10. Ghazi, B., Pagh, R., Velingker, A.: Scalable and differentially-private distributed aggregation in the shuffled model (2019). arXiv preprint. https://arxiv.org/abs/1906.08320
11. Hosseini, H., Yun, S., Park, H., Louizos, C., Soriaga, J., Welling, M.: Federated learning of user authentication models (2020). arXiv preprint. https://arxiv.org/abs/2007.04618
12. Konečný, J., McMahan, H.B., Ramage, D., Richtárik, P.: Federated optimization: distributed machine learning for on-device intelligence (2016). arXiv preprint. https://arxiv.org/abs/1610.02527
13. Konečný, J., McMahan, H.B., Yu, F.X., Richtárik, P., Suresh, A.T., Bacon, D.: Federated learning: Strategies for improving communication efficiency. In: NIPS Workshop on Private Multi-Party Machine Learning, Barcelona, Spain, pp. 1–10 (2016)
14. Kang, J., Xiong, Z., Niyato, D., Xie, S., Zhang, J.: Incentive mechanism for reliable federated learning: a joint optimization approach to combining reputation and contract theory. IEEE Internet Things J. **6**(6), 10700–10714 (2019)
15. Li, T., Sahu, A.K., Talwalkar, A., Smith, V.: Federated learning: challenges, methods, and future directions. IEEE Signal Process. Mag. **37**(3), 50–60 (2020). https://doi.org/10.1109/MSP.2020.2975750
16. Lim, W.Y.B., Luong, N.C., Hoang, D.T., Jiao, Y., Liang, Y.-C., Yang, Q., Niyato, D., Miao, C.: Federated learning in mobile edge networks: a comprehensive survey. IEEE Commun. Surv. Tutor. **22**(3), 2031–2063 (2020). https://doi.org/10.1109/COMST.2020.2972708
17. Li, B., Wu, Y., Song, J., Lu, R., Li, T., Zhao, L.: DeepFed: federated deep learning for intrusion detection in industrial cyber–physical systems. IEEE Trans. Indus. Inf. **17**(8), 5615–5624 (2021)
18. Liu, Y., Garg, S., Nie, J., Zhang, Y., Xiong, Z., Kang, J., Hossain, M.S.: Deep anomaly detection for time-series data in industrial IoT: a communication-efficient on-device federated learning approach. IEEE Internet Things J. **8**(8), 6348–6358 (2021). https://doi.org/10.1109/JIOT.2020.3042147
19. McMahan, H.B., Ramage, D., Talwar, K., Zhang, L.: Communication-efficient learning of deep networks from decentralized data. AISTATS **30**(1), 1273–1282 (2017)

20. Ma, C., Li, J., Ding, M., Yang, H.H., Shu, F., Quek, T.Q., Poor, H.V.: On safeguarding privacy and security in the framework of federated learning. IEEE Netw. **34**(4), 242–248 (2020)
21. Nayak, S.K., Ojha, A.C.: Data leakage detection and prevention: Review and research directions. In: Machine Learning and Information Processing, pp. 203–212. Springer (2020)
22. Papadopoulos, P., Abramson, W., Hall, A.J., Pitropakis, N., Buchanan, W.J.: Privacy and trust redefined in federated machine learning. Mach. Learn. Knowl. (2021)
23. Parimala, M., Swarna Priya, R.M., Pham, Q.-V., Dev, K., Maddikunta, P.K.R., Gadekallu, T.R., Huynh-The, T.: Fusion of federated learning and industrial Internet of Things: a survey (2021). arXiv preprint. https://arxiv.org/abs/2101.00798
24. Peng, Z., Xu, J., Chu, X., Gao, S., Yao, Y., Gu, R., Tang, Y.: VFChain: enabling verifiable and auditable federated learning via blockchain systems. IEEE Trans. Netw. Sci. Eng. (2021). https://doi.org/10.1109/TNSE.2021.3095756
25. Riedl, M.O.: Human-centered artificial intelligence and machine learning. Human Behav. Emerg. Technol. **1**(1), 33–36 (2019). https://doi.org/10.1002/hbe2.158
26. Rønn, K.V., Søe, S.O.: Is social media intelligence private? Privacy in public and the nature of social media intelligence. Intell. Natl. Secur. **34**(3), 362–378 (2019). https://doi.org/10.1080/02684527.2018.1520703
27. Rahman, S.A., Tout, H., Talhi, C., Mourad, A.: Internet of things intrusion detection: centralized, on-device, or federated learning? IEEE Netw. **34**(6), 310–317 (2020)
28. Shokri, R., Stronati, M., Song, C., Shmatikov, V.: Membership inference attacks against machine learning models. In: 2017 IEEE Symposium on Security and Privacy (2017). https://doi.org/10.1109/SP.2017.41
29. Suresh Babu, C.V.: Introduction to IoT Applications. Anniyappa Publications (2023)
30. Suresh Babu, C.V., Srisakthi, S.: Cyber physical systems and network security: The present scenarios and its applications. In: Thanigaivelan, R., Kaliappan, S., Jegadheesan, C. (eds.) Cyber-Physical Systems and Supporting Technologies for Industrial Automation, pp. 104–130. IGI Global (2023). https://doi.org/10.4018/978-1-6684-9267-3.ch006
31. Suresh Babu, C.V., Yadavamuthiah, K.: Cyber physical systems design challenges in the areas of mobility, healthcare, energy, and manufacturing. In: Thanigaivelan, R., Kaliappan, S., Jegadheesan, C. (eds.) Cyber-Physical Systems and Supporting Technologies for Industrial Automation, pp. 131–151. IGI Global (2023). https://doi.org/10.4018/978-1-6684-9267-3.ch007
32. Srisakthi, S., Suresh Babu, C.V.: Cybersecurity: protecting information in a digital world. In: Saeed, S., Azizi, N., Tahir, S., Ahmad, M., Almuhaideb, A. (eds.) Strengthening Industrial Cybersecurity to Protect Business Intelligence, pp. 1–25. IGI Global (2024). https://doi.org/10.4018/979-8-3693-0839-4.ch001
33. Suresh Babu, C.V., Abhinaba Pal, A., Vinith, A., Muralirajan, V., Gunasekaran, S.: Enhancing cloud and IoT security: leveraging IoT technology for multi-factor user authentication. In: Ahmed Nacer, A., Abdmeziem, M. (eds.) Emerging Technologies for Securing the Cloud and IoT, pp. 258–282. IGI Global (2024). https://doi.org/10.4018/979-8-3693-0766-3.ch011
34. Suresh Babu, C.V., Auroshaa, A., Sowmi Saltonya, M., Sathyanarayanan, A.S.: Cloud-enabled fire safety in Industry 5.0 smart factories: leveraging IoT and sensor networks for real-time monitoring and proactive prevention. In: Hassan, A., Dutta, P., Gupta, S., Mattar, E., Singh, S. (eds.) Emerging Technologies in Digital Manufacturing and Smart Factories, pp. 150–166. IGI Global (2024). https://doi.org/10.4018/979-8-3693-0920-9.ch009
35. Truex, S., Baracaldo, N., Anwar, A., Steinke, T., Ludwig, H., Zhang, R., Zhou, Y.: A hybrid approach to privacy-preserving federated learning. In: Proceedings of the 12th ACM Workshop on Artificial Intelligence and Security, pp. 1–11. ACM (2019). https://doi.org/10.1145/3314415.3314420
36. Zhang, T., He, C., Ma, T., Ma, M., Avestimehr, S.: Federated learning for internet of things: a federated learning framework for on-device anomaly data detection (2021). arXiv preprint https://arxiv.org/abs/2106.07976
37. Zhang, Z., Li, J., Yu, S., Makaya, C.: SAFELearning: enable backdoor detectability in federated learning with secure aggregation (2021). arXiv preprint https://arxiv.org/abs/2102.02402

38. Zhang, J., Chen, B., Cheng, X., Binh, H.T.T., Yu, S.: PoisonGAN: generative poisoning attacks against federated learning in edge computing systems. IEEE Internet Things J. **8**(5), 3310–3322 (2021)
39. Zhao, Y., Zhao, J., Yang, M., Wang, T., Wang, N., Lyu, L., Niyato, D., Lam, K.-Y.: Local differential privacy based federated learning for internet of things. IEEE Internet Things J. **8**(11), 8836–8853 (2021). https://doi.org/10.1109/JIOT.2021.3087238

Challenges and Opportunities in Federated Learning for the Internet of Everything (IoE)

Sumit Bansal, Piyush Kumar, Tanmay Raj, Saurabh Mishra, and Pankaj Goyal

Abstract Today, the introduction of the Internet of Everything (IoE) which is marked by the interconnectivity of numerous gadgets and devices will result to the creation of a large amount of data. The decentralised approach of the federated learning expands to address these data by using them while circumventing the issues of privacy and security. The chapter illustrates the relationship between federated learning and IoE and also discusses the problems it faces. The chapter further discusses and elucidate the benefits of federated learning in the context of IoE while highlighting decentralized data ownership, edge computing, reinforced privacy preservation methods, and the aspect of cooperation in FL. Various applications that are relevant to real life in medical field, smart cities, industrial applications or IoT, and self-driven vehicles give a clear meaning to the impact of federated learning. In addition, the chapter discusses the latest trends and further research prospective for development of federated learning in context of IoE environments. Overall, the chapter is a synthesis of federated learning as a key facet of the IoE combining the benefits and difficulties of the technology innovation.

Keywords Federated learning · Internet of Everything (IoE) · Decentralized data ownership · Edge computing

S. Bansal (✉) · P. Kumar · T. Raj · S. Mishra
Department of Computer Science and Engineering, Gurukul Kangri (DU), Haridwar, India
e-mail: sumitbansal.cse@gkv.ac.in

P. Goyal
Infosys Technologies, Feltham, Greater London, UK
e-mail: pankaj.goyal09@infosys.com

R. Mohana et al. (eds.), *Federated Learning Applications in the Industrial Internet of Everything (IoE)*, Studies in Systems, Decision and Control 611,
https://doi.org/10.1007/978-3-031-99270-4_7

1 Introduction to Federated Learning and the Internet of Everything (IoE)

IoE and Federated Learning are two very important keys that make a significant shift in the fast-growing trend of modern technology [1]. This is a shift in the processing and utilization of data of connected devices and systems, which is illustrated by this convergence. It is therefore important to provide a brief background of the topic so as to be able to grasp the various systems and the life changing consequences involved. Google even came up with federated learning where several devices contribute to learning from data that is downloaded on each device while maintaining user privacy and decentralizing the conventional paradigm of data processing [2]. In addition to this, the IoT has taken the concept of connectivity to the next level with Internet of Everything ranging from smart homes, industrial control and many more and not restricted to gadgets alone. These two perspectives create the previously unimaginable opportunities for optimizing and inventing at the same time. However, there are a lot of problems with this convergence, including, but not limited to the issues of security and privacy. Because federated learning is distributed, and since the data may be the sensitive type that circulates from one network and device to another, privacy and confidentiality concerns arise [3]. Likewise, the high density of connections of the IoE turns it into a weakness in which enemies can pose threats against the security of data exchange and preservation. Based on the discussed challenges, it is crucial to implement these transformational technologies responsibly and securely as we advance further deep into federated learning and IoE. An example of IoE is illustrated in the following Fig. 1.

The Internet of Everything (IoE) is a concept that attempts to enhance a lot of sectors such as healthcare, smart cities, and transportation by participating in an intelligent environment full of connected systems and devices [4, 5] that will be operated autonomously. Federated learning is a technique of machine learning in which data never leaves the device and yet contributes to worldwide models [6, 7]. FL provides a use case in which a data scientist serves as one party while a wristwatch user is the other to explain how collaborative intelligence can be made possible without surrendering privacy [8, 9]. The chapter situates federated learning in the context of the Internet of Everything focusing on the general questions and particularities of the learning process and also future possibilities of the approach.

1.1 *Objectives of the Chapter*

The objectives of the chapter are:

1. To enlighten the basics of the federated learning within the context of the Internet of Everything (IoE);
2. To illustrate the strengths of security and privacy achieved by federated learning;

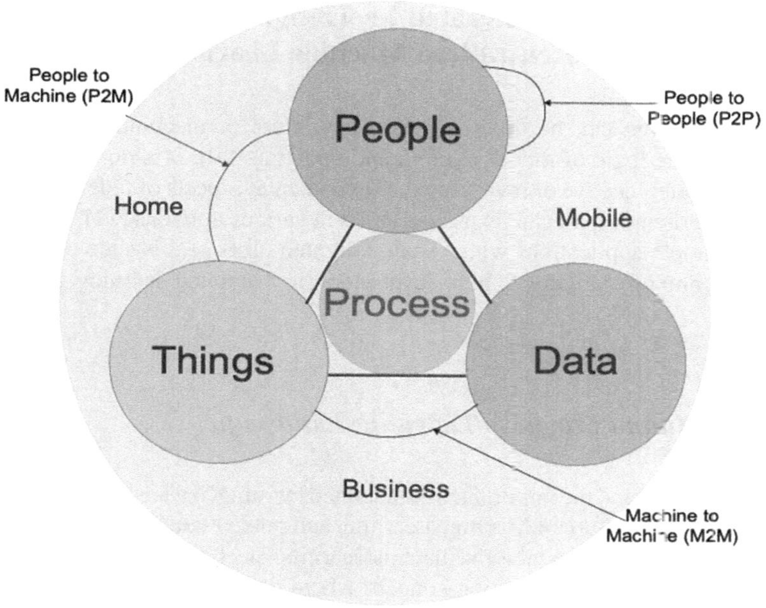

Fig. 1 An illustration of the Internet of Everything (IoE)

3. To assess the use cases as implementing federated learning in the different contexts of IoE;
4. And, to find out directions of further research and overview of the future trends in the field of federated learning.

1.2 Organization of the chapter

The rest of the chapter is organized as:

Section 2 discusses the demarcations of federated learning such as the learning procedure and application. The integration of IoE and federated learning is explained in detail in Sect. 3 of this research. There are several challenges that are encountered in federated learning, these challenges are; privacy concerns, security concerns, and network constraints, as discussed in Sect. 4. Section 5 involves a discussion of the promise of FL, FL's advantages, measures that have been adopted for preserving privacy while using FL, and the concept of collaborative learning in FL. The use case and various application of federated learning in the Internet of Everything is discussed in Sect. 6. Section 7 provides an analysis of the potential future trends and directions for this ground's study. And, finally, Sect.8 concludes the chapter with future scope.

2 Fundamentals of Federated Learning: Unleashing the Power of Decentralized Machine Learning

Federated learning can be described as a new class of machine learning which does rewrite the logic of model training and updating [10]. It is done in a detailed manner with an objective of reviewing the fundamental aspects of federated learning concerning principles, mechanisms, and shifts in various domains [11]. We will also include example applications which shall enlighten users of this slice of paradigm-shifting technology. Figure 2 high level steps in federated learning in machine learning.

2.1 *Traditional Machine Learning Landscape*

In order to understand the importance of federated learning, one needs to first acquaint with the standard Picture of Machine Learning and related issues [12]. By definition, in their traditional implementation, machine learning mechanisms are built according to the centralized data acquisition concept where data is transferred to the model repository from different sources to enable training. While this method is effective in certain scenarios, it gives rise to several concerns: While this method is effective in certain scenarios, it gives rise to several concerns.

Data Privacy: Here centralization means inclusion of specific data in a central place that creates risks to privacy and limit the compliance of regulations.

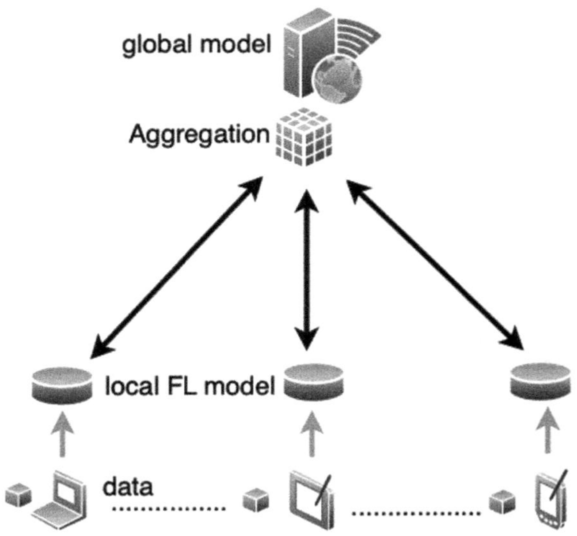

Fig. 2 Fundamental steps of machine learning

Data Imbalance: The centralized systems create an environment in which some data sources contribute data in disproportional amounts and thus the models turn out to be 'biased'.

Scalability: When there are many data sources, centralized systems face problems of scalability efficiently.

2.2 The Birth of Federated Learning

FL stands for the decentralized approach to machine learning as a counter to the disadvantages of the centralized scheme; in the federated learning framework, model training takes place on data sources without transferring raw data [13]. Rather than sharing raw data to a central hub, only the models' updates are asynchronously shared, saving data privacy and security. The key components of federated learning are as follows:

Local Models: Every data source has its local machine learning model in the local database of its machine.

Global Model: An initial condition is defined though the system is global, and parameters are usually random.

Model Updates: Local models communicate by reporting to a master server their detector updates.

Aggregation: The parameters are updated and sent to the central server which helps to update the global model.

2.3 Federated Learning Process

A fundamental prerequisite of comprehending the operating nature of the federated learning is recognition of the process. The step-by-step breakdown is as follows:

Step 1: Initialization—The global model is set up randomly or using any pre-selected values and then shared on all sub-models employed by the contributors of data.

Step 2: Local training: The global model is trained by each data source through the local data; thus, it maintains data privacy.

Step 3: Model Update—Local training results in each data source calculating the disparity between the global model and its issued model, to form a model update.

Step 4: Update transmission—The updates of models are sent to the central server when no raw data leaves the data source.

Step 5: Aggregation—the central server gathers the obtained several model updates and optimizes the global model using methods like averaging.

Step 6: Iteration—steps 2 to 5 are repeated until there has been an improvement of the global model.

2.4 Real-World Examples of Federated Learning

Federated learning's power becomes palpable when real-world applications across diverse domains are examined.

2.4.1 Healthcare

In this case, many hospitals work together to enhance a machine learning model for disease prediction without passing patient data [14].

2.4.2 Mobile Keyboards

Mobile keyboard applications can learn typing habits of users without compromising on data of each keystroke through federated learning [15].

2.4.3 Autonomous Vehicles

In the case of autonomous vehicles, federated learning enables the navigation and safety models to get updated continuously using the data from the various sensors while optimising the fleet's performance without revealing the data of any single auto [16].

2.4.4 Smart Grids

The IoT is also being integrated with electricity grids transforming them into smart electricity grids. This way, through federated learning, it is possible to lower the energy consumption by training models based on the data of several sensors and appliances [17]. This approach makes sure that the energy usage pattern of users is not revealed to other people and also makes the grid to function well.

2.5 *Advantages of Federated Learning*

Federated learning offers several advantages that make it a transformative paradigm in machine learning.

2.5.1 Data Privacy

The first major advantage is protection of data privacy to a greater extent than usually offered. FL maintains that the personal data do not leave the devices but only the model update that is converged to all the participants hence affording individual privacy and meeting regulations such as the GDPR [18].

2.5.2 Scalability

The last advantage of FL highlights its high scalability that is a feature allowing the model's adaptation when more data sources are introduced.

2.5.3 Decentralization

It is also favourable for data sources as it maintains their ownership over the data they contribute.

2.5.4 Reduced Latency

Due to the nature of its application, federated learning is especially useful in situations where a decision is needed in real-time. Local model updates put less strain on the cloud and allow edge devices to work faster and give quick responses.

2.5.5 Global Model Improvement

Due to accumulating the information from local datasets and using them to train a global model, federated learning takes advantages of the wisdom of the crowd; thus, creating a more accurate and resilient model.

Federated learning signifies a significant shift in machine learning, where decentralized intelligence aligns with data privacy imperatives. Real-world examples showcase its applications across industries, fostering innovation and safeguarding sensitive information [19]. Ongoing research and development are rapidly advancing its capabilities amidst persistent challenges. In today's data-driven world, federated learning stands as a beacon for collaborative, privacy-preserving, and intelligent machine learning.

3 Convergence of the IoE and Federated Learning

The combination of federated learning and the Internet of Everything (IoE marks a significant milestone the evolution of technology data-informed decision-making [20]. We explore how these two influential frameworks, which began as separate concepts, now work together to create new avenues for innovation, intelligence. Below, Fig. 3 depicts elements that make up Internet of Everything (IoE).

3.1 A Marital of Scale and Intelligence

The Internet of Everything (IoE) represents a vision for a world that is completely connected. In this world, devices, sensors, people, processes, and data link together. This vast web is defined by an ever-increasing number of connected entities. It generates a wealth of data and opportunities for insights [21]. However, it also brings forth intricate challenges concerning data privacy, security, and scalability.

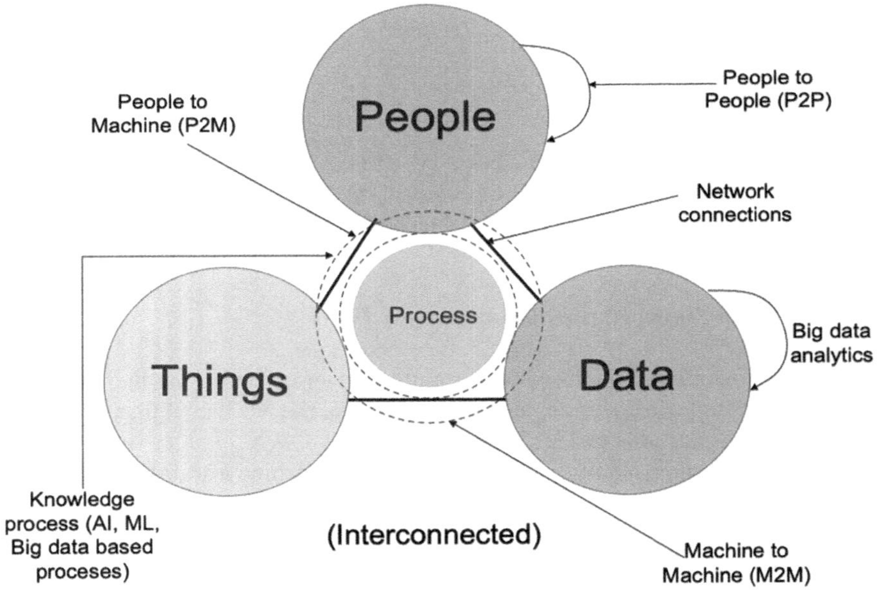

Fig. 3 The components of the Internet of Everything (IoE)

- **Vision of the IoE**: The IoE imagines a hyperconnected experience where nearly every part of our lives—and industries—interacts with technology. This includes a wide range of connectivity among people, processes, data, and things.
- **Challenges of Scale**: As more devices connect within IoE ecosystems, the amount of produced data reaches new heights. This surge demands innovative strategies for managing and analyzing the data effectively.
- **Data Privacy and Security**: The vast array of interconnected entities raises important questions about privacy and safety. Personal information, confidential business records, and sensitive medical information are all integral to this interconnected framework.
- **Scalability challenges**: Scalability is a fundamental challenge in the IoE. Ensuring that systems can handle the ever-increasing number of connected devices while maintaining performance and security is a complex problem.

3.2 *Privacy-Preserving Intelligence*

Both IoE and federated learning are not alien to the issue of privacy. The data gathered by IoE devices, sensors, and consumers' interactions are generally of delicate nature, from individual health statistics to organizational mercantile secrets. Thus, strong privacy guarantees are required because of the huge amount and diversity of information produced by things linked to the IoE. Data users and organizational expectancies are that their data should not be infringed and accessible to anybody. The distributed nature of the training and learning of the model in federated learning means that it can meet expected IoE privacy. The basic concept of federated learning is that data never leave their owner devices and the main focus is to share the updates of the received models in order to enhance the global model [22, 23]. This approach enables the data owners to own their data, which is important for the build-up of trust in the IoE environment [24]. Still, as the system of federated learning is privacy-oriented the system is vulnerable to such privacy threats as adversarial attacks and usage of model inversion techniques to extract certain information from model updates.

3.3 *Edge Intelligence and Real-Time Processing*

Edge computing is one of the defining concepts of the IoE in which computations are carried closer to the source of the data. Federated learning complements this paradigm well as the training of local models can be conducted on clients with computing capacity [20]. The distribution of intelligence reduces delay and prompts quick decision-making, a necessity in uses such as self-driving cars and smart cities.

- **Edge Computing in the IoE**: Edge computing minimizes the time in which data need to travel from where they are generated to a central data center and then back through actual computation. This is important especially when low latency is required in the specific application.

- **Federated Learning at the Edge**: Subsequently, federated learning flexibility places it in a perfect position for edge computing solutions. Local devices or edge servers can fine-tune and increase the efficiency of the AI models by their data.
- **The Role of Federated Learning**: Federated learning empowers edge devices with the ability to learn and adapt autonomously. For example, an autonomous vehicle can continuously improve its navigation model on the basis of its own data and the collective knowledge of other vehicles while preserving the privacy of individual drivers.

3.4 Diverse IoE Applications

We have also seen that IoE and federated learning can be integrated not necessarily for the manufacturing sector or any domain of the industry. It does not have a specific domain; rather it is used in a broad spectrum of processes in different fields. Let us explore some of the prominent domains where this convergence has a substantial impact. Let us explore some of the prominent domains where this convergence has a substantial impact:

- **Healthcare**: In the context of the healthcare delivery, federated learning can lead to individualized health assessment and identification of ailments without compromising on patients' rights to privacy. Smart garments, sensors and telemedicine create abundance of data that can help in improving the state of healthcare [25].
- **Smart Cities**: The IoE and federated learning's primary application is in managing the traffic patterns, energy usage, and public security within smart cities. Real time traffic data from different intersection can be helpful in synchronizing the traffic light so as to ease traffic jam and enhance safety [26].
- **Industrial IoT (IIoT)**: Originally, the discrete production systems utilized federated learning in the areas of predictive maintenance and product quality assurance in industries with extensive interconnectivity of production processes. On manufacturing floors, it is estimated that there is constant data gathering through sensors and IoT devices that are used in training the maintenance prediction machine learning algorithms [27].
- **Autonomous Vehicles**: Federated learning has a very significant role to play in the process of infusing autonomous features in vehicles. Every automobile gathers information about the environment, which is empowered to improve models used in navigation and protection [28]. Cooperative learning is critical to enabling the accomplishment of independent transportation systems [29].

4 Challenges in Federated Learning for the IoE

With federated learning converging with the Internet of Everything (IoE), various unique challenges have arisen. On one hand, federated learning provides significant benefits for privacy and efficiency in handling data. However, applying it within the IoE context brings complexities that require careful thought and attention [30]. This section delves deeply into these challenges, highlighting the intricacies of preserving privacy, managing diverse data sources, overcoming communication and network constraints, and ensuring scalability and efficient resource utilization. Figure 4 illustrates the challenges present within the domain of the Internet of Things (IoT).

4.1 *Privacy and Security Concerns*

Below are the privacy and security concerns discussed in association with federated learning.

Fig. 4 A depiction of the challenges within the realm of the Internet of Things (IoT)

4.1.1 Preserving Individual Privacy

Security remains a paramount issue in the IoE given that data are produced incessantly from various instruments, gadgets, and structures. People expect their data to be personal, and federated learning is built on that very conceptualization [31]. But it presents particular privacy issues. Thus, the federated learning approach guarantees that the raw data do not escape local devices and edge servers. However, only the model updates which are the aggregated summary of the knowledge gets transmitted to a central server. The above approach reduces the chances of exposing data especially in transmission [32]. However, it is quite important to understand that federated learning is not completely protected from privacy threats. The major threats are adversarial attacks and model inversion. That is why it may become possible for adversaries to reconstruct the sensitive data by analyzing the updates of the model. Future studies will aim at designing effective countermeasures for such attacks, methods such as secure summation and DP.

4.1.2 Differential Privacy

Differential privacy is a method of measuring the distortion that one person's data can cause within a model. It added some noise into the model updates so as not to be able to reconstruct a specific data point. Privacy is improved by its implementation but it comes at the expense of model accuracy as well as privacy. In the IoE, where data correctness is a prominent factor since it relates to proper diagnosis in the case of healthcare or correct navigation in the case of automated vehicles, it becomes harder to strike this balance. Despite the fact that it is possible to add noise for privacy preserving, researchers will study how to make the added noise useful for models.

4.1.3 Securing Model Updates

The primary motivation of federated learning is to protect the privacy of data, which, however, brings new threats to the model updates. Some of the possible threats are that the participants of the process may try to control the updates in such a way that the new information in the global model contains a significant portion of harmful and biased data. The authenticity and update of any model is very important hence the need to ensure the model was not tampered with. Security of communication and cryptography are thus among the crucial factors enabling the solution to this problem [33]. Secure multiparty computation (SMPC) methods allow the computation of a function with several parties' inputs without revealing the data to other parties. While the leveraging of the SMPC and related approaches into federated learning is to guarantee that the updated models are proof and respectable.

4.2 *Data Heterogeneity and Distribution*

This section discusses the issues of data heterogeneity and distribution.

4.2.1 Diverse Data Sources

IoE environments encompass a vast array of data sources, ranging from consumer smartphones and wearables to industrial sensors and infrastructure. These data sources differ not only in their types, including structured and unstructured data but also in their data formats, quality, and distribution. Federated learning must adapt to this data heterogeneity to build accurate and reliable global models. Traditional centralized machine learning approaches are less affected by data diversity since data are consolidated into a central repository [3]. In federated learning, each local device or edge server has its own data distribution, making it crucial to handle this diversity effectively. One of the defining characteristics of federated learning is the nonidentically and independently distributed (non-IID) nature of data across devices or edge servers. In non-IID scenarios, data from different devices or servers may follow distinct data distributions. This can lead to challenges in model convergence and overall performance. Federated learning algorithms need to be adapted to handle non-IID data effectively. Researchers are exploring techniques to mitigate the impact of data diversity, such as personalized federated learning and data augmentation. These approaches aim to improve model convergence and accuracy in non-IID settings.

4.3 *Communication and Network Constraints*

This section discusses the issues related to communication and network constraints in federated systems.

4.3.1 Bandwidth Limitations

IoE scenarios may encompass devices in the areas that are physically remote or lack resources to share huge amounts of data. Distribution of model updates from local devices to the central servers is often costly and sometimes impossible especially when resources are limited. To resolve this issue, federated learning relies on a communication-efficient procedure. An example of Model quantization is compressing the updates of a model before transmission with only some amount of data getting through the network [34]. Other optimisation methods with quantisation methods become pivotal to keep using the IoE Fed-HeB bandwidth to the minimal requirement.

4.3.2 Unreliable Networks

IoE environments encompass a vast array of data sources, ranging from consumer smartphones and wearables to industrial sensors and infrastructure. These data sources differ not only in their types, including structured and unstructured data but also in their data formats, quality, and distribution. Federated learning must adapt to this data heterogeneity to build accurate and reliable global models. Traditional centralized machine learning approaches are less affected by data diversity since data are consolidated into a central repository. In federated learning, each local device or edge server has its own data distribution, making it crucial to handle this diversity effectively.

One of the defining characteristics of federated learning is the nonidentically and independently distributed (non-IID) nature of data across devices or edge servers. In non-IID scenarios, data from different devices or servers may follow distinct data distributions. This can lead to challenges in model convergence and overall performance [35]. Federated learning algorithms need to be adapted to handle non-IID data effectively. Researchers are exploring techniques to mitigate the impact of data diversity, such as personalized federated learning and data augmentation. These approaches aim to improve model convergence and accuracy in non-IID settings.

4.4 Scalability and Resource Management

Below, a subsection discusses the issues of scalability and resource management.

4.4.1 Scalability Challenges

IoE scenarios can involve device located in areas of unpredictable connectivity or for that matter in areas with limited bandwidth. Model updates between locally implemented devices and central server can be computationally expensive and where bandwidth is limited may not be easily done. To that end, federated learning uses communication-efficient methods and procedures with the members of the network. Some of these optimizations include Model quantization for instance entails that the model update is compressed before being sent over the network [36, 37]. As other optimization procedures, practising techniques involve quantization so that the usage of bandwidth in the federated learning of IoE can be reduced.

4.4.2 Resource-Efficient Federated Learning

IoE devices may often work in scenarios where they can temporarily or sometimes even permanently lose the network connection. There are different reasons why continuity of the network may be affected; this may be as a result of signal interferences, congestion of the network or mobility of the devices. For the federated learning to be effective under such conditions, then the algorithm has to be very strong in the areas of network interruption. This line of research deals mainly with communication paradigms that are suitable for unreliable networks. Such measures afford trending processes the perspective to proceed unhampered under adverse conceivable network situations. Even in the case of network instability issues, smart practices like choosing adaptive learning rates and performing the updates asynchronously are beneficial. The following sub section offers information concerning the difficulties of scale and resource control.

5 Opportunities and Advantages of Federated Learning in the IoE

With The convergence of federated learning and the Internet of Everything (IoE) marks a transformative moment in the realm of technology and data-driven decision-making [38]. This fusion presents a multitude of opportunities and advantages that have the potential to redefine how we harness the power of data in interconnected ecosystems. These advantages span various dimensions, encompassing technological innovation, data privacy, and the cultivation of collaborative intelligence.

5.1 Decentralized Data Ownership

One of the fundamental principles underlying federated learning in the IoE is the restoration of data ownership and control of the entities that generate it. Whether these entities are individuals or organizations, federated learning ensures that they retain full control over their data, an aspect that has significant implications for privacy, security, and the autonomy of data creators.

5.1.1 Empowering Data Owners

The current environment characterized by the Internet of Everything signifies that data is created in voluminous amounts and in various settings. These data sources may as well include the health vitals of an individual captured through wearable technologies, data generated from industrial sensors, among other data sources. Federated learning can be observed in this context as a shift in data ownership where the power lays with the Data Producers [39]. These fears are due to federated learning that performs the model training locally in devices and edge servers thereby empowering the end-users. Therefore, data are equally owned and managed by their owners, such as people observing their health condition or companies that oversee industrial processes. Who owns the data now also reflects significant changes that occur with data collection rules which states that people must have more control over what data is being collected from them. Federated learning allows users to participate in the management of data initiatives to choose specific circumstances and ways of using their data for creating models.

5.1.2 Privacy by Design

Security and privacy are fundamental issues in the environment of IoE since most of the data that are produced in the system are sensitive. Privacy issues, federated learning takes a 'privacy by design' solution to it. In its design structure, federated learning guarantees that the data are never transferred from the end devices or edge servers in raw form. However, model updates which contain improved knowledge as opposed to individual data points are sent to a central server or aggregator. In the IoE applications it is often the case that the exchanged datasets include parameters containing people's health records, manufacturing method of a product, or other details that could be of value to third party interlopers. Not only does it preserve information privacy at the individual and organizational level but also assure on the safety of passed information throughout the FL procedure. Such a strict privacy preserving mechanism is arguably the key enabler for making federated learning valuable in the IoE.

5.2 Edge Computing and Real-Time Processing

Another significant advantage of the convergence of federated learning and the IoE is the seamless integration of edge computing and real-time data processing. This synergy enhances the efficiency, responsiveness, and scalability of IoE applications. A visual representation of the advantages of Edge computing is provided in Fig. 5.

5.2.1 Localized Intelligence

Edge computing suggests dispersing computing assets and decision-making to the edge of the network, that is, closer to the data. As seen in the IoE environment, data are produced in real time both from numerous nodes and sensors, thus localized intelligence is critical. This style of training is perfect with federated learning since model training and model inference occur on the devices or edge servers that are computational. It provides localized processing and sound 'response time' which is a key issue when it comes to self-driven cars, smart cities and manufacturing industries. For instance, in the instance of self-driving cars, federated learning allows cars to decide on the spot on the basis of raw data as well as other collaborating conclusions, which significantly cuts short the time necessary for information processing and

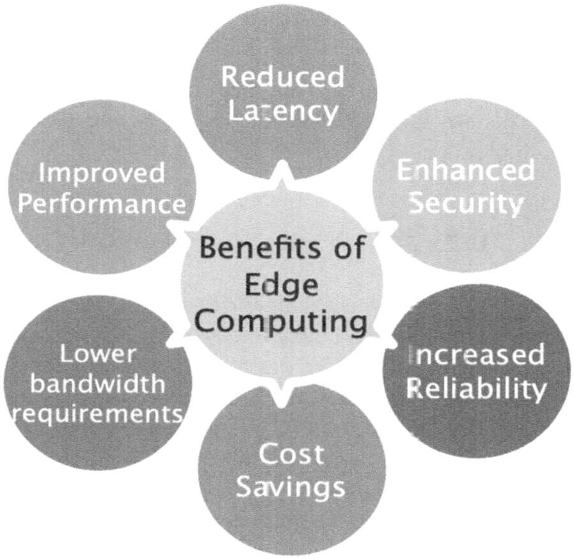

Fig. 5 A visual representation highlighting the advantages of Edge computing

consequent decision making. The outcome is the enhanced and more secure means of transport systems.

5.2.2 Bandwidth Efficiency

IoE environments are frequently heterogonous in data sources that are available and that have restricted bandwidths. A key aspect of federated learning is the fact that data and model training are performed locally thus enhancing the bandwidth. Due to the lesser amount of raw data that needs to be shared in the network, federated learning helps to offload substantial loads from central servers and the network. As much as this is the case the ability to transmit data within the limited bandwidth is a plus in situations where the network resources are either scarce or expensive, this being so if the setting is in an area that is geographically isolated or in populated areas such as urban centers. Not only does it lower the expenses of a business but also improves the sustainability of IoE systems and their abilities to grow rapidly. Thus, federated learning helps achieve the goal of using the data where they are located without transferring them when it is unnecessary and can burden the network with the flow of data.

5.3 *Enhanced Privacy-Preserving Techniques*

Privacy preservation is an ongoing concern in IoE applications, and federated learning incorporates advanced techniques to ensure data privacy throughout the model training process.

5.3.1 Differential Privacy

Differential privacy is a mathematical framework that quantifies how individual data points can influence the parameters of a machine learning model. This is achieved by introducing controlled noise into the model updates. The primary goal of differential privacy is to prevent the inference of individual data points from the model updates [40]. In the context of federated learning within the IoE, where data privacy is paramount, differential privacy plays a critical role. This ensures that even when participating devices or edge servers collaborate on model training, the privacy of individual data sources is preserved. This protection extends to highly sensitive data, such as personal health metrics and proprietary business data, making federated learning a robust solution for privacy-conscious applications.

5.3.2 Homomorphic Encryption

Homomorphic encryption is another cornerstone of privacy preservation in federated learning. This cryptographic technique allows computations to be performed on encrypted data without the need for decryption [41]. In the context of federated learning, homomorphic encryption ensures that data remain confidential throughout the model aggregation process. The importance of homomorphic encryption is evident in applications where data must be protected at all costs. Financial transactions, personal records, and intellectual property are examples of data that require the utmost confidentiality. Federated learning, powered by homomorphic encryption, enables organizations to collaborate on model training without exposing their sensitive data to risk.

5.4 *Collaborative Learning in IoE Ecosystems*

One of the most compelling advantages of federated learning in the IoE landscape is its capacity to foster collaborative learning within ecosystems that span diverse domains and sectors.

5.4.1 Collective Intelligence

Federated learning empowers devices, sensors, and machines to collaborate and collectively improve their machine learning models on the basis of insights gained from the entire network. In practical terms, this means that data-driven decisions and improvements become a collaborative effort. For example, in a smart city, traffic data from various intersections, along with insights from sensors and cameras, can be used to optimize traffic flow. This collaborative intelligence results in reduced congestion, enhanced safety, and improved urban mobility. Collaborative learning is particularly valuable in IoE applications where the complexity of challenges often transcends the boundaries of individual domains. By pooling insights and knowledge across sectors such as healthcare, transportation, energy management, and environmental monitoring, federated learning enables holistic solutions to emerge [42]. For example, the lessons learned from optimizing energy consumption in smart homes can be applied to industrial processes, promoting sustainability and resource efficiency.

5.4.2 Cross-Domain Insights

IoE ecosystems are often connected with several domains where each domain has different factors and prospects. Through federated learning, there're capabilities of combining the findings and knowledge in these areas enhancing innovation and sharing of ideas amongst the domains. This kind of cross-domain design has the advantage of discovering new solutions and improvements that might be kept internal in domain-specific systems. For instance, let us discuss the healthcare option in combination with the transportation option within the smart city environment. Therefore, federated learning in the analysis of traffic patterns' data might yield insights of how traffic patterns are related to certain health conditions and such data could assist in the planning of cities, as well as, where to allocate health resources. This is a challenge for the federated learning as to how it can effectively solve problems in different domains that intersect in real-life.

In other words, it is not just a keeper of technology achievements but rather a wide-scope area which includes federated learning on the Internet of Everything benefits. They have stunning societal and economic consequences as it enables people or companies to have more decision-wise control on the data, facilitate real-time, safe and speedy data processing, and advance cooperative wisdom among multiple fields. Continuing on, we describe real-world cases and scenarios with practical examples speaking to how and where federated learning makes a difference to industries and communities globally.

6 Use Cases and Applications of Federated Learning in the IoE

With The combination of federated learning with the Internet of Everything (IoE) paves the way to new field of data-oriented applications and findings. In its quintessential form the biggest advantage of federated learning is the use of the data drawn from distributed sources while not violating data security. In the IoE context, this capability boils down to the following prolific application areas across different domains of interest.

6.1 Healthcare and the IoE

In the healthcare sector, the IoE represents a paradigm shift, with the potential to significantly improve patient care, reduce costs, and enhance medical research [43]. Federated learning plays a pivotal role in realizing these goals while addressing the critical concerns of data privacy and security. A graphical representation of the use cases and applications of federated learning in the IoE is shown in Fig. 6.

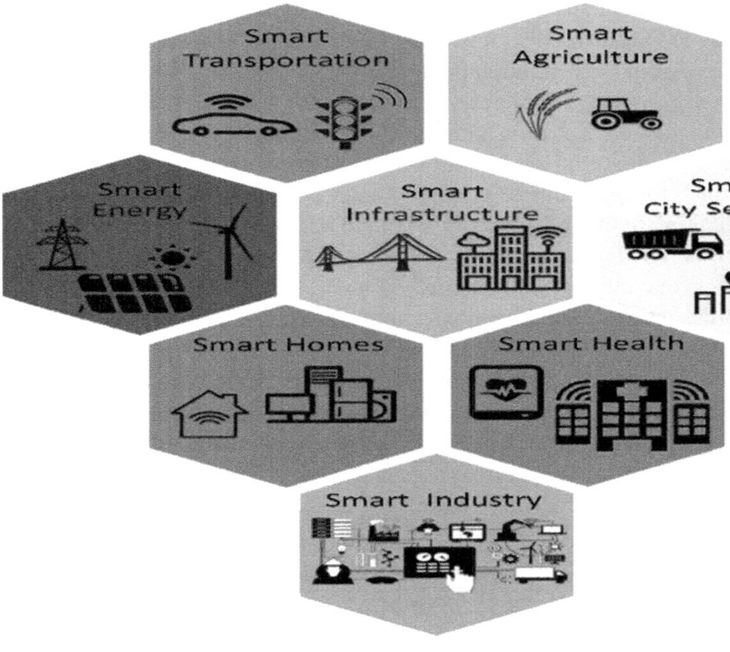

Fig. 6 Visualizing the use cases and applications of federated learning in the Internet of Everything (IoE)

For example, consider a scenario where federated learning is employed to analyse the cardiac health data of thousands of patients, each monitored through wearable devices. By training machine learning models collaboratively across these devices while keeping the data decentralized, federated learning can identify subtle patterns and risk factors associated with heart conditions. The result is a system that can provide early warnings and personalized interventions, potentially saving lives and reducing healthcare costs.

6.1.1 Drug Discovery and Research

In the fields of pharmaceuticals and medical research, federated learning is a game changer. Drug discovery and development involve the analysis of extensive datasets, often containing proprietary information. Federated learning enables pharmaceutical companies and research institutions to collaborate on model training without sharing sensitive data. For example, organizations researching a new drug can collectively train models on their datasets to identify potential candidates more efficiently. By keeping their data within their domains, they protect intellectual property and adhere to regulatory requirements. This collaborative approach expedites drug discovery processes and promotes the development of innovative treatments.

6.2 Smart Cities and Urban Planning

Smart cities are at the forefront of IoE adoption, and federated learning is poised to revolutionize how cities operate, from optimizing traffic flow to enhancing environmental sustainability.

6.2.1 Traffic Management and Optimization

This is a familiar problem known to the people living in large cities. By using federated learning, it is possible to gather and analyse traffic data coming from various sources such as traffic cameras, sensors placed in the roads, and vehicles data. When models that are trained in the collaborative manner are used it can control traffic signals, reroute traffic to avoid congestion and optimize transportation. Suppose a smart city environment in which, based on traffic conditions, lighting systems and signals of intersection, public transport systems, and even the cars themselves adjust their levels accordingly. This is made possible federated learning that operates at the edge, processing data and making split-second decision on traffic flow. Therefore, traffic becomes less of an issue, travel time to work is cut down, and pollution is kept at bay.

6.2.2 Environmental Monitoring and Sustainability

Environmental management is becoming the major concept of smart cities. Federated learning plays this role because it allows for the accumulation and processing of data from a vast number of sensors and devices with regard to the environment. Starting with monitoring the air we breathe to optimising the way we dispose our waste; federated learning enables well informed decision making to improve the quality of life in cities. For example, residents of a city can use sensors to get environmental data including pollution, temperature, and humidity. These sensors are using federated learning models, where models can be learned across these sensors about the air quality and can predict future trends and intimate preventive measures like flow of traffic and energy consumption in buildings. Hence, the following conclusions enhance a cleaner and healthier living environment in urban areas.

6.3 Industrial IoT (IIoT) and Manufacturing

In the realms of the Industrial IoT (IIoT) and manufacturing, federated learning holds the promise of increased efficiency, reduced downtime, and enhanced safety.

6.3.1 Predictive Maintenance

Manufacturing facilities rely heavily on the continuous operation of machinery and equipment. Unplanned downtime can be costly and disruptive. Federated learning enables predictive maintenance by analysing data from sensors and equipment to detect anomalies and patterns indicative of potential failures. For example, in a factory with hundreds of machines, each equipped with sensors, federated learning models can collaboratively analyse data from these sensors. By identifying early warning signs of equipment degradation or failure, maintenance teams can be alerted to perform preemptive maintenance. This approach minimizes downtime, extends the lifespan of equipment, and optimizes production processes.

6.3.2 Quality Control and Defect Detection

Pervasive are the views on timely consistency of the product quality in manufacturing. It is common to note that federated learning improves quality control procedures since the data collected at several stages in the production process is combined to identify problematic areas while maintaining the data's confidentiality. Suppose a manufacturer is involved in the production of intricate electronic parts. This is because federated learning enables the combination of data from the sensors, cameras, as well as the quality control stations. The internet-supported machine learning models allow for the simultaneous recognition of defects or deviation from the target state and thus can guarantee that all products released into the market correspond to the highest quality. Not only from the perspectives of waste and rework but also from the viewpoints of customer satisfaction, this is also the case.

To sum it up, the application opportunities and the application area of federated learning in the Internet of Everything (IoE) are as diverse as the areas of healthcare or smart city, Industry 4.0, and autonomous transportation systems. As a decentralized machine learning technique, federated learning empowers an organization with an efficient method to accrue valuable insights from sources of data that are devolved by decentralization while maintaining privacy and security in the IoE environment. This is true because with the expansion of the IoE in the society, federated learning is going to be a center stage of most of the solutions required to drive engagements in the interconnected world.

7 Future Trends and Research Directions

With having its basis on machine learning as well as utilizing decentralized data, federated learning is a constantly developing area that adapts to new issues and explores fresh opportunities [44]. Several interesting trends and research directions in the context of the Internet of Everything (IoE) and are depicted in Fig. 7 that are influencing the future of federated learning.

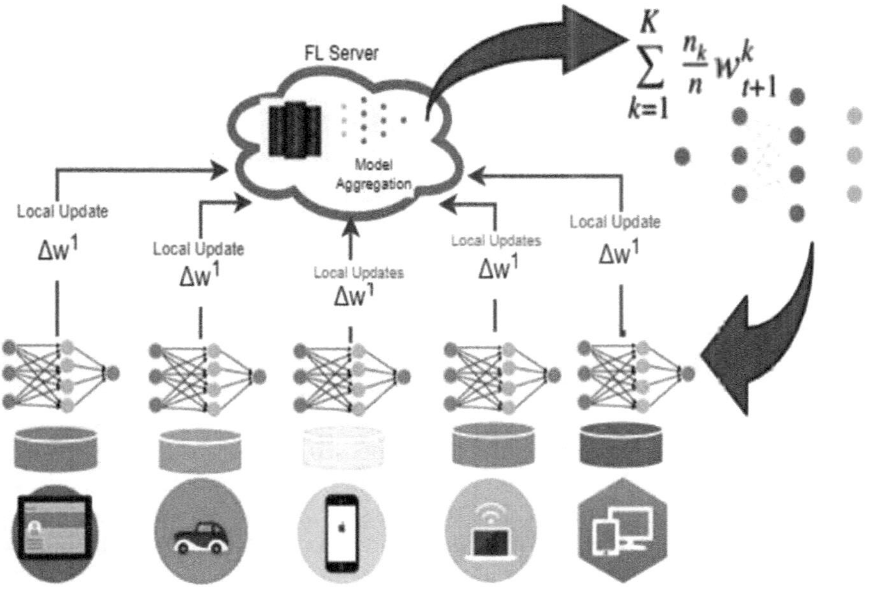

Fig. 7 Illustrating future trends and research directions in the IoE landscape

7.1 Federated Learning for Edge Devices

One prominent trend is the integration of federated learning with edge computing. Edge devices, such as smartphones, IoT sensors, and autonomous vehicles, generate vast amounts of data locally. Federated learning allows these devices to collaboratively learn and make predictions without sending data to centralized servers. This trend is poised to have a profound impact on various IoE applications.

- **Edge AI and real-time inference**: Edge devices are becoming increasingly capable of running complex machine learning models locally. Federated learning enhances this ability by enabling edge devices to train and fine-tune models collectively. This leads to real-time inference for applications such as autonomous vehicles, smart home automation, and predictive maintenance in industrial settings [45].
- **Bandwidth and Latency Reduction**: By performing model updates locally, federated learning reduces the need for continuous data transmission to the cloud. This not only conserves bandwidth but also minimizes latency, which is critical for applications where split-second decisions are necessary, such as augmented reality (AR) and remote robotic control.
- **Privacy at the edge**: Edge devices often handle sensitive data, from personal health information to surveillance footage. The emphasis of federated learning on decentralized training ensures that data remain on the device, preserving user privacy. This aligns with evolving privacy regulations and consumer expectations.

7.2 Federated Learning for Unstructured Data

Typically, federated learning has primarily been used with fairly clean and ordered formats of shared data, including numerical data accumulated by sensors or data kept in tables in a database. However, the IoE creates a significantly large volume of unstructured data inclusive of images, audio and text. Thus, future research will lie in applying federated learning approaches and modifications to optimally deal with such data types.

- **Multimodal Federated Learning**: This is the case because, depending on the adopted technology, IoE devices are in most cases capable of capturing data from more than one sensor at the same time. For instance, a smart city camera may contain a feature to capture or record images and at the same time record sound. Multimodal federated learning has been proposed to learn from data of multiple modalities to enhance the performance. This could result in new IoE applications like audio visual event detection or multisensory anomalous event detecting.
- **Natural Language Processing (NLP) and Text Data**: Social media posts or chat transcripts are examples of text data that are important for the IoE business. The use of federated learning for NLP makes it possible to perform sentiment analysis, acquire contextual knowledge, build chatbots, and many others without

infringing user privacy. However, new approaches for the generation of text at the federation level can be developed, which will open new opportunities for creating and recommending content.

7.3 Federated Learning for Federated Learning

As federated learning ecosystems grow increasingly complex, managing the training process itself becomes a challenge. Future research will focus on federated learning methods to improve the efficiency, scalability, and robustness of federated setups.

- **Communication-Efficient Algorithms**: Developing communication-efficient federated learning algorithms is critical for reducing the overhead of transmitting model updates between devices. Techniques such as quantization, sparsification, and compression are refined to minimize communication costs.
- **Cross-Device Transfer Learning**: Federated learning models trained on one set of devices could benefit from the knowledge acquired on other devices. Cross-device transfer learning mechanisms enable models to generalize better, even when trained on highly diverse data sources within the IoE.

7.4 Federated Learning and Security

Security remains a paramount concern in federated learning, particularly in the IoE, where the data come from diverse, potentially untrusted sources. Future research directions will focus on enhancing the security and privacy guarantees of federated learning.

- **Differential Privacy for Federated Learning**: Federated learning can benefit from advanced privacy-preserving techniques such as differential privacy. Researchers will work on integrating robust differential privacy mechanisms into federated learning frameworks to provide stronger privacy guarantees.
- **Blockchain and Federated Learning**: Blockchain technology offers a decentralized and secure ledger for tracking and validating model updates in federated learning. Integrating blockchain with federated learning can enhance the traceability and integrity of the training process, which is critical for applications such as supply chain monitoring within the IoE.

Finally, the combination of federated learning and Internet of Everything (IoE) becomes one of the most emerging and highly developable domains. As for the future trends, more decentralization, efficiency, and security will predetermine further development of federated learning. All these trends complement one another and, together with the growing emphasis on compliance and ethics, will define the IoE's data-oriented future. The desire for advancements in federated learning, as well as its practical implementation, falls on everyone in the role of the researcher, the practitioner, and the policymaker to ensure that this technology is not only sustaining and functional in enabling the world's interconnectivity but is also responsible in its application.

8 Conclusion and Future Scope

The chapter dissected the interconnection between federated learning and Internet of Everything (IoE) by analysing some of the challenges, taking new opportunities, and envisioning the prosperous future in this field. Other methods, such as federated learning with differential privacy, enhance the confidentiality of the data, protecting data value while honouring data subjects' privacy. Integration in the IoE ecosystems promote cooperation which enhances collective intelligence as the models are improved and predictive techniques enhanced. In various applications in healthcare, smart city, industrial IoT and autonomous transportation, we looked at how these applications can make federated learning a more efficient, secure and scalable solution. It's entry into fields like agriculture, education, and finance is believed to have the potential to revolutionaries those industries too.

The process of federated learning within the context of the Internet of Everything has faced some objective difficulties, yet the result is promising. By solving for privacy and security, by capitalizing on the potential of edge learnings, and fostering collaboration with the members within the IoE platform the principles behind federated learning will quite soon redraw how data and machine learnings can be utilised in this interconnected world. It is eagerly anticipated that the future perspective will have infinite prospects, and federated learning will lead the world towards a higher level of intelligence, security, and integration.

References

1. Allioui, H., Mourdi, Y.: Exploring the full potentials of IoT for better financial growth and stability: a comprehensive survey. Sensors **23**(19), 8015 (2023)
2. Bonawitz, K., Kairouz, P., McMahan, B., Ramage, D.: Federated learning and privacy: building privacy-preserving systems for machine learning and data science on decentralized data. Queue **19**(5), 87–114 (2021)
3. Nguyen, T.V., Dakka, M.A., Diakiw, S.M., VerMilyea, M.D., Perugini, M., Hall, J.M.M., Perugini, D.: A novel decentralized federated learning approach to train on globally distributed, poor quality, and protected private medical data. Sci. Rep. **12**(1), 8888 (2022)
4. Reddy, K.S., Agarwal, K., Tyagi, A.K.: Beyond things: a systematic study of internet of everything. In: Innovations in Bio-Inspired Computing and Applications: Proceedings of the 10th International Conference on Innovations in Bio-Inspired Computing and Applications (IBICA 2019) Held in Gunupur, Odisha, India, pp. 226–242. Springer International Publishing (2021)
5. Tyagi, A.K., Nair, M.M.: Internet of Everything (IoE) and Internet of Things (IoTs): threat analyses, possible opportunities for future. J Inform Assur Secur **15**(5) (2020)

6. Dodda, S.B., Maruthi, S., Yellu, R.R., Thuniki, P., Reddy, S.R.B.: Federated learning for privacy-preserving collaborative AI: exploring federated learning techniques for training AI models collaboratively while preserving data privacy. Austr. J. Mach. Learn. Res. Appl. **2**(1), 13–23 (2022)
7. Wittkopp, T., Acker, A.: Decentralized federated learning preserves model and data privacy. In: International Conference on Service-Oriented Computing, pp. 176–187. Springer International Publishing, Cham (2020)
8. Gadekallu, T.R., Pham, Q.V., Huynh-The, T., Bhattacharya, S., Maddikunta, P.K.R., Liyanage, M.: Federated learning for big data: a survey on opportunities, applications, and future directions (2021). arXiv preprint arXiv:2110.04160
9. Agrawal, S., Sarkar, S., Aouedi, O., Yenduri, G., Piamrat, K., Alazab, M., Gadekallu, T.R.: Federated learning for intrusion detection system: concepts, challenges and future directions. Comput. Commun. **195**, 346–361 (2022)
10. Farahani, B., Monsefi, A.K.: Smart and collaborative industrial IoT: a federated learning and data space approach. Digit. Commun. Netw. **9**(2), 436–447 (2023)
11. Rauniyar, A., Hagos, D.H., Jha, D., Håkegård, J.E., Bagci, U., Rawat, D.B., Vlassov, V.: Federated learning for medical applications: a taxonomy, current trends, challenges, and future research directions. IEEE Internet Things J.
12. Tan, A. Z., Yu, H., Cui, L., Yang, Q.: Towards personalized federated learning. IEEE Trans. Neural Netw. Learn. Syst.
13. Kairouz, P., McMahan, H.B., Avent, B., Bellet, A., Bennis, M., Bhagoji, A.N., Zhao, S.: Advances and open problems in federated learning. Found. Trends® Mach. Learn. **14**(1–2), 1–210 (2021)
14. Reddy, K.D., Gadekallu, T.R.: A comprehensive survey on federated learning techniques for healthcare informatics. Comput. Intell. Neurosci. (2023)
15. Xu, X., Peng, H., Bhuiyan, M.Z.A., Hao, Z., Liu, L., Sun, L., He, L.: Privacy-preserving federated depression detection from multisource mobile health data. IEEE Trans. Indus. Inf. **18**(7), 4788–4797 (2021)
16. Chellapandi, V.P., Yuan, L., Brinton, C.G., Żak, S.H., Wang, Z.: Federated learning for connected and automated vehicles: a survey of existing approaches and challenges. IEEE Trans. Intell. Vehicles (2023)
17. Varlamis, I., Sardianos, C., Chronis, C., Dimitrakopoulos, G., Himeur, Y., Alsalemi, A., Amira, A.: Using big data and federated learning for generating energy efficiency recommendations. Int. J. Data Sci. Anal. **16**(3), 353–369 (2023)
18. Truong, N., Sun, K., Wang, S., Guitton, F., Guo, Y.: Privacy preservation in federated learning: An insightful survey from the GDPR perspective. Comput. Secur. **110**, 102402 (2021)
19. Chaudhary, P.S., Khurana, M.R., Ayalasomayajula, M.: Real-world applications of data analytics, big data, and machine learning. In: Data Analytics and Machine Learning: Navigating the Big Data Landscape, pp. 237–263. Springer Nature Singapore, Singapore
20. Al-Quraan, M., Mohjazi, L., Bariah, L., Centeno, A., Zoha, A., Arshad, K., Imran, M.A.: Edge-native intelligence for 6G communications driven by federated learning: a survey of trends and challenges. IEEE Trans. Emerg. Top. Comput. Intell. (2023)
21. Ma, L., Sun, B.: Machine learning and AI in marketing–connecting computing power to human insights. Int. J. Res. Mark. **37**(3), 481–504 (2020)
22. Ettredge, M., Guo, F., Li, Y.: Trade secrets and cyber security breaches. J. Account. Public Policy **37**(6), 564–585 (2018)
23. Liang, P.P., Liu, T., Ziyin, L., Allen, N.B., Auerbach, R.P. Brent, D., Morency, L.P.: Think locally, act globally: Federated learning with local and global representations (2020). arXiv preprint arXiv:2001.01523
24. Hiwale, M., Walambe, R., Potdar, V., Kotecha, K.: A systematic review of privacy-preserving methods deployed with blockchain and federated learning for the telemedicine. Healthc. Anal. 100192 (2023)
25. Rudin, R.S., Friedberg, M.W., Shekelle, P., Shah, N., Bates, D.W.: Getting value from electronic health records: research needed to improve practice. Ann. Internal Med. **172**(11_Supplement), S130–S136

26. Guo, Q., Li, L., Ban, X.J.: Urban traffic signal control with connected and automated vehicles: a survey. Transp. Res. Part C: Emerg. Technol. **101**, 313–334 (2019)
27. Teoh, Y.K., Gill, S.S., Parlikad, A.K.: IoT and fog-computing-based predictive maintenance model for effective asset management in Industry 4.0 using machine learning. IEEE Internet Things J. **10**(3), 2087–2094 (2021)
28. Muhammad, K., Ullah, A., Lloret, J., Del Ser, J., de Albuquerque, V.H.C.: Deep learning for safe autonomous driving: Current challenges and future directions. IEEE Trans. Intell. Transp. Syst. **22**(7), 4316–4336 (2020)
29. Bellini, E., Iraqi, Y., Damiani, E.: Blockchain-based distributed trust and reputation management systems: a survey. IEEE Access **8**, 21127–21151 (2020)
30. Duan, Q., Huang, J., Hu, S., Deng, R., Lu, Z., Yu, S.: Combining federated learning and edge computing toward ubiquitous intelligence in 6G network: Challenges, recent advances, and future directions. IEEE Commun. Surv. Tutor. (2023)
31. Hafi, H., Brik, B., Frangoudis, P.A., Ksentini, A., Bagaa, M.: Split federated learning for 6G enabled-networks: requirements, challenges and future directions. IEEE Access (2024)
32. Alazab, M., Rm, S.P., Parimala, M., Maddikunta, P.K.R., Gadekallu, T.R., Pham, Q.V.: Federated learning for cybersecurity: concepts, challenges, and future directions. IEEE Trans. Indus. Inform. **18**(5), 3501–3509 (2021)
33. Samaila, M.G., Neto, M., Fernandes, D.A., Freire, M.M., Inácio, P.R.: Challenges of securing Internet of Things devices: a survey. Secur. Privacy **1**(2), e20 (2018)
34. Deng, L., Li, G., Han, S., Shi, L., Xie, Y.: Model compression and hardware acceleration for neural networks: a comprehensive survey. Proc. IEEE **108**(4), 485–532 (2020)
35. Wen, Z., Yang, R., Garraghan, P., Lin, T., Xu, J., Rovatsos, M.: Fog orchestration for IoT services: issues, challenges and directions. IEEE Internet Comput. **21**(2), 16–24 (2017)
36. Imteaj, A., Thakker, U., Wang, S., Li, J., Amini, M.H.: A survey on federated learning for resource-constrained IoT devices. IEEE Internet Things J. **9**(1), 1–24 (2021)
37. Shuvo, M.M.H., Islam, S.K., Cheng, J., Morshed, B.I.: Efficient acceleration of deep learning inference on resource-constrained edge devices: a review. Proc. IEEE **111**(1), 42–91 (2022)
38. Himeur, Y., Sayed, A., Alsalemi, A., Bensaali, F., Amira, A.: Edge AI for internet of energy: challenges and perspectives. Internet of Things 101035 (2023)
39. Feng, T., Ramakrishna, A., Majmudar, J., Peris, C., Wang, J., Chung, C., Gupta, R.: Partial federated learning: unlocking non-biometric text information sharing for federated learning (2024)
40. Wei, K., Li, J., Ding, M., Ma, C., Yang, H.H., Farokhi, F., Poor, H.V.: Federated learning with differential privacy: algorithms and performance analysis. IEEE Trans. Inf. Forensics Secur. **15**, 3454–3469 (2020)
41. Sharma, D.K., Singh, N.C., Noola, D.A., Doss, A.N., Sivakumar, J.: A review on various cryptographic techniques and algorithms. Mater. Today: Proc. **51**, 104–109 (2022)
42. Rahman, A., Hossain, M.S., Muhammad, G., Kundu, D., Debnath, T., Rahman, M., Band, S.S.: Federated learning-based AI approaches in smart healthcare: concepts, taxonomies, challenges and open issues. Clust. Comput. **26**(4), 2271–2311 (2023)
43. Darwish, A., Hassanien, A.E., Elhoseny, M., Sangaiah, A.K., Muhammad, K.: The impact of the hybrid platform of internet of things and cloud computing on healthcare systems: opportunities, challenges, and open problems. J. Ambient. Intell. Humaniz. Comput. **10**, 4151–4166 (2019)
44. Singh, B.: Federated learning for envision future trajectory smart transport system for climate preservation and smart green planet: insights into global governance and SDG-9 (industry, innovation and infrastructure). Natl. J. Environ. Law **6**(2), 6–17 (2023)
45. Bansal, S., Sindhi, V., Singla, B.S.: Exploration of deep learning and transfer learning techniques in bioinformatics. In: Applying Machine Learning Techniques to Bioinformatics: Few-Shot and Zero-Shot Methods, pp. 238–257. IGI Global (2024)

Revolutionizing Industry 4.0: Multi-level Federated Learning for a Dynamic Ecosystem

Rashmi Sharma, Bharat Singh, Monali Gulhane, Piyush Chauhan, and Nitin Rakesh

Abstract Federated learning is becoming a prominent area of research in computer science, showing significant promise in various applications. Nowadays, we can see innovative methods where millions or even billions of IoT devices can collaborate to achieve a shared machine learning objective through federation. However, current methodologies mainly focus on single-task learning, with one objective and a single task owner, and assume that most devices involved in federated learning share similar designs, types, or limitations. It is believed that federated learning's full potential can only be realized in a dynamic and open ecosystem where a diverse range of devices, industrial units, machine manufacturers, non-governmental organizations, and governmental entities can all contribute to learning across multiple tasks and objectives in a crowdsourced fashion. In this chapter, a multilevel framework is introduced that demonstrates how federated learning, IoT, and crowdsourcing can synergize to create a robust ecosystem for multi-level federated learning tailored for Industry 4.0. This integration supports the development of future intelligent applications for Industry 4.0, such as predictive maintenance and fault detection in smart manufacturing environments. Additionally, various use cases are explored where multi-level federated learning can be applied within Industry 4.0. If successfully implemented, this approach will not only improve performance but also contribute to broader goals like the UN Sustainable Development Goal No. 13, which focuses on reducing the carbon footprint.

Keywords Federated learning · Industry 4.0 · crowdsourcing · Smart manufacturing · UN SDG 13 · Predictive maintenance

---

R. Sharma · B. Singh · M. Gulhane (✉) · P. Chauhan · N. Rakesh
Department of CSE, Symbiosis Institute of Technology, Symbiosis International (Deemed University), Pune, India
e-mail: monaligulhane4@gmail.com

R. Mohana et al. (eds.), *Federated Learning Applications in the Industrial Internet of Everything (IoE)*, Studies in Systems, Decision and Control 611,
https://doi.org/10.1007/978-3-031-99270-4_8

1 Introduction

The fourth industrial revolution, commonly referred to as Industry 4.0, marks a monumental shift in manufacturing and industrial practices through the adoption of advanced digital technologies. Making its way in the early twenty-first century, Industry 4.0 integrates the physical and digital globe, blending cutting-edge innovations such as the Internet of Things (IoT), Artificial Intelligence (AI), Big Data Analytics (BDA), cloud computing, and robotics to create smart automated factories. These factories are characterized by their ability to simultaneous collect, analyze, and respond to data in real-time, leading to unprecedented stages of efficiency, flexibility, and customization in construction processes [1].

The central idea of Industry 4.0 is connectedness, as the industrial sector is transformed into an active and dynamic data-driven environment by empowering machines, devices or sensors to interact with each other. This data is collected and then processed to drive action, enable decisions, enhance productivity and predict maintenance requirement of machines which in turn provides operational resilience. Industry 4.0 facilitates sustainability and energy efficiency, which makes it possible for industries to reduce waste, emissions or meet environmental regulations [2]. It is in this sense that it overall enables the change from traditional production lines to a more automatically operating, flexible and decentralized process. Figure 1 gives a broad illustration about Industry 4.0.

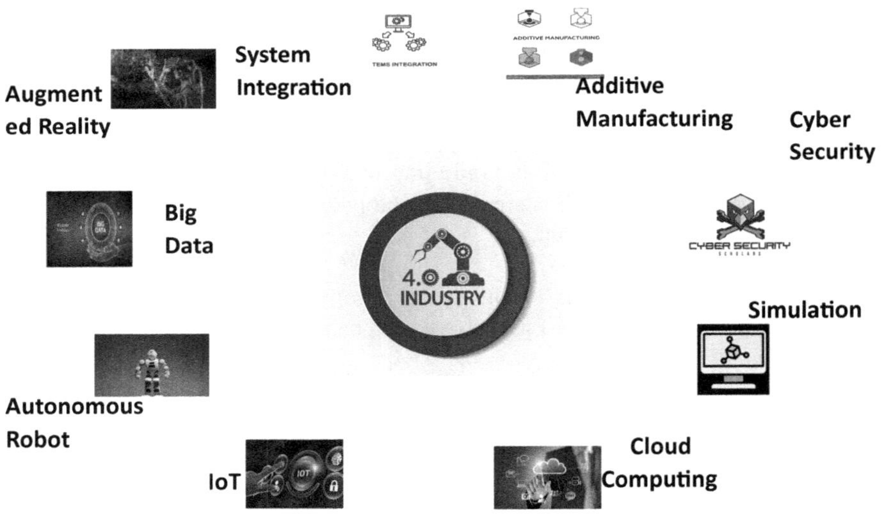

Fig. 1 Industry 4.0 that focuses in inter connection, automated machine learning, and real-time data

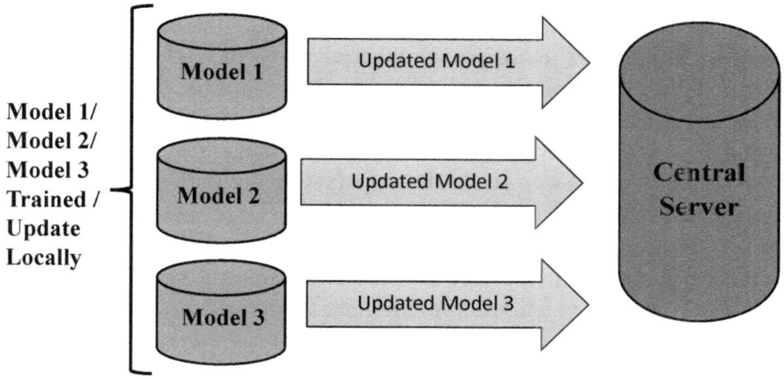

Fig. 2 Federated learning approach

1.1 About Federated Learning

In parallel with the progress of Industry 4.0, federated learning is emerging as an effective approach to machine learning in distributed environments. Unlike traditional machine learning paradigms where data is centralized in one place for training, federated learning allows data to remain decentralized across devices or local servers [3]. Models are trained locally on devices (such as smartphones, IoT devices or industrial machines) and only updates (model parameters) are sent to a central server. The central server aggregates these updates to improve the global model and then sends the improved model back to the device for further training [4]. Figure 2 represents the federated learning approach.

Federated learning addresses some of the key issues that arise in distributed systems, especially when it comes to data privacy and security. Sensitive data such as patient health records or financial transactions do not leave the local facility, reducing the risk of data breaches and ensuring compliance with data protection regulations (e.g. GDPR in Europe). Federated learning is therefore particularly valuable in privacy-sensitive sectors such as healthcare, finance and industrial operations.

In addition, federated learning is bandwidth efficient. Since only model updates are shared, not the raw data itself, the communication overhead is significantly reduced. This makes it an ideal solution for large-scale distributed networks where communication costs are an issue, such as in the connected ecosystems of Industry 4.0.

1.2 Federated Learning in Industry 4.0

Federated learning can be used in various aspects of Industry 4.0 [5–9]. Few of them are as follows:

1. **Smart Manufacturing**: In smart manufacturing, federated learning allows machines in different factories or industrial plants to work together to improve their predictive maintenance algorithms. Each machine locally collects data about its operating status, trains a maintenance model, and sends model updates to a central server that collects information from all machines. The improved model is then distributed back to all machines, enabling predictive insights without the need for centralized data collection.
2. **Smart Cities**: In smart cities, where devices in different sectors (such as energy management or traffic control) contribute to a common model without sharing raw data, ensuring privacy and minimizing risks associated with data centralization.

1.3 Levels of Federated Learning

Federated learning can be partitioned into single level and multilevel as per their applicability.

1.3.1 Single-Level Federated Learning

Traditional federated learning typically works at a single level where local devices communicate directly with a central server. In this architecture, each device performs local training on its own data and periodically sends updated model parameters (rather than raw data) to a central server [10]. The server then gathers information from all the connected devices and forms a global model and sends it to the device for the next level of a procedure. This cycle is repeated until the global model is acceptable according to a chosen level of global model accuracy.

Although this method is efficient for relative uniform tasks and smaller scale networks, it possesses constraints when used for large scale, variant systems in Industry 4.0. Such constraints include communication constraints which include delays in data flow from one device to another, scalability which is a challenge in most swarm systems since the computing capabilities of different devices can greatly vary as well as the network conditions and the distribution of computing power between devices [11].

1.3.2 Multi-level Federated Learning

To handle with the issues with single level federated learning which was defined as traditional approach, Multi-Level Federated Learning (MLFL) presents a hierarchy structure for decision making. In MLFL, the training and aggregation process is done at different levels of aggregation points some of which include local, regional and global servers. This kind of structure is suitable in complex systems that have several devices scattered in diverse areas or departments [12]. Several layers in a federated learning hierarchy work in parallel; devices at the local level process parameters locally on the local data and then send the newly adjusted model parameters to a higher level, such as a regional server. All the devices in the region send their updates to the regional server and it in return sends the averaged model to the global server. Updates from all regions are then consolidated at the global server to make a global model then the procedure is reversed up through the hierarchy to the regional server and possibly even local devices [13].

This hierarchical structure has several advantages over single-level federated learning [14–17]:

- **Scalability**: By distributing the aggregation process across multiple levels, MLFL can scale to accommodate millions or even billions of devices. Group servers for local and regional are used for aggregation of smaller sets of devices to minimize the flow of traffic to global server.
- **Communication Efficiency**: MLFL cuts down communication overhead since it prevents most of the messages that are exchanged between devices and the global server. Local devices mainly communicate with local servers and these local servers compile and then sync updated with the global server. This decreases the overall communication which takes place and effectively utilize the network resources.
- **Heterogeneity Management**: Multi-level federated learning is beneficial to improve the heterogeneity of device and data scenario because intermediate levels for model aggregation and updating are possible. For instance, devices in a given region or in a given industry will likely have some similar characteristics with regards to their data, so a regional model can then be enhanced to fit the region before making contributions to the global model.

2 Relationship Between Each Involved Body

The effectiveness of multi-level federated learning in the context of Industry 4.0 is the result of several parties that are involved in the learning process, each of them contributing their part by following specific important tasks.

2.1 Devices and Edge Nodes

On the local tier, the devices and edge nodes involved in federated learning are the objects (or things), i.e., IoT sensors, smart industrial robots, or smart machines. These devices gather data from their surroundings, update the local machine learning model, and transmit model coefficients to the regional centre. The operation of the devices is in the background and the devices do not exchange their raw data with a third party, thereby respecting privacy and minimizing the possibility of attacks [18]. In real-time environment, these devices can be machines on a production line, energy consumption sensors or self-driving vehicles in a smart factory [19]. This is because each device offers different information depending on the data collected locally and aids in enhancing the performance of the FL system.

2.2 *Intermediate Aggregators (for Example, Regional Servers)*

The regional servers are situated in the middle of the hierarchical structure within the multi-tiered federated learning architecture. The devices within a certain region send new model parameters to their respective regional servers and the servers combine the parameters to form a regional model. The collected data is then forwarded to the global server for more aggregation from this regional model. It is required to have regional servers that will help to minimize the communication bottleneck problem and also will help to create a system that can work with a large number of devices. This makes the system to aggregate updates at the regional level consequently, avoid congestion of the network from being flooded with updates that congest the global server [20].

In Industry 4.0, regional servers may be placed in different factories, departments or zones to organize updates that have been gathered from the devices that are in their domain. For example, there will be a regional server that will collect the machine updates of all the machines of a production line in a factory; a regional server that will collect the updates of all machines in a different factory.

2.3 *Global Aggregators*

At the global level, there is Global Server that is an integration center for the multi-level federated learning system. When the global server receives the total update from all the serve regions, it then integrates all the regional models to make the global model to incorporate all the knowledge of all the apparatuses. It is then disseminated back down the hierarchy to the regional servers and then to the local devices so that they also can implement the learning that has been accomplished by the group [21].

In the federated learning environment, the global server has a central role of making sure that the system converges a high-quality global model. Through pulling updates from all around the globe, the global server makes it possible to ensure that the model will work suitably in different devices and contexts [22].

3 Collaboration Between Entities

In general, the operational prospects of the multilevel federated learning in Industry 4 spotlight the following. The policies and its implementation depends on efficient teamwork of the various actors. Communication between the various devices, local servers and global ones must follow and enable the learning process to occur in a proper and efficient manner [23–25]. This is an important area that calls for proper coordination of communication time frames, aggregation plans and updates to models.

One of the problems of using the federated learning is the heterogeneity of the devices and data. Various devices may possess rather diverse computational power, communication environment, and data distribution, which complicates the process of the learning process synchronization. For instance, the devices may be in a position where their numerical capabilities are insufficient for the training process; the devices also may have weak network connections with the regional servers.

In order to overcome these problems, federated learning systems usually use some proactive methods that allow optimizing the learning process based on the device features. For instance, low-end devices should be able to perform a lower number of local updates than hi-end devices due to the computational capacity as well as devices with low network ability to connect with the regional servers less than hi-end devices.

However, a number of other issues are challenging to implement federated learning systems; two of the most significant of these issues are related to heterogeneity and data privacy and security. The underlying concept of federated learning is that instead of sharing the raw data, model updates are exchanged, and hence it becomes necessary to guarantee that the updates cannot expose information that may be regarded as sensitive by the system. This may be done through differential privacy, secure aggregation and similar techniques, whereby the model updates are kept in such a manner that it will be impossible for anybody to reverse engineer the information from it.

- **Cross-Silo Learning**

In Industry 4.0 another significant characteristic of federated learning is worth to be mentioned the next important aspect of federated learning in Industry 4.0 is cross-silo learning whereby different organisations or silos cooperate in the improvement of their own models in Machine learning without sharing their data. In the cross silo learning context, the local model is trained using data resident locally and the model update information is sent to a central server for combining [24, 26]. The changes of parameter estimates are then communicated from each of the organizations to the central server where a global model is created which is then disseminated to the organizations.

Cross-silo learning is particularly suitable in industries where one or more organizations require improving on their machine learning models with data from other organizations, but cannot do so because of data sensitivity. For instance, in the manufacturing industry, it is common that factories and or suppliers work together in enhancing their predative maintenance models but cannot share data as this is deemed strapping due to competition [27, 28]. Federated learning makes it possible for all these organizations to train together yet independently instead of sharing data hence helping them get the benefits of shared learning without the drawbacks of sharing the data.

4 Current State of Federated Learning

Federated Learning is nothing short of a sea change in the world of model training as far as machine learning goes. Firstly, conventional machine learning frameworks entail one-dimensional data collection—information gathered into a single location and subsequently fed for training to initialize the model. However, this process can be difficult to perform because of privacy concerns as well as regulatory and logistical challenges in moving large amounts data [8].

Federated learning addresses these issues by allowing models to be trained locally on devices or servers where the data resides. Instead of sending data to a central server, only the model updates (e.g., weights and gradients) are shared. These updates are aggregated to form a global model that benefits from the diverse data spread across multiple sources [29–31].

4.1 The Progress of Federated Learning

Federated learning has advanced rapidly since its inception, driven by the need for privacy-preserving machine learning and the growing capabilities of edge devices. The current state of federated learning can be understood through several key developments:

1. **Broadening Applications**:
   - **Healthcare**: FL now powers new models trained on patient data that can be executed outside hospitals and research institutions around the world. An example is federated learning, which allows for the creation of predictive models to diagnose diseases without sharing actual patient data from a hospital with another.
   - **Mobile Devices**: Further, companies such as Google have been utilizing federated learning in mobile devices where tasks like the next-word prediction of keyboards are offloaded to user sensitive data. This means we can train our model on the text written onto millions of smart phones without also memorizing their corresponding keyboards without the need to store or share personal data.
   - **Finance**: In this sector, federated learning helps institutions collaborate on fraud detection models without exposing sensitive transactional data, ensuring both privacy and enhanced security.
2. **Technological Improvements**:
   - **Algorithmic Advances**: Significant progress has been made in designing federated learning algorithms that are robust to challenges like non-IID (non-independent and identically distributed) data, where different participants have

data distributions that vary significantly from one another. Techniques like FedAvg (Federated Averaging) have become foundational in improving the efficiency and effectiveness of federated learning.
- **Privacy Enhancements**: To further protect the privacy of data during the federated learning process, methods such as differential privacy and secure multi-party computation (SMPC) are being integrated. These techniques ensure that the shared ensuring that the shared updates of model do not inadvertently disclose any sensitive data regarding to local.
- **Communication Efficiency**: Various studies around federated learning focus on communication efficient solutions. In scenarios with low bandwidth, there are strategies to reduce the required amount of transferred data—for instance model compression, sparsification and asynchronous updates.

3. **Challenges Being Addressed**:

- **Scaling**: The efficiency cost of scaling the system across devices and different organizations will increase as federated learning is used in more companies. To address such concerns, researchers have been investigating hierarchical or multi-level federated learning methods that support large-scale coordination across thousands of devices.
- **Personalized Models**: As federated learning traditionally tries to achieve a single global model, more recently there has been interest in creating personalized or group-specific models. This method overcomes the drawback of a universal model which can handle specific domain based applications efficiently like healthcare and personalized services.
- **Fairness and Bias**: An active area of research is also on making sure that models data remains fair when distributed as per federated learning. Decentralized data means there is a potential for the global model to not accurately represent minority groups or less-represented data distributions. The methods to detect and reduce bias are being developed so that federated learning benefits all users equally.

4. **Real-World Implementations**:

- **Federated Learning Industry Adoption**: Automotive, telecom and energy industries are starting to develop federated learning instead of classic industry practice. This allows a predictive maintenance model to optimise federated learning by combining knowledge from machines in multiple rooms, which ensures better and faster performance of production systems at smart factories.
- **Cross-Silo Collaboration**: For cases where data is not allowed to be aggregated (illegal or proof of competitive conflict, such as between different companies or departments) and centralized storage becomes impossible federated learning may produce model training. In this model, institutions have the ability to improve models by collaborating despite not sharing sensitive data—very useful in sectors like finance and healthcare.

5 Challenges Federated Learning Methods

These are some challenges of federated learning method with the aspect of case studies.

5.1 Data Heterogeneity (Non-IID Data)

Case Study: Personalized Healthcare Applications

- **Scenario**: Used in personalized healthcare to train models on data from different hospitals using federated learning. However, because each hospital serves different patient populations and uses different equipment or protocols in care delivery there are diverse data distributions.
- **Major Limitation**: One of the main assumptions federated learning makes is that data in all clients are independent and identically distributed (IID). But in reality, data is often non-IID because the locally available training a customer‘s source may come from different distributions. However, this creates the problem of heterogeneity—which further breaks down model generalization at a global scale.
- **Example**: Let's say, a distributed learning model is developed over hospitals in diverse locations. Perhaps one hospital is notorious for cardiology, and the other rule last due to its oncology department. Since it is non-IID, the federated model struggles to perform well for both categories of patients.

Communication Bottlenecks

Case Study: IoT Networks in Smart Cities

- **Convoy**: IoT devices are thousands of miles apart in a smart city and used to measure traffic, air quality, energy consumption. This model operates city-wide on the devices and works together with other in forming to an intelligent resource allocation strategy using federated learning.
- **Downsides**: Requires constant communication between devices and central server, which can be taxing on models where updates must be periodically aggregated. That being said, IoT devices historically tended to have low bandwidth communication capabilities which can make the network congested and unreliable dictating delays that drain model performances.
- **Example**: A citywide network of air quality sensors required to constantly feed their data; the data infrastructure needs featured processing complex queries and exploratory analysis. Congestion in the network or too many devices going offline can cause the global model to become stale, which might give wrong results.

5.2 *Stragglers and System Heterogeneity*

Case Study: Collaborative Autonomous Vehicle Networks

- **Use Case**: A fleet where autonomous cars and trucks cooperate to learn driving models using federated learning. Each vehicle is trained using its own driving data, which it then sends updates to a central server.
- **Straggler**: In federated learning, the communication bottleneck comes from stragglers–those devices that are slow to return their local updates due to weak computation or networks, limited battery life. These stragglers could slow down the whole training process. Furthermore, the heterogeneous system (the variety of devices having various hardware configurations, different OS and network conditions) would make it worse by providing an unbalanced learning environment.
- **Example**: Imagine a fleet of autonomous vehicles, where the cars with older hardware and slower processors. As a result, this will cause the vehicles to finish their training updates slower than the others and slow down the overall model convergence of whole fleet. In time-dependent applications, such as autonomous driving this inconsistency is especially problematic.

5.3 *Privacy and Security Concerns*

Case Study: Financial Services Across Banks

- **Scenario**: Banks use federated learning to jointly train models for fraud detection, without exchanging customer data across institutions.
- **Limitation**: Federated learning models designed to preserve privacy of each client's raw data, but the model update still leaks information. Finally, adversarial attacks such as model inversion or poisoning can compromise the privacy and integrity of the system.
- **Example**: Suppose multiple banks use federated learning to detect fraudulent transactions. Even though they are not sharing raw transaction data, a malicious entity could potentially infer private information from the shared model updates or corrupt the model by introducing poisoned data.

5.4 *Scalability Issues*

Case Study: Global Collaboration in Pharmaceutical Research

- **Scenario**: Pharmaceutical companies across the globe collaborate to train models that predict the efficacy of new drug compounds using federated learning. Each company contributes its data while keeping proprietary information confidential.
- **Limitation**: As the number of participating entities (e.g., research labs, companies) increases, the complexity of coordinating updates and aggregating models

grows. The central server must manage the synchronization of updates from a large number of participants, leading to scalability challenges in terms of computation and communication.

- **Example**: In a scenario where hundreds of pharmaceutical companies are involved, the central server may struggle to efficiently aggregate updates from all participants within a reasonable timeframe, slowing down the overall research process.

5.5 Lack of Personalization

Case Study: Mobile Keyboard Predictions

- **Scenario**: Federated learning for a mobile keyboard that learns local typing habits while improving predictive text, but without using personal data.
- **Disadvantage**: The federated model only grants user-level granularity to a small extent, as it is trained on all users globally and cannot be optimized for individual preferences for any user, there are unique ways to type the same words that may not be well-covered in a model trained across all user's data causing less-effective predictions for some of them.
- **Use Case: e.g.,** A user that types in Spanish slang—the global model is too generalized and does not fully capture this specific typing habit so it provides inaccurate predictions for a federated learning-model example.

5.6 Model Convergence

Case Study: Predictive Maintenance in Smart Factories

- **Scenario**: Federated learning is used to build predictive maintenance models in multiple factories for smart manufacturing. Every factory trains a model on their machinery data to help build the globalized predictive maintenance model.
- **Drawback**: Model Convergence is a challenge in Federated Learning, especially when Data and Systems are being Heterogeneous. The global model may not converge or need a lot of constructs after multiple iterations, especially in the case when there is discrepancy between the data or out-of-date device dropping off from network.
- **Examples**: Factories may provide different updates to the global model depending on machinery type or operating conditions, leading to slow convergence and oscillations between suboptimal states. As for what happens next…this could in turn lead to a range of bad scenarios from false predictions, which runs the gamut Either in avoidable persistent maintenance or sudden equipment failures.

6 Envisioning a Dynamic and Open Ecosystem for Federated Learning

The federated learning (FL) landscape has been changing fast, and it is a future that will take place within Pelion with an open dynamic ecosystem. This ecosystem would enable innovation and growth without boundaries, permeating different stakeholders which could be facilities, industries or research institutes or government bodies coming together to work in a flexible and inclusive manner. This ecosystem would increase the power of federated learning, allowing models to be crowd-sourced and constantly updated using collective intelligence without compromising privacy or security.

6.1 A Vision for the Future

So instead of this boring world, how about a planet where Federated Learning can actually be used fully open and dynamic? This future was one in which federated learning did not mean isolated silos of data or the most powerful actors spying on everyone except that now instead it allowed collaboration across industries, regions and domains. Such a partnership would involve multi-part interactions leading up to a common goal of developing an intelligent, dynamic and individualized system.

This system could be designed with an interface interoperability layer that would allow different devices and platforms to federate their learning cycles, without concern for the tech underneath or how exactly data is structured. It would also argue in favour of on-going innovation and models that can be updated as they learn at speed, just as you demand as additional data and insights are produced.

6.2 Real-World Applications in a Dynamic Ecosystem

- **Google's Gboard (Mobile Keyboard Predictions)**

Google Federated Learning is the technology that makes Gboard's predictive text features as good as possible without burning your super private information on Google Cloud. With the help of nearby millions who speak keyboard, Gboard is using on-device learning to make for a more personalized and faster suggestion experience. It can even be generalized to others and more applications for a tailored user experience on various devices with different platforms in an agile ecosystem.

- **Healthcare Collaboration**

This is changing what healthcare can do and allowing hospitals to stop sending patient data elsewhere for models, meaning patients are not only more secure but now you too. In such a vibrant and open ecosystem, this collaboration has the ability to grow worldwide as well as spark predictions of future modelson the disease diagnostic and treatment front through disparate datasets from around the world. That could result in better targeted and effective health care interventions that protect patient privacy.

- **Apple's Siri and Dictation**

Federated learning is used by apple in Siri and dictation to make improvements with the help of on-device machine learning. Over the time, Siri becomes more adaptive to individual user preferences by analyzing how they interact with it on device and sharing only model updates. This new form of multimodal speech recognition system could be an effective approach to making more natural and intuitive the voice-activated systems for smart home appliances, car interfaces or other applications in a global ecosystem.

- **Edge Computing in Autonomous Vehicles**

It allows maintaining the driving model in autonomous vehicle, which can be used to training with data when collected from different vehicles. It is possible to evolve safer and more adaptable driving algorithms, while still keeping personal vehicle data confidential. In a so-called locky, but also open ecosystem this cooperative concept could be one of the methods that will establish in related way with drones as well as intelligent local public transportation systems network edge to increase safety and efficiency on whole transport field.

- **Financial Fraud Detection**

Federated learning enables financial institutions to collaboratively train fraud detection models without revealing customers' sensitive data. In an open ecosystem, this type of collaboration could eventually be initiated by multiple banks and also insurance companies—even if one day what we experience is the beauty of Kickbox—keeping a fraud protection block constantly updated with current events. Such solutions would make financial systems safer and also retain the privacy of individual users

- **Medical Imaging**

For example, federated learning allows hospitals or research centers in medical imaging to train their diagnostic models on different sets of (anonymous) medical images and reach a common model agreement without exchanging patient data. In a dynamic environment, these models could be sponsored among different healthcare facilities providing speedier diagnosis of higher accuracy on global scale. This would allow identification of the disease sooner and hence a better prognosis for patients.

- **IoT and Smart Homes**

Federated learning is critical for smart homes and devices, enabling them to learn from user behavior and get better with time. Smart thermostats, for instance, can be trained to use less power by understanding household patterns. In an ecosystem that was distributed, such insights might be shared across disparate smart home systems and in noe so doing help create a more seamless and intelligent environmentoids and then molding to fit the needs of those who reside in it.

- **Telecommunications Industry**

In telecommunications, federated learning is used to analyze data from a large number of interconnected devices in order to optimize network management and ensure the best quality of service without privacy breaches for users. These companies could cooperate to build better, more redundant and efficient networks where there is massive data flow in a flexible ecosystem. This would enhance connectivity and service delivery, especially in urban slums or rural settings.

- **Federated Learning in Transportation**

For transportation, federated learning can inform on the improvement of operational traffic signals and optimize traffic control systems across vehicles, infrastructure. This data helps to predict traffic flow and how we can reduce congestion. Those insights can then be shared across cities and regions, theoretically allowing for a civic innovation akin to the trial-and-error experimentation of Silicon Valley to more sustainable, faster and efficient transport networks.

7 Application of Multi-level Federated Learning in Industry 4.0

- **Scaling Up for Large Industrial Systems**

Industry 4.0 means huge networks of devices and machines that are distributed all over the place to address this challenge, Multi-Level Federated Learning (MLFL) manages the learning process in a hierarchical manner which is detailed as three levels of computations: local computation on devices; regional global servers collaborate with each other to learn high-level feature representation from multiples regions and finally infer about privacy-sensitive tasks.

In layman terms, MLFL is simply shifting the load from one centralized system to many mini systems. First, heads read and act on their data locally before adding aggregated metrics to the regional server where they talk with a global above. Efficient and less efficient this approach more efficient and bottleneck-resistant.

- **Dealing with Unstructured Data from Various Sources**

In Industry 4.0 scenarios, the data depending on its location in the system and even by their tasks are generated with a whole universe about how to change these marvelous world driving model development at scale can be solved through MLFL creating an incremental proposition able gather tremendous amounts of local data first before being inserted as partent/part/son (whatever metaphor) from other sources

Reasoning: For example, every single factory or department may do things differently thus have different sources of data. This allows each region to create bespoke models that are finely tuned for their environment and the best global model, E.g., in MLFL based, which performs well across regions.

- **Efficient Communication and Faster Responses**

To keep equipment running or to enhance production in an Industry 4.0 environment, real-time decision-making is vital for improving operations and reduce costs. It keeps the amount of data that needs to be sent back and forth with MLFL, minimizing processing delays.

Explanation: Instead of every machine constantly sending data to a central server, they send updates to a nearby regional server. This server processes the data locally, only passing on the essential information to the global server. This reduces traffic and speeds up the whole process.

- **Better Security and Data Privacy**

Protecting sensitive data is a top priority in industrial settings. MLFL helps by keeping data within local or regional servers, reducing the risk of breaches.

Explanation: Since data doesn't need to travel far, there's less chance it will be intercepted or leaked. Only model updates, which are much less sensitive than raw data, are shared with higher-level servers. This makes MLFL a more secure option for industries handling confidential information.

- **Encouraging Collaboration Across Organizations**

Many industries involve multiple companies working together, like suppliers, manufacturers, and distributors. MLFL supports this collaboration without forcing anyone to share their raw data.

Explanation: Imagine a group of factories that want to improve their maintenance systems. With MLFL, they can work together by sharing model updates, not the actual data. This way, they can benefit from each other's insights while keeping their own data private.

- **Adapting to Continuous Changes**

Industrial environments are always changing—new machines, updated processes, and shifting market demands are the norm. MLFL supports continuous learning, allowing models to be updated in real-time as conditions change.

Explanation: Local models are updated with new scans and use the currents scan to fix old errors. Thereby staying relevant & adaptable to serve in a far more underpowered manner.

- **Contributing to Sustainability Goals**

Industry 4.0 also has a large sustainability component: less waste, more energy efficiency MLFL is a powerful platform that can help to drive industries toward all these objectives by enabling optimization at each process level.

MLFL allows industries to more easily operate; for example, reducing energy use, waste. For instance, smart factories to consume energy and resources more efficiently with MLFL.

- **Real-World Applications**

MLFL is in practice while MLFL remains an abstract concept, it is not entirely theoretical frameworks are already being employed in a number of different Industry 4 variations such as smart factories right through to the supply chain Management.

E.g. In smart factory MLFL can help in enhancing predictive maintenance by getting machines services before they breakdown and thereby reducing down time said Ashish Ghildiyal. For supply chains, it can aggregate learnings integrated from a breadth of sources aid with improved demand forecasting and inventory management.

8 Alignment with UN Sustainable Development Goal No. 13 (Climate Action)

Climate Action: Take urgent action to combat climate change and its impacts. Federated learning enables industries to adopt sustainable processes due to on-device data processing, low computation cost efficient model training and no/less data movement.

Inefficient processes and energy management increase carbon footprints, to optimize the use of specific strategies allowing lower energy consumption [31–33], resulting in higher efficiency and less greenhouse gas emission. Here are some important strategies to achieve this:

- **Decentralized Data Processing**

Federated Learning acts as a building block of sustainability by providing decentralized data processing. The basic idea is that rather than collecting gobs of data in centralised data centres, cogitating it (often using vast amounts of energy) and then tweaking models with the updated weights or parameters—all within a severe privacy accord setup which quickly becomes large headache to administer devices or regional servers. It eliminates the need for complete data transfers and decreases energy requirements due to centralized storage. Decentralized processing also allows for quick decisions directly at the data station and thus potential immediate effective action against climate change (e.g., by better utilization energy in industrial processes or changed production schedules to reduce emissions).

- **Energy-Efficient Model Training**

Federated learning further helps achieve sustainability goals by improving energy efficiency in model training. Centralized machine learning as traditionally possible requires resource-heavy compute operations in vast data centers, taxes energy consumption and thus is responsible for carbon emissions. However, federated learning spreads the computational burden across multiple devices; each device performs local model updates at a fraction of the total cost lower energy requirements. By doing so, federated learning reduces computational complexity and makes the training process ecologically smarter with known methods like model compression/ quantization or even specification. And because the distributed approach results in

AI models that require less carbon to train, it is also consistent with ongoing initiatives across regions and industries focused on reducing energy consumption and supporting a sustainable environment.

- **Minimized data movement**

Reducing the carbon cost of data-intensive operations is now a new environmental concern, and minimized movement between processor resources directly affects this. Traditional machine learning frameworks require the downloading of large data sets to a central location for processing, in many cases utilizing extensive bandwidth and energy. Localized versus Globalized: Federated learning addresses the above problem by maintaining data on device and sharing only model updates (typically much smaller than training data.) This way greatly reduces the data you need to load for each page and hence, improves performance transferred which ultimately leads to improved energy efficiency and less carbon-emitting data transmission. Furthermore, by reducing the amount of data moving around and thus cleaning up network infrastructure it also leads to more sustainable industrial operations.

8.1 Strategy and Challenge Deployment

Despite the sustainability advantages of Federated Learning, there have been significant shortcomings that need to be resolved in real-world deployments [32, 33].

- **Scalability**: When federated learning scales across devices and regions, energy efficiency is hard to maintain and there will be a lot of data movement. Strategies for leveraging better communication protocols and managing decentralized processing to accommodate scale.
- **Heterogeneous environment**: The device capabilities in a federated learning network are different, differences in power consumption by all devices. To solve this problem to adapt the training to be run in each device under its own energy constraints and computational power (adaptive algorithms).
- **Privacy and security**: Decentralized data processing allows for higher sustainability and also brings new privacy and security concerns the effectiveness of the training process has to be balanced with ensuring model updates do not reveal private information, and hence requires strong encryption as well as secure aggregation protocols.

9 Conclusion

To sum up our findings of federated learning in Industry 4.0, it is evident however this ground breaking technology has the capability to change industries through operations and innovation. The chapter went deeper into federated learning on everything from the basics to actually using in a multi-tiered system; and how it can face the challenges facing modern industrial systems. We think federated learning is a very promising solution for the large-scale, inter-connected networks characteristic of Industry 4.0 developments. It helps industries to scale, process a wide variety of data and also enhances the real-time decision making by providing decentralization in both processing the data as well as collaborating simultaneously at multiple levels of operations. Being able to work with these capabilities is imperative in a world where industries are becoming more and more intricate and data-oriented. Furthermore, federal education is also tied to international sustainability aims—in particular UN SDG Goal 13 (Climate Action). Federated learning helps enable a more sustainable industry ecosystem by eliminating the necessity for centralized data processing, reducing overhead and resources needed to move massive amounts of data back-and-forth, support training with energy-efficient models. It is more than compliance, it is a real carbon reduction tracker of industry to satisfy wider ambitions on the environment of course, the strategy implementation comes with some issues as well. Energy efficiency, preserving data privacy and effective management of device peripherals are some issues to be addressed with the increasing scale and complexity of federated learning systems. But the track we have covered to date says these are overcomable problems. Finally, the study over federated learning also outlines the disruptive role of this technology in Industry 4.0 era. In addition to operational efficiency and data security, it enables a more sustainable and connected industrial future. Placing federated learning at the forefront of our next phase into increasingly evolved and mature versions of these technologies.

References

1. Kairouz, P., McMahan, H.B., Avent, B., Bellet, A., Bennis, M., Bhagoji, A. N., Zhao, S.: Advances and open problems in federated learning. Foundations and trends® in machine learning. **14**(1–2), 1–210 (2021)
2. Li, T., Sahu, A.K., Talwalkar, A., Smith, V.: Federated learning: challenges, methods, and future directions. IEEE Signal Process. Mag. **37**(3), 50–60 (2020). https://doi.org/10.1109/MSP.2020.2975749
3. Bonawitz, K., Eichner, H., Grieskamp, W., et al.: Towards federated learning at scale: system design. In: Proceedings of the 2nd SysML Conference, Palo Alto, CA, USA (2019)
4. McMahan, H.B., Moore, E., Ramage, D., Hampson, S., Arcas, B.A.: Communication-efficient learning of deep networks from decentralized data. In: Proceedings of the 20th International Conference on Artificial Intelligence and Statistics (AISTATS), pp. 1273–1282. PMLR (2017)
5. Yang, Q., Liu, Y., Cheng, Y., Kang, Y., Chen, T., Yu, H.: Federated learning. In: Zhou, Z.H. (eds) Advances in Neural Information Processing Systems. Springer, Singapore (2019). https://doi.org/10.1007/978-981-15-4952-6_7
6. Lu, Y., Huang, G., Dai, Y., Maharjan, S., Zhang, Y.: Federated learning for data privacy preservation in industrial internet of things. IEEE Netw. **34**(5), 36–41 (2020). https://doi.org/10.1109/MNET.001.1900656
7. Zhang, C., Xie, Y., Bai, H., Yu, B., Wu, Y., Gao, F.: Federated learning for industrial internet of things in Industry 4.0: concerns, methods, and challenges. IEEE Commun. Surv. Tutor. **23**(3), 1622–1678 (2021). https://doi.org/10.1109/COMST.2021.3085591

8. Zhang, H., Bosch, J., Olsson, H.H.: Engineering federated learning systems: a literature review. In: International Conference on Software Business, pp. 210–218. Springer, Cham, Switzerland (2020). https://doi.org/10.1007/978-3-030-67292-8_17
9. Kulkarni, V., Kulkarni, M., Pant, A.: Survey of personalization techniques for federated learning. In: 2020 Fourth World Conference on Smart Trends in Systems, Security and Sustainability (WorldS4), pp. 794–797 (2020). https://doi.org/10.1109/WorldS450073.2020.9210355
10. Nguyen, D.C., Pathirana, P.N., Ding, M., Seneviratne, A.: Federated learning for smart healthcare in Industry 4.0: a survey. IEEE Trans. Indus. Inf. **17**(11), 7600–7613 (2021). https://doi.org/10.1109/TII.2021.3073834
11. Smith, V., Chiang, C.-K., Sanjabi, M., Talwalkar, A.S.: Federated multi-task learning. Adv. Neural Inform. Process. Syst. (NeurIPS) **30**, 4424–4434 (2017)
12. Lyu, L., Yu, H., Yang, Q.: Threats to federated learning: a survey. IEEE Trans. Knowl. Data Eng. (2020). https://doi.org/10.1109/TKDE.2020.2991380
13. Brisimi, T.S., Chen, R., Mela, T., Olshevsky, A., Paschalidis, I.C., Shi, W.: Federated learning of predictive models from federated electronic health records. Int. J. Med. Informatics **112**, 59–67 (2018). https://doi.org/10.1016/j.ijmedinf.2018.01.007
14. Yang, Q., Liu, Y., Chen, T., Tong, Y.: Federated machine learning: concept and applications. ACM Trans. Intell. Syst. Technol. **10**(2), 1–19 (2019). https://doi.org/10.1145/3298981
15. Reis, J.R., Gonçalves, R.F., Martinho, R., Cota, M.P.: Federated learning for smart manufacturing in Industry 4.0. In: Advances in Intelligent Systems and Computing, vol. 1386, pp. 591–599. Springer, Cham (2022). https://doi.org/10.1007/978-3-031-00602-2_52
16. Li, T., Sahu, A.K., Zaheer, M., Talwalkar, A., Smith, V.: Federated optimization in heterogeneous networks. In: Proceedings of Machine Learning and Systems (MLSys), pp. 429–450 (2020)
17. Zhang, Y., Luo, T., Zhang, J., Liu, X.: Multilevel federated learning with heterogeneous data for human activity recognition. In: Advances in Neural Information Processing Systems, vol. 34, pp. 28002–28015 (2021). https://doi.org/10.48550/arXiv.2107.07595
18. Xu, J., Hu, L., Sun, J., Gu, X., Li, P.: Multilevel federated learning for smart grid load forecasting. IEEE Trans. Industr. Inf. **18**(2), 1392–1400 (2022). https://doi.org/10.1109/TII.2021.3091938
19. Ghosh, A., Chung, S., Yin, D., Ramchandran, K.: An efficient framework for clustered federated learning. Adv. Neural Inf. Process. Syst. (NeurIPS) **33**, 19586–19597 (2020)
20. Lepoint, T., Patel, S., Raykova, M., Seth, K., Trieu, N.: Private join and compute from PIR with default. IACR Cryptol. ePrint Arch. **2020**, 1011 (2020)
21. Li, X., Huang, K., Yang, W., Wang, S., Zhang, Z.: On the convergence of FedAvg on non-IID data. In: International Conference on Learning Representations (ICLR) (2021). https://doi.org/10.48550/arXiv.1907.02189
22. Li, Q., Diao, Y., Chen, Q., He, B.: Federated learning on non-iid data silos: An experimental study. In 2022 IEEE 38th international conference on data engineering (ICDE). IEEE, pp. 965–978 (2022)
23. Shi, R., Wang, X., Wu, Q., Yuan, K., Liu, Q.: Communication-efficient cross-silo federated learning via decentralized updates. In: Proceedings of the 24th International Conference on Artificial Intelligence and Statistics (AISTATS), pp. 728–737 (2021)
24. Jiang, Y., Shen, Z., Jin, Y., Ren, Y.: Cross-silo federated learning for healthcare: a case study in medical imaging. IEEE Trans. Indus. Inf. **18**(5), 3167–3176 (2022). https://doi.org/10.1109/TII.2021.3089160
25. Li, J., Cui, T., Yang, K., Yuan, R., He, L., Li, M.: Demand forecasting of E-commerce enterprises based on horizontal federated learning from the perspective of sustainable development. Sustainability **13**(23), 13050 (2021). https://doi.org/10.3390/su132313050
26. Maheswaran, J., Jackowitz, D., Zhai, E., Wolinsky, D.I., Ford, B.: Building privacy-preserving cryptographic credentials from federated online identities. In: Proceedings of the Sixth ACM Conference on Data and Application Security and Privacy, New Orleans, LA, USA, pp. 3–13 (2016). https://doi.org/10.1145/2857705.2857711

27. Wang, Y., Bennani, I.L., Liu, X., Sun, M., Zhou, Y.: Electricity consumer characteristics identification: a federated learning approach. IEEE Trans. Smart Grid **12**(4), 3637–3647 (2021). https://doi.org/10.1109/TSG.2021.3060625
28. Li, X., Zhou, Y., Guo, W., Huang, X.: (2021). Multi-level federated learning system for industrial intelligence. In: 2021 IEEE International Conference on Computer Science and Automation Engineering (ICCSAE), pp. 372–376. https://doi.org/10.1109/ICCSAE52496.2021.9524829
29. Xu, G., Li, H., Liu, S., Yang, K., Lin, X.: VerifyNet: secure and verifiable federated learning. IEEE Trans. Inf. Forensics Secur. **15**, 911–926 (2020). https://doi.org/10.1109/TIFS.2019.2938887
30. Deshpande, N.M., Gite, S.S.: A brief bibliometric survey of explainable AI in the medical field. Library Philosophy and Practice, pp. 1–27 (2021)
31. Liu, Y., Kang, J., Niyato, D., Zhang, J., Liang, Y.C.: A joint federated learning and edge computing framework for AI-driven smart cities. IEEE Netw. **34**(3), 76–83 (2020). https://doi.org/10.1109/MNET.011.1900575
32. Zulfiqar, M., Kamran, M., Rasheed, M.B., Alquthami, T., Milyani, A.H.: A hybrid framework for short-term load forecasting with a novel feature engineering and adaptive grasshopper optimization in smart grid. Energy Rep. **7**, 6262–6273 (2021). https://doi.org/10.1016/j.egyr.2021.09.119
33. Zhao, Z., Zheng, P., Xu, S., Wu, X.: Federated learning with blockchain for autonomous vehicles: a decentralized learning approach to privacy-preserving energy efficiency. IEEE Trans. Intell. Transp. Syst. **22**(3), 1678–1689 (2021). https://doi.org/10.1109/TITS.2020.2992437

Federated Learning for Energy Optimization in the Industrial Internet of Things: Applications and Challenges

Ankita Tiwari, Nitin Rakesh, and Monali Gulhane

Abstract The Industrial Internet of Things (IIoT) provides an unprecedented opportunity to address the issue on a macro scale. However traditional centralized learning approaches cannot manage the enormous amount of data from anywhere, like IIoE devicesCreateTime. Therefore, federated learning began to be considered as a solution for decentralized collaborative model training that will maintain the confidentiality of data. This chapter investigates the effectiveness of federated learning on energy-efficient optimization within IIoE and provides insight into different federated learning algorithms and architectures for energy-efficient operations, exploring the resource-constrained characteristics of IIoE devices. Additionally, the chapter discusses how federated learning would benefit intelligent energy management systems, predictive maintenance, and resource configuration.

Keywords Internet of Energy (IoE) · Smart grid · Energy management · Digitalization · Decarbonization · Renewable energy

1 Introduction

The IIoT is revolutionizing industries, with its multitude of machines, sensors, and systems generating vast data streams. This new, information-centric ecosystem offers significant opportunities for companies to enhance their industrial operations, reduce energy consumption, and promote sustainability. However, current centralized methods have limitations when it comes to energy optimization within IIoT, given the large-scale production and widespread sensitive data. Federated learning, with its

A. Tiwari
Department of Engineering Mathematics, College of Engineering Koneru Lakshmaih Education Foundation, Guntur, India

N. Rakesh · M. Gulhane (✉)
Department of Computer Science and Engineering, Symbiosis Institute of Technology, Symbiosis International (Deemed University), Pune, India
e-mail: monali.gulhane4@gmail.com

R. Mohana et al. (eds.), *Federated Learning Applications in the Industrial Internet of Everything (IoE)*, Studies in Systems, Decision and Control 611,
https://doi.org/10.1007/978-3-031-99270-4_9

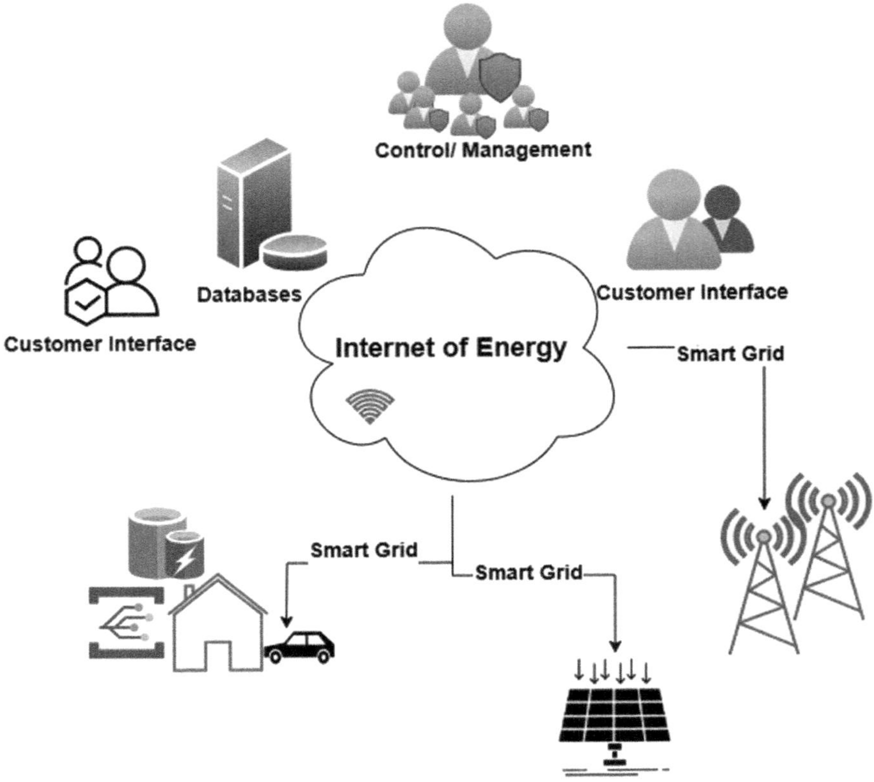

Fig. 1 The structure of the Internet of Energy (IoE)

potential to transform industrial operations, emerges as a potential solution to these challenges.

One potential answer to these problems is federated learning [1]. It enables devices to download models and train them by themselves in a privacy-preserving manner without sending their raw data to any centralized server. It ensures that data is private, reduces the communication bandwidth, and processes edge using the device. As a result, FL can help industrial AI developers use different edge-sensitive data to train various models (e.g., energy intelligence, energy management, and predictive maintenance) for intelligent IIoT in collaboration [1]. This is a good idea as this system has been used in some upcoming technologies and applications in the current power distribution system to improve the overall efficiency of electrical supply practices, including renewable resources across decarbonized smart grids with IoT-based digital Smart Grids at IoE through Electrical Practical Framework Media Subset. IoE Encompasses Infrastructure (The energy network), Electronics Devices, and mobility. As we see in Fig. 1, all the digitalization processes are present.

IoT based modern era Smart Grids with Internet of Energy (IoE) on Decarbonised Electrical Practical Framework Media Subset incorporating Renewable standard and Latest Technologies Among current system, few novel technologies and applications have been created to enhance more efficiency in electrical power systems. From (With the energy network, electronic devices and mobility shown in Fig. 1) what is all around IoE all its process of digitalization everything is that encompassed Fig. 1.

Dynge et al. [1] constructed a price corridor for different range values regarding Levelized Cost of Electricity LCOE so that prosumers are competitive while having equal pricing with any market counterpart. The system considered the entire market interaction as a Cyber-Physical-Social-System (CPSS), which will be deployed in the retails of an eCommerce environment and is built on top of a Distributed Ledger Technology (DLT)-based platform. We evaluate the case relative to conventional Feed-in Tariffs (FiT) and recommended alternatives, assessing the strengths or weaknesses of both fixed vs variable contracts with the wholesale supplier. Which reveals the energy consumer pays less, but this is traded with the prosumer receives almost slightly lower revenue compared with FiT scheme. However, in this study, the cost mechanism proposed has a much larger welfare improvement than FiT other things being equal (for both prosumers and consumers).

Ali and Zorlu Partal [2] proposed ZigBee with Low-Power Wide-Area Network (LPWAN) based wireless sensor network for smart-building energy management system. The framework was illustrated by formalizing a set of rules for basic automatic room temperature and humidity control in smart spaces, supported with commonly available home lighting systems. This makes the system more scalable, allowing for additional capabilities to be programmed in alternatively for some internet things (IoT) applications with a more customized embedded code. Technically it is an intermediate through which LoRa transceiver module and ZigBee transceiver modules exchange data The performance parameters that were evaluated based on which included wireless transceiver module communication range and throughput data.

Nkuriyingoma et al. [3], Tahir et al. [4] and Chauhan et al. [5] analyzed the integration of renewable energies in the distribution systems of Dynge et al [1]. Given this context, the present chapter conducts a techno-economic analysis of battery energy storage for grid-connected Solar PV systems at a small community scale in Rwanda.. The research proves that it accomplishes the desired ratio. SR = 89.2 kWh BESS) 33 kWp solar PV, (Annual load self-sufficiency level 64.38%, Performance Ratio 86.05%. The proposed solution, applied to the address of power outages in developing countries, proved technically and economically viable: the numbers support economic analysis reinforced by a return on asset of 9.14% with payback in nine years and five months.

As per the research conducted by Tahir et al. [4] no energy is wasted and individuals can sell their excess electricity to someone who needs it. Blockchain bolsters the efficiency of energy savings and IoT in smart grids. Here, a model had been presented where authors build microgrids on their own blockchain, and the users transacted in energy with smart contracts.

1.1 Objectives of the Chapter

The objectives of the chapter are:

To articulate the critical challenge of high energy consumption within the Industrial Internet of Things (IIoT) and establish Federated Learning (FL) as a viable, privacy-preserving framework for developing intelligent energy optimization models without centralizing sensitive operational data from distributed industrial assets;

To investigate and detail the diverse applications of Federated Learning for energy management in IIoT, including its use in predictive maintenance to reduce energy waste, smart grid load balancing, and the optimization of manufacturing processes to enhance overall energy efficiency.

And, to critically evaluate the inherent technical challenges and practical trade-offs in deploying FL for energy optimization, specifically addressing the complexities of data heterogeneity across different industrial machines, the energy cost of communication overhead, the computational constraints of edge devices, and the scalability of the FL architecture in dynamic industrial environments.

1.2 Organization of Chapter

The rest of the chapter is organized as: Sect. 2 enlightens the roles of federated learning in energy optimization. Section 3 discusses the methods of energy optimization. Sections 4, 5 and 6 highlights meta-heuristic optimization, reinforcement learning and game theory approaches. Section 7 highlights case studies for federated learning in IIoT Energy optimization, while, Sect. 8 illustrates federated learning algorithms, followed by federated averaging framework in Sect. 9. Section 10 illustrates Fedprox, followed by Sects. 11 and 12 discussing agnostic federated learning and renewable energy applications in federated learning. And, finally, Sects. 13 and 14 concludes the chapter with challenges and future directions.

2 Roles of Federated Learning in Energy Optimization

Federated learning offers several key features for the energy-efficient operation of the Industrial IoT:

1. **Distributed Intelligence**: FL enables intelligent energy management systems operating directly on IIoT devices. This allows for instantaneous decision-making and control, minimizing the need for constant communication with central servers, thereby reducing data transfer delays and energy use.
2. **Predictive Maintenance**: FL facilitates maintenance before actual breakdowns by analyzing diverse device data and recognizing patterns that indicate potential malfunctions. This proactive approach reduces downtime and saves energy by preventing equipment failure and optimizing maintenance schedules.

3. **Redistribution of Resources**: FL algorithms can predict energy usage trends based on operational data, optimizing electric power distribution across systems. This enhances efficiency and minimizes waste.
4. **Improved Renewable Energy Integration**: As industries increasingly adopt renewable energy sources, FL can assist in their efficient utilization. FL models can analyze data from weather sensors, energy storage systems, and grid infrastructure to predict energy production and adapt consumption patterns for maximum use of green energy.
5. **Collaborative Optimization**: FL fosters collaboration between various IIoT stakeholders–equipment manufacturers, energy providers, and facility operators. FL-trained models using data from diverse sources can design comprehensive energy optimization strategies across the entire IIoT ecosystem.

With these capabilities, federated learning can enhance energy efficiency, reduce operational costs, and promote the creation of a sustainable industrial infrastructure.

3 Methods for Energy Optimization

3.1 Rule-Based Optimization

Rule-based optimization is used in scenarios where decision-making is driven by predefined logic expressions. Unlike mathematical models, this technique uses "if-then" conditions or expert knowledge for the decision-making process.

Example: A Rule-Based System for a Thermostat Controller:

- If the temperature exceeds 25 °C, turn on the air conditioning.

This method does not rely on a specific mathematical formula, but instead follows a heuristic or logical approach.

3.2 Mathematical Optimization

Mathematical optimization concerns problems in which the objective is to optimize (maximize or minimize) a mathematical function subject to constraints. It can be divided into two categories: deterministic and uncertainty-based optimization.

Deterministic Optimization: Deterministic optimization involves no uncertainty in the parameters or variables of the model. All variables are known with certainty.

- **Dynamic Programming**: Dynamic Programming (DP) is a technique that solves problems by breaking them into smaller sub-problems. A solution is constructed from the solutions to sub-problems, which can be formulated as a recursive equation, commonly referred to as the **Bellman equation**:

$$V(x) = \min\{C(x, u) + V(f(x, u))\} \tag{1}$$

where:

- $V(x)$ is the value function for the optimal cost starting from state x,
- $C(x, u)$ is the cost incurred by taking action u from state x,
- $f(x, u)$ is the transition function representing the next state.

– **Linear Programming (LP)**: Linear programming involves optimizing a linear objective function subject to linear constraints. The general form is:

$$\text{Maximize/Minimize} \quad c^T x \tag{2}$$

Subject to:

$$Ax \leq b \tag{3}$$

where:

- c is the vector of coefficients for the objective function,
- x is the vector of decision variables,
- A is the matrix of constraint coefficients,
- b is the vector of constraints.

– **Non-Linear Programming (NLP)**: Non-linear programming deals with problems where the objective function or the constraints are non-linear. The general form is:

$$\text{Minimize/Maximize } f(x) \tag{4}$$

Subject to:

$$g_i(x) \leq 0, \quad i = 1, \ldots, m \tag{5}$$

where $f(x)$ is the non-linear objective function and $g_i(x)$ are the non-linear constraints.

3.3 *Uncertainty-Based Optimization*

Uncertainty-based optimization is used when the parameters or variables are uncertain, and the models must account for these uncertainties.

Fuzzy Programming: Fuzzy programming applies fuzzy logic to deal with imprecise data. It's used when the objective or constraints involve fuzzy variables. An example fuzzy optimization problem can be:

$$\text{Maximize } \tilde{c}^T x$$

Subject to:

$$Ax \leq \tilde{b}$$

where $\tilde{c}$ and $\tilde{b}$ are fuzzy numbers.

Stochastic Programming: Stochastic programming is used when some of the parameters follow a probability distribution. A typical formulation is the two-stage stochastic programming model:

$$\min_{x \in X} \left\{ f(x) + \mathbb{E}_{\xi}[Q(x, \xi)] \right\}$$

where:

- $f(x)$ is the first-stage cost,
- $Q(x, \xi)$ is the second-stage cost involving the random variable ξ,
- $\mathbb{E}_{\xi}$ is the expected value over the distribution of ξ.

4 Meta-Heuristic Optimization

Meta-heuristic algorithms are high-level frameworks designed to find near-optimal solutions for complex optimization problems. These algorithms are inspired by natural processes and are suitable for large-scale problems, though they don't guarantee finding the global optimum.

4.1 Nature-Inspired Optimization

Nature-inspired optimization algorithms mimic biological processes to find optimal solutions.

Examples:

- **Genetic Algorithm (GA)**: GA is based on natural selection and evolves a population of candidate solutions to find the best one.
- **Particle Swarm Optimization (PSO)**: Inspired by the movement of swarms, PSO finds the optimal solution through the collective behavior of particles.

4.2 Hybrid Optimization

Hybrid optimization combines different techniques, often blending heuristic and deterministic methods to improve performance. An example is combining genetic algorithms with gradient-based methods to optimize solutions.

5 Reinforcement Learning

In Reinforcement Learning (RL) an agent interacts with an environment and learns to perform actions by receiving rewards. We want to maximize the reward we accumulate over time.

The RL problem is usually formulated as:

$$J(\pi) = \mathbb{E}\left[\sum_{t=0}^{\infty} \gamma^t r(s_t, a_t)\right]$$

where:

- $J(\pi)$ is the expected total reward,
- γ is the discount factor ($0 < \gamma \leq 1$),
- $r(s_t, a_t)$ is the reward received at time t for taking action a_t in state s_t.

Q-learning and deep reinforcement learning are frequently used reinforcement learning algorithms for solving complex optimization problems.

6 Game Theory Approaches

Game theory studies models that describe the interactions between different decision-makers (agents) whose behavior is determined by those of all involved. The idea is to see if there are strategies that end in a sort of Nash equilibrium, where none of the n players can improve their payoff by changing randomization alone.

Nash Equilibrium in a two-player game is defined as:

$$u_1(s_1^*, s_2^*) \geq u_1(s_1, s_2^*), \quad u_2(s_1^*, s_2^*) \geq u_2(s_1^*, s_2) \tag{6}$$

where:

Let $u_1(s_1, s_2)$ and $u_2(s_1, s_2)$ be the utility functions for Players 1 and 2, respectively, where s_1 and s_2 represent their strategies.

The optimal strategies for Player 1 and Player 2 are denoted by s_1^* and s_2^*, respectively. These strategies maximize each player's utility function:

$$s_1^* = \arg\max_{s_1} u_1(s_1, s_2^*)$$

$$s_2^* = \arg\max_{s_2} u_2(s_1^*, s_2)$$

where s_1^* and s_2^* are the Nash equilibrium strategies that provide the best response for both players.

It is a widely used approach in various fields and a good way to understand economics, political sciences, and multi-agent systems in which agents have to play strategically given the moves of the others (Fig. 2).

7 Case Studies for Federated Learning in IIoT Energy Optimization

7.1 *Optimizing Federated Learning in Distributed IIoT: A Multi-agent Approach*

A three-tier collaborative FL structure for aiding deep neural network (DNN) training in decentralized IoT networks is proposed. The method uses deep multi-agent reinforcement learning to enhance device selection and resource allocation [6].

7.2 *Adaptive Federated Learning and Digital Twin for IIoT*

This chapter proposes an IIoT architecture enabled by digital twins (DT). DTs assist FL by understanding the features of industrial devices and mitigating estimation deviations using trust-based aggregation [7].

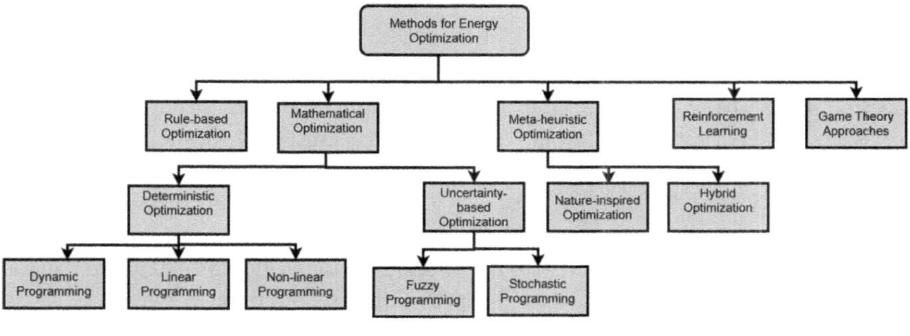

Fig. 2 Methods for energy optimization

8 Federated Learning Algorithms

This section outlines various federated learning algorithms for energy optimization:

8.1 Aggregation Algorithms

In standard machine learning, neural network training minimizes an objective function. In FL, this objective is extended across participating clients. The standard objective function is:

$$\min_{\omega} \sum_{i=1}^{N} \frac{n_i}{n} \ell_i(x_i, y_i; \omega) \tag{7}$$

8.2 Federated Stochastic Gradient Descent (FedSGD)

FedSGD minimizes the objective function in Eq. 1 by averaging gradients sent from clients in each round, calculating one epoch per batch. The updated model parameters are:

$$\omega^{t+1} = \omega^t - \eta \sum_{i=1}^{N} \frac{n_i}{n} \nabla \ell_i(\omega^t) \tag{8}$$

FedSGD guarantees model convergence but introduces significant communication overhead at each iteration.

9 Federated Averaging Framework

As a tool, this work uses FedAvg, which is the premier framework fi whitepaper minutecluster.orgin distributed learning (DL) and from which many extensions spawned for Federated Learning (FL). It relatively speeds up FedSGD by updating client models many times before averaging (multiple mini-batches per epoch). That is why the model training does not take place at every client; it happens in several batches or several repeats of an epoch, where clients update their models and send them to the server. Thus, FedAvg does not distribute gradient averages, but instead utilizes model weights directly to obtain the updated server weights ω^{t+1} on aggregation:

$$\omega^{t+1} = \sum_{i=1}^{N} \frac{n_i}{n} \omega_i^{t+1} \tag{9}$$

where ω_i^{t+1} is the updated model calculated by client C_i at epoch t.

This based approach allows faster training and a decrease of communication costs. While convergence of the global model's optimization cannot be guaranteed, we have enjoyed strong empirical success with this setup using FedAvg. The simplicity and flexibility of FedAvg have also promoted its use, and in most studies it has become the most used FL framework for renewable energy applications. In the case of strong statistical heterogeneity, however, the weights of locally trained models will tend to pull the server's weights in opposite directions, leading to a decrease in performance for FedAvg.

10 FedProx

Non-IID data aggravates the problem of client drift, leading to reduced average performance in terms of convergence and accuracy by FedAvg. FedProx trains local client models so that they do not diverge from the global model. Each client's loss function, per distance to local client weights, is also affected by the regularisation term, ω_i, and the global model weights, ω^t:

$$h_i(\omega_i, \theta) = \ell_i(\omega_i) + \frac{\mu}{2} \|\omega_i - \theta^t\|^2 \tag{10}$$

where μ is the regularization parameter (analogous to lasso or ridge regularization, depending on the model) that scales the importance of the term in the loss function.

Next, the server aggregates all updated parameters ω_i^{t+1} to get the updated global model parameters analogous to FedAvg.

Regularization compels the local models to remain close to the initial model, mitigating data consistency. Nevertheless, convergence may be slow as the models are encouraged to match the global model parameters, leading to longer training rounds. Results: The empirical results suggest improved accuracy over FedAvg on highly heterogenous data, but slower convergence (relative to FedAvg) than IID datasets.

The regularization parameter μ can be dynamically adjusted to balance training speed and accuracy. FedProx is a moderately complex extension of FedAvg, but remains as versatile as FedAvg and can be combined with many other federated learning algorithms.

11 Agnostic Federated Learning

Agnostic Federated Learning (AFL) is a novel paradigm that guarantees model fairness by eliminating biased data distributions on the client's side. This issue arises when there are strongly correlated client clusters that share an extremely similar data distribution. In such cases, a few clients have more influence on the learned federated model, while those with different data distributions are not well-represented.

As an example, imagine a model learning to control clients who are residential buildings (e.g., for heating control). The model could favor an ensemble where all buildings house large families (sharing similar consumption patterns), over one where the distribution includes single households, night shift workers, mobility providers, or holiday travelers.

For client datasets $\mathcal{D}_i$ of size n_i, according to distributions $\pi_{\mathcal{D}_i}$, the biases are caused by FedAvg learning a target distribution that is the average over all clients, weighted by their dataset size ratios:

$$\pi_t = \sum_{i=1}^{N} \frac{n_i}{n} \pi_{\mathcal{D}_i} \tag{11}$$

The target distribution is not be the true objective distribution and it is by biased being closer to 1 than 0. AFL solves this by running through all possible target distributions that could have formed the optimal combination of client data distributions:

$$\pi_\lambda = \sum_{i=1}^{N} \lambda_i \pi_{\mathcal{D}_i} \tag{12}$$

where $\lambda_i$is the weights of the mixture rather than dataset sizes. The weights on the incoming edges are optimized according to this objective in the aggregation step on the server AFL is a mechanism by which the global model generalizes well to all clients, not just most of them.

12 Renewable Energy Applications in Federated Learning

Federated Learning (FL) is becoming increasingly popular with the rise of renewable energy adoption. Fleet members in renewable energy systems are usually geographically distributed, and data is often located on different ends of the fleet. This presents challenges in practical applications such as condition monitoring and power forecasting.

12.1 Condition Monitoring Model

A condition monitoring system for a wind turbine (WT) fleet was built as a typical application by Zhao et al. [8]. The results clearly demonstrate that the proposed personalized federated approach performs better than, or at least as well as, local training methods (i.e., models trained only on private datasets). However, it is far stronger in terms of preserving privacy compared to traditional centralized learning systems.

Wind turbines (WTs) can share their incident status information privately and securely through federated learning. This highlights the importance and benefits of collaborative learning, improving local models' accuracy without sacrificing data privacy, using only locally stored datasets. We describe a range of applications of FL for various tasks.

12.2 Forecasting of Resources, Production and Loads

Training data are needed for any accurate machine learning forecast model, which can be compatible with PVFactory production forecasts. The strategic importance of the estimates to renewable energy system integration with variable energy sources (solar and wind). Accuracy of a power forecast varies depending on the length of horizon For experienced developer who may not need an integrated platform Price Free Plans available Functionality VariableRanges from globe-level historical to individual plant real-time and cross tool management processes Training Deep Learning development timesPower BlocksA summary of types is—SupervisedRespective prototype... The model will not perform well without a good amount of data.

This will be possible respectively if the two sides include extensive information about observed properties of what can be predicted and, hence, how reliable its long-term predictions are. There are exceptions, such as wind farms, PV systems, and residential building sites having sparsity of specific datasets or localized weather, but these won't be the case for every other site. This keeps the clients' local data private while allowing a model of what all clients have learned from their own datasets to be trained and shared without any dataset actually getting exchanged among clients. As an example, clients (wind turbines or wind farms) work together to model power output and wind speeds in the context of wind power forecasting. This is proven by the various empirical studies [6], which continuously reach exact same conclusion that FL provides higher performance when compared to models trained in locally. Most of the time FL approaches outperform centralized models that require sharing at least some privacy-violating data.

FedAvg remains the most common baseline in the literature. Nonetheless, [9] delves into Federated-Proximal (FedProx) and an adversarial deep domain adaptation framework used with FedAvg. Some other researchers like Lu et al. [10] talked about the improvement it models to be privacy and security through model parameter encryption, while some others like Zhang et al. [10] focused on the improvement at security by avoiding model data injection attack.

To tackle the cold-start problemin machine learning, FL can also be of help because it could find the way to save computation time. This might be applicable to solve the data scarcity challenge, for example a new power grid where there are no features available in order to train any model like condition monitoring of that newly installed power grid is particularly common use cases in case of usage of renewable energy. The work by Salh et al. [12] was based on developing a unified model for forecasting wind power and testing the trained model with new wind farms. This shows an additional advantage of FL in renewable energy use-cases.

13 Challenges and Future Directions

From the above sections, it is concluded that the Federated Learning (FL) [13–16] paradigm has its own specific characteristics. In this section, we will present the drawbacks, unanswered questions and potential future research directions related to federated learning in renewable energy.

Nearly all of the studies we looked at have a central server network and only a few clients. A common issue in the literature with this FL method is scalability, specially in scenarios where there are many participating clients at large scale. For the most part, differences exist in configuration; asynchronous/decentralized (server-less) and peer-to-peer network structures among other approaches. This fact is demonstrated to be the case for many custom kernels and a large communication cost reduction was made out of it, so this point has not been covered very well in earlier works.

Furthermore, both the impact of client computation capacity and related heterogeneity has been little studied. The exploration of computational complexities, benefits and detriments of different network configurations could help in leaps forward for theoretically more approachable and productive applications of FL in real-world renewable energy scenarios. More extensive scale trials, which consider many clients or working with more powerful capabilities of computation, could corroborate the way proposed options operate in an operational setting.

Statistical heterogeneity is still a major challenge in the practical application of renewable energy. Non-IID data arises in many settings, across a range of fields. Case studies show that state of the art FL frameworks perform even worse—sometimes even worse than local models—unless specific mitigations are used due to the non-IID issue. The positive news is there a relatively new set of specific techniques called Personalized FL to help us address this. Altogether, a range of methods—predominantly clustering-based techniques in the case of personalization layers (and despite some fine-tuning-based solutions) are vastly superior to non-personalized centralized methods.

It seems that personalization is a great way to deal with the non-disjoint data one often sees in renewable energy use cases. In contrast, the majority of existing studies assume a certain personalized approach without empirically comparing to (other) methods. Not one method of personalization is undefeated in all use cases but comparing them can uncover key differences. The largest differences in distribution appear among different applications i.e. individual household consumption vs power generation of a fleet of PV systems. So how does one choose the right method of personalization, making it a difficult consideration to make and hold bigger implications?

The structure in the form of clustering or personalization layers exacerbates the cold-start problem since new clients do not get a global model. Additionally, some methods need modifications to the framework and consensus among all nodes which fine-tuning instead of being elective is a post-FL step. For instance, clustering-based personalization is still suspected of boosting FL performance under a large number of wind farms and different types of turbines.

The non-IID data issue continues to drive the operation of FedAvg. Recent novel algorithms such as FedProx and FUALA have been developed to improve both convergence speed and performance with non-IID data, but their application in renewable energy remains limited. These results suggest that personalized FL is a remarkably useful solution for renewable energy applications despite the presence of significant statistical diversity.

Given the common nature of this problem, it might be worth exploring other aggregation algorithms that could yield better results. Weighted variants of FedAvg have been proposed in several studies to ensure fairness. However, many issues persist regarding systematic bias and fairness in renewable energy applications. The effects of dominating factors, such as weather-related operating conditions, are often insufficiently described. There is also a lack of case studies on how well the models perform.

The federated learning framework design of 'tensorflow' can extend a flexible aggregation operation, communication client sampling, or training stop configuration. Some more novel FL frameworks for renewable energy applications or other types of FLs have also been presented, but a comprehensive comparison among these is still not possible. Results: These results can facilitate the app developer's choice of a better-performing framework objectively and bound to data as well as a domain-dependent perspective, showing that there may not exist an absolute recommendation for the right choice of framework. Instead, they assert that—in all the aggregation types—a FULLY NEURAL ALGORITHMS (i.e., average weights unit-by-unit distribution of u itself in our review neighofuando widely used for renewable energy neurons mouse) In contrast to federated learning methods for dynamical network architectures, like Federated Match Averaging, Probabilistic Federta Neural Matching or Federated Distance: which distinctly regard neural networks behavior differentiated by each unit. Federated Learning is a rapidly new field, and many new flavors are still emerging to make it more efficient or robust. These were the latest algorithms that often appeared to be derived from older technologies—for instance; the recent FedMA and PFNM projects are "state-of-the-art methods" that cannot be applied in sufficient fields, such as renewable energy. We demonstrate in this paper that we can bring the training time of a model with similar or better accuracy down to the order of magnitude well above classical algorithms by having only a small amount of data transferred compared to what would have classically been picked as less appropriate. It will also be interesting to look at whether more recent FL methods such as FedMA, PFNM, and the like provide further benefits in terms of performance enhancement that can be explored in future works. Our research also discovered that because federated learning is a classic case of transfer learning setting (heterogeneous data), research on applying a few adaptation learning methods would be interesting for future work—hyper-parameter tuning. The model architecture, and to a lesser extent maybe hyperparameter tuning, are other complex parts of federated learning that you must take care of. In the more sumptuous feel, model architecture and hyper-parameter tuning aren't federated (if we now need to assume on a reasonable channel hyper-parameters random one device data here in each wish of global choice local client counts as 1 decision which may not best global choice). Like a neural network architecture for instance? Could very well overfit extremely fast on little, low-dimensional, or sparse data from any one particular local client but not solve the global level complexity over all of these clients. The only other method we will consider here to study a federated learning problem is hyper-parameter tuning epoch by epoch, and this, too, must be carefully designed/optimized as one of two subtasks in the renewable energy application. Implementation challenges. However, lock-in at details in implementation level in processes and policies are still to be ironed out to make Federated Learning practical. In other words, What is anyone participating down the FL path to do? That is just one factor that plays a role in configuring and deploying roles. This takes us to what can be called the regulatory part regarding who should be able to join one federated learning process, and that even decides how the neural network is. For instance, What if there is enough resistance so that a client opposes some parts of the federated training or participates ineligibly but wants to join, or an operator can handle and schedule federated learning processes, ensuring continuous pool-level data distribution? The manufacturer might even offer it as an alternative for anyone who does not want to give the maker access to their data using federated learning. Further, in making a case of research needs for federated learning in renewable power systems, it can be seen that there is a multitude of question marks on its pragmatisms.

14 Conclusion and Future Scope

The chapter delves into the integration of Federated Learning (FL) for energy optimization in the Industrial Internet of Things (IIoT). It discusses how traditional centralized learning methods are inadequate for managing the vast amount of data generated by IIoT devices, and how a decentralized solution like FL is crucial. FL not only resolves issues like privacy and communication overhead, but also enhances the efficiency of applications like predictive maintenance (PM), intelligent energy management (IEM), and dynamic resource allocation, thereby playing a pivotal role in the IIoT landscape.

The case studies reviewed highlighted FL's capability to enhance the performance of local models and its potential to optimize energy efficiency in industrial systems. Various algorithms such as FedAvg and FedProx were examined, offering insights into their strengths and limitations when applied to non-IID data in IIoT networks. While promising, FL still faces challenges in scaling, handling heterogeneous data, and improving personalization.

Future research will focus on addressing the challenges of scalability, personalization, and heterogeneity, along with exploring newer aggregation algorithms and advanced variants like FedMA and PFNM. As renewable energy systems continue to grow, the application of FL in optimizing energy usage and enhancing collaboration between IIoT devices presents a sustainable and efficient path forward.

Overall, FL presents a compelling solution for industries seeking to implement energy-efficient operations while preserving data privacy. However, more work is required to refine these methods for practical, large-scale deployment, particularly in renewable energy applications.

References

1. Dynge M.F., Halden U., Klæboe G., Cali U.: Lcoe-based pricing for DLT-enabled local energy trading platforms. Front. Energ. Res. **10**, 901009 (2022)
2. Ali, A.I., Zorlu Partal, S.: Development and performance analysis of a Zigbee and Lora-based smart building sensor network. Front. Energy Res. **10**, 933743 (2022)
3. Nkuriyingoma O., Ozdemir E., Sezen S.: Techno-economic analysis of a PV system with a battery energy storage system for small households: a case study in Rwanda. Front. Energ. Res. **10**, 957564 (2022)
4. Tahir M., Ismat N., Rizvi H.H., Zaffar A., Nabeel Mustafa S.M., Khan A.A.: Implementation of a smart energy meter using blockchain and internet of things: a step toward energy conservation. Front. Energ. Res. **10**, 1029113 (2022)
5. Chauhan, A., Sharma, A., Mohana, R.: A pre-trained model for aspect-based sentiment analysis task: using online social networking. Procedia Comput. Sci. **233**, 35–44 (2024). In: 5th International Conference on Innovative Data Communication Technologies and Application (ICIDCA 2024). [Online]. Available: https://www.sciencedirect.com/science/article/pii/S1877050924005520
6. Zhang, W., Yang, D., Wu, W., Peng, H., Zhang, N., Zhang, H., Shen, X.: Optimizing federated learning in distributed industrial IoT: a multi-agent approach. IEEE J. Select Areas Commun. 1–1 (2021)
7. Feng, J., Zhang, W., Pei, Q., Wu, J., Lin, X.: Heterogeneous computation and resource allocation for wireless powered federated edge learning systems. IEEE Trans. Commun. **70**, 3220–3233 (2022)
8. Zhao, T., Chen, X., Sun, Q., Zhang, J.: Energy-efficient federated learning over cell-free IoT networks: modeling and optimization. IEEE Internet Things J. **10**, 17436–17449 (2023)
9. Sun, W., Lei, S., Wang, L., Liu, Z., Zhang, Y.: Adaptive federated learning and digital twin for industrial internet of things. IEEE Trans. Indus. Inform. **17**, 5605–5614 (2020)
10. Lu, Y., Huang, X., Zhang, K., Maharjan, S., Zhang, Y.: Communication-efficient federated learning for digital twin edge networks in industrial IoT. IEEE Trans. Indus. Inform. **17**, 5709–5718 (2021)
11. Zhang, T., Mao, S.: Energy-efficient federated learning with intelligent reflecting surface. IEEE Trans. Green Commun. Netw. **6**, 845–858 (2022)

12. Salh, A., Ngah, R., Audah, L., Kim, K.S., Abdullah, Q., Al-Moliki, Y.M., Aljaloud, K.A., Talib, H.N.: Energy-efficient federated learning with resource allocation for green IoT edge intelligence in b5g. IEEE Access **11**, 16353–16367 (2023)
13. Su, Z., Wang, Y., Luan, T., Zhang, N., Li, F., Chen, T., Cao, H.: Secure and efficient federated learning for smart grid with edge-cloud collaboration. IEEE Trans. Indus. Inf. **18**, 1333–1344 (2021)
14. Tang, Y., Du, G., Long, J., Zhao, Y.: Integration of online deep reinforcement learning and federated learning: a dynamic energy management strategy for industrial applications. IET Commun. (2023)
15. Yang, W., Xiang, W., Yang, Y., Cheng, P.: Optimizing federated learning with deep reinforcement learning for digital twin empowered industrial IoT. IEEE Trans. Indus. Inf. **19**, 1884–1893 (2023)
16. Chen, X., Liu, G.: Federated deep reinforcement learning-based task offloading and resource allocation for smart cities in a mobile edge network. Sensors (Basel, Switzerland) **22** (2022)

Revolutionizing Smart Transportation with Federated Learning: Applications, Challenges, and Future Directions

Rashmi Sharma, Shilpa Choudhary, Monali Gulhane, and Ankita Tiwari

Abstract Smart transportation systems are rapidly evolving, driven by advancements in technology and the growing demand for efficient, safe, and sustainable mobility solutions. Federated learning, a novel approach that enables decentralized machine learning across multiple devices without compromising data privacy, is emerging as a powerful tool to revolutionize smart transportation. This chapter explores the applications of federated learning in various facets of smart transportation, including traffic management, autonomous vehicle navigation, predictive maintenance, and public transport optimization. By leveraging data from diverse sources while maintaining data security and privacy, federated learning can enhance decision-making processes, improve operational efficiency, and reduce costs. However, implementing federated learning in smart transportation presents several challenges, such as data heterogeneity, scalability, communication costs, and regulatory concerns. This chapter discusses these challenges in detail and proposes potential solutions and strategies to overcome them. Furthermore, it outlines future directions for research and development, highlighting the integration of federated learning with other emerging technologies, such as the Internet of Things (IoT) and edge computing, to create more robust and adaptive smart transportation systems. Ultimately, this chapter aims to provide a comprehensive overview of how federated learning can transform the future of smart transportation, addressing current limitations and paving the way for innovative applications and improved urban mobility.

R. Sharma · M. Gulhane (✉)
Department of CSE, Symbiosis Institute of Technology, Nagpur Campus, Symbiosis International (Deemed University), Pune, India
e-mail: monaligulhane4@gmail.com

S. Choudhary · M. Gulhane
Department of CSE, Neil Gogte Institute of Technology, CSE (AIML), Hyderabad, India

A. Tiwari
Department of Engineering Mathematics, College of Engineering, Koneru Laksamaih Education Foundation, Guntur, India

R. Mohana et al. (eds.), *Federated Learning Applications in the Industrial Internet of Everything (IoE)*, Studies in Systems, Decision and Control 611,
https://doi.org/10.1007/978-3-031-99270-4_10

1 Overview of Smart Transportation Systems

Connected Vehicle (CV): A CV is a vehicle that can communicate with the surrounding world, in particular other vehicles and infrastructure elements like traffic lights, road signs or weather services over data communication networks. With cutting edge communications technologies, such as 5G and V2X (Vehicle-to-Everything), these vehicles will be able to share data amongst themselves in order to give real-time updates and provide for a safer, more efficient, intelligent transportation system.

Connected vehicles ecosystem components (Fig.1):

Connected Vehicles:

- Sensor, communication and ODB (on-board computer) integrated vehicles
- Sharing data with other vehicles (V2V), infrastructure (V2I), pedestrians (V2P) or the cloud (V2C).
- They can involve GPS, telematics, real-time vehicle diagnostics and even autonomous driving features.

Traffic Management Systems:

- The operations associated with the monitoring, directing and improving of traffic through data gathered from connected vehicles and other sensors.
- FeaturesTDACResourceTraffic signal controlYesDrakeOReal-time traffic congestion monitoringYesATRAMAdaptive traffic routingYesABORIncident managementNoTVehicle_LocationPrediction of vehicle paths based on their recent location updates.

Infrastructure:

- Connected infrastructure: Traffic lights, sensors, and signage that can communicate with vehicles; a.k.a. smart road infrastructure,
- High Speed Communication: Gives real time date transfer between the vehicles and the infrastructure, through communication networks like 5G.
- Data Processing Centers: Centralized or cloud-based systems that process the vast amounts of data gathered from connected vehicles and infrastructure to improve traffic flow and safety.
- Stations would have V2G interaction, allowing vehicles to communicate with the grid for optimized charging and energy management (EV-only).

These parts work on optimizing the flow of traffic, preventing congestion, and reducing pollution and in addition to ensuring safety in smart cities and future transportation [1, 2] scenarios.

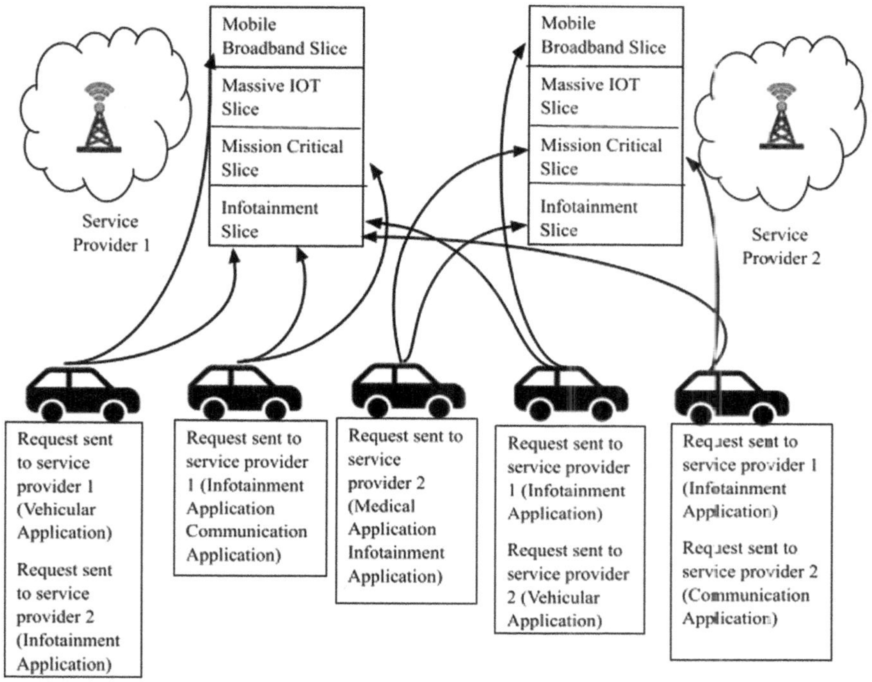

Fig. 1 Overview of Smart Transportation System

1.1 Objectives of the Chapter

The objectives of the chapter are:

- To establish the transformative potential of Federated Learning (FL) in addressing the critical data privacy, latency, and scalability challenges prevalent in modern intelligent transportation systems, positioning it as a key enabler for collaborative intelligence among vehicles, infrastructure, and traffic management centers;
- To systematically survey the diverse applications of FL within the smart transportation ecosystem, detailing its implementation for real-time traffic flow prediction, cooperative perception for autonomous vehicles, intelligent parking management, and enhancing the safety and efficiency of vehicular networks;
- And, to critically analyze the significant technical and operational hurdles to deploying FL in transportation, including communication bottlenecks in high-mobility environments, heterogeneity of on-board units and roadside sensors, and security vulnerabilities like model poisoning, while also outlining future research directions to create more robust, scalable, and secure FL frameworks for the next generation of smart mobility.

1.2 Organization of the Chapter

The rest of the chapter is organized as: Sect. 2 enlightens the role of AI in Smart Transportation. Section 3 illustrates the challenges of centralized AI in smart transportation, followed by discussion on federated learning in smart transportation. Section 5 highlights diverse applications of federated learning in smart transportation. And, finally, Sect. 6 concludes the chapter.

2 Role of AI in Smart Transportation

Leveraging artificial intelligence (AI), smart transportation [3–5] systems are transformed utilizing predictive analytics, real-time decision-making, and automation. Such features enhance safety, reduces congestion and deliver a better overall experience for urban transportation.

- **Predictive Analytics**:

Traffic prediction: AI can analyze historical data traffic and real-time data to predict the future of the car market, so that authorities and travelers have a clearer outlook. By using predictive models, they can detect bottlenecks before they happen.

Prediction for Maintenance: AI supports predictive maintenance in vehicles and infrastructure (like streets, bridges and traffic lights) by monitoring conditions that can lead to faults or breakdowns. This cuts outside time and dogging costs.

Predicting Demand: Using time, weather conditions and events, respectively AI is able to predict the demand of such services as public transport or ride-sharing enabling effective resource allocation and management of transport fleets.

- **Real-Time Decision-Making**: Real-time traffic data from sensors, cameras, connected vehicles AI systems are capable of analyzing real-time traffic data and using it to make instantaneous decisions on how long a traffic signal should remain green, re-routing of traffic or even how to manage congestion. This results in a smooth flow of traffic and better delay times.

Optimized Emergency Response: AI can assist emergency services by measuring traffic and real-time data to find the quickest routes for ambulances, fire trucks or police cars. It can even give priority to an emergency vehicle in traffic management systems.

Autonomous driving: For real-time decision making in complex traffic environments, detecting obstacles (e.g., humans), vehicles and decide on braking, acceleration or lane change.

- **Automation**:

Autonomous Driving: At the center of self-driving vehicles is AI that allows cars to use the existing roads and objects while driving without human control. Using data fed by cameras, sensors, radar and LIDAR systems, AI algorithms enable the vehicle to make safe decisions.

AI can do everything from automating scheduling, fleet management and route planning to optimising public transport using an intelligent network. In practice, this means automated buses or trains able to modify their speeds, routes and schedules based on actual passenger demand and traffic.

Logistics and Freight Management: In logistics, automation is utilized in managing and controlling the movement of goods from point to point, helps in optimizing the routes for transporting goods form one place to another and thus minimizes the inventory management cost and reduces fuel consumption. And right now, it is autonomous trucks and drones that are the emerging technologies in this space.

AI-Powered Toll Collection and Parking Automation: AI will enable automated toll booths and parking systems that identify vehicle information, accept payment, and assign parking spots—significantly reducing delays and making driving more consumable for drivers.

These improvements serve to bolster contemporary transportation systems, making them safe, efficient and sustainable by optimising infrastructure while enabling greener urban mobility through lower emissions both from vehicles but also for public transport options as well as finally breathing life into the dream of autonomous driving [6–9].

3 Challenges of Centralized AI in Smart Transportation

While centralized AI in smart transportation may have a huge benefit, it does not come without its share of challenges.

- **Data Volume**:

Great Data Production: Smart transportation systems produce large amounts of data generated by connected vehicles, traffic signals, sensors, cameras and other infrastructure. Central AI systems may suffer under the weight of this volume overall, which can compromise their ability to handle and process it all in real-time fashion.

Inferior storage and processing of data: While centralized software may not be efficient in storing and processing all the data coming in, which means that there would either be a bottleneck, inefficiencies or the data might also get lost.

Data Transmission Overload: Transferring huge amounts of data from a large number of thousands or millions of connected vehicles and infrastructure to be processed by centralized AI system can overwhelm communication networks input/output which may result impacts in performance and accuracy.

- **Latency**:

And especially Real-Time Requirements: Instant decisions are mostly required in smart transportation (e.g., autonomous driving, traffic signal adjustments) However, a centralized AI system will result in lags as information is transferred to the central processing unit.

Network Delays: In this, Network latency due to traffic congestion or outages can introduce further delays raising the net time. Ultimately making systems less effective and possibly increase safety issues in some cases.

Decision-Making Lag: In critical systems such as accident prevention, emergency services being able to act in real time centralized systems cant provide fast decisions.

- **Scalability**:

System Complexity: With smart cities growing, the connected devices and sources of data grow exponentially as well. The centralized AI system may not be able to cope with the increased levels of computational power, storage, and bandwidth that are going to be in demand.

Infrastructure Constraints: Turning a centralized system into something that can scale as the number of vehicles and communications grows and grows, may be expensive and convoluted. Deploying physical infrastructure like servers and data center can slow down the pace of scaling out.

If this happens, it could lead to a performance degradation at the centralized AI system where processing time lags and poor decision making features.

- **Privacy**:

Sensitive Data Collection: Smart transportation systems collect a huge amount of sensitive data such as personal information about drivers, passengers, and their routes. In case of a centralized AI system, this data might find itself in one repository (the central place) and get opened up as a prime area for cyberattacks and loss of data.

Risks of Data Centralization: A central data distribution point will lead to occasional leakages on a massive scale. If the central AI system were to be hacked, there would be a huge chunk of probable personal and important data on display.

Lack of Transparency: The opacity in centralized AI decision-making processes heightens concerns around the privacy of personal data and makes it challenging for individuals to understand how their data is being used, who has access to it, whether it is being used for surveillance or commercial purposes.

Typical challenges in this context can only be met with decentralized or edge-based AI, where the processing can be moved to where the data is created and required to handle, reducing latency, improving scalability and increasing privacy by limiting needed transmission of sensitive information back to a central place.

4 Federated Learning in Smart Transportation

In smart transportation systems [10], Federated Learning (FL) play an evolutionary role where inter connected vehicles, traffic management systems with smart infrastructure produce and process large amount of data. In the context of smart transportation multiple, vehicles, traffic cameras, and infrastructure sensors collaborate with each other to develop more intelligent and adaptive models without sharing delicate data. In view of smart transportation, improvements in technical frameworks with its real world implementations will enhance the model efficiency, data privacy and collaborations with other transportation systems like road, rail, water, air and many other transportation systems. Further some advancement in FL relevant to smart transportation is discussed.

4.1 Recent Advances in Federated Learning

In recent years, many developments, techniques, frameworks, and applications of Federated Learning (FL) have emerged. Few developments in terms of technical are discussed here in smart transportation field [11].

- Latest algorithms and techniques relevant to transportation.

In terms of Federated Learning (FL) literature focused on transportation, we review the latest algorithms and methods that are introduced to maintain data privacy preserving, model accuracy improving and real-time decision-making in this sector [12–14]. These algorithms are important for intelligent transportation systems such as autonomous vehicles, traffic management and logistics optimization, connected vehicle network [13]. Here are a few of the central algorithms and methods [15–20].

(1) **Enhanced Privacy-Preserving Technique**

Differential Privacy (DP): Progress on incorporating DP guarantees to protect individual data while exchanging model updates. Differential Privacy (DP) is the property of a mechanism that ensures no single output can reveal information about, or identify an individual user in the dataset.

Homomorphic Encryption (HE): Improvements in the domain of HE have permitted operations on encrypted data, allowing FL models to participate aggregation without needing raw data decryption.

Secure Aggregation: Techniques such as Google's Federated Secure Aggregation can now perform privacy-preserving aggregation even for thousands of shards, preventing an adversary from learning anything about the global model update via traffic analysis.

(2) **Scalability and Communication Efficiency**

Compression Techniques: Gradient Sparsification, quantization and pruning have been used as innovative methods to shrink the model update sizes in FL so that it can be effectively utilized over limited bandwidth environments such.

Federated Learning Aggregate Data on hierarchical grouping to have parts of models synchronize at different rates and aggregation tiers before updates are sent from local servers (decentralized or proximate) to a central one, effectively scaling FL solutions for higher population sizes.

(3) **Federated Learning for Heterogeneous Data**

Federated Transfer Learning (FTL): FTL enables models trained on one domain or data distribution to be adapted for another, which is useful especially in cases where different devices/organizations have disparate types of data e.g., urban traffic versus rural traffic.

Personalized Federated Learning (PFL): PFL is designed to adapt the aggregation strategy based on individual user or device by considering heterogeneity in terms of data distributions. This is especially crucial for example, in healthcare and personalized driver assist systems where individual differences translate to changes.

(4) **Federated Reinforcement Learning**

FRL = reinforcement learning + federated training: FRL is a new approach to combining RL with FL, in which decentralized agents (e.g., autonomous vehicles or robots) can share their learned policies without sharing raw data based on an improved actor-critic method. This is very handy, especially for real-time decision-making applications such as autonomous driving, smart grid optimization and robotics.

(5) **Federated Learning Across Silos and Devices**

Cross-Silo FL: Cross-Silo FL requires only collaboration between a small number of organizations (e.g., hospitals, banks or transportation agencies); each organization will locally train models on its private data. There have been recent advances in developing more robust models and allowing disparate data silos to work securely together.

Cross-Device FL: Consist of models trained in millions of devices (smartphones, IoT devices etc…) An increasing number of new techniques have been developed to address the challenges imposed by device churn, intermittent availability and non-IID (independent and identically distributed) data.

(6) **Blockchain Empowered Federated Learning**

Looking at FL, blockchain has arisen to secure model updates by using a tamper-evident and decentralized ledger in such systems. It also serves as the basis for maintaining and tracking all of the contributions from various participants, which secures that no bad actor could poison our aggregated models.

In cases where a federated healthcare system or intelligent transportation systems are considered, then blockchain-FL can be most useful.

(7) **Edge and IoT Federated Learning**

Edge AI becomes Edge Computing: FL are widely used at the edge computing environment where data is processed near generation place for example in IoT devices. This includes minimizing the computational and communication load on edge devices, which is one among new set of FL algorithms that are known as energy-efficient FL techniques.

IoT Networks: Federated learning allows distributed machine learning on huge IoT networks for better applications (e.g., smart homes, industrial IoT) benefitting from data generated by the connected sensors and devices.

(8) **Improvements that Are Specific to Application**

Healthcare—FL has been applied to healthcare data, enabling hospitals and clinics to train disease diagnosis/patient risk score prediction models together without sharing any sensitive patient information.

Autonomous vehicles and transportation: With the progress in FL, autonomous driving has allowed sharing of models across an entire fleet of self-driving cars to power collaborative learning for not only improved driving performance but also better traffic predictions along with advancements around vehicle-to-vehicle (V2g) communication without compromising on patients privacy.

Finance: Financial institutions have started using FL for fraud detection, credit scoring and risk assessment—allowing banks to work together on models without revealing sensitive data.

(9) **Federated Learning Libraries**

FedAvg (Federated Averaging) is the most fundamental FL algorithm. With the recent variants such as FedProx, FedNova or SCAFFOLD are able to solve these challenges by addressing heterogeneity in data and models faster convergence better handling of noisy updates.

FL Platforms—Some of the open-source FL platforms like TensorFlow Federated (TFF) and PySyft are adding advanced features to support larger-scale federated learning deployments and different application scenarios.

4.2 Integration of FL with Smart Transportation Systems

Employing FL methods to decentralize the train infrastructure entities, allowing data processing collectively without violating privacy by reducing the latency and enhancing system-wide decision-making. This is especially important to many existing operations whereby the data across vehicles, sensors and Urban infrastructure. Further some methods uses to deploy FL in distributed network of vehicles and infrastructure is discussed.

4.2.1 Methods for Deploying FL in Distributed Networks of Vehicles and Infrastructure

1. **Local Model Training via Edge Computing**

Method: Edge devices, such as vehicles' onboard computers, roadside units or traffic sensors train models using their own data in a distributed vehicle and infrastructure network.

Application:

Models for detecting obstacles, optimizing routes or managing energy can be trained using local data (e.g. camera feeds, LiDAR and GPS) which may not leave the vicinity of autonomous vehicles as often.

At intersections, traffic sensors can locally train models to predict congestion based on local vehicle flow.

Edge Hardware: Edge should have processing power include GPU, FPGA or TPU in order to handle model training at edge.

2. **The Hierarchically Federated Learning (HFL)**

Method: Hierarchical FL optimizes both locally at each region (e.g. city, traffic district) and globally over the full network gathered across all regions

Application:

In a city-wide traffic management system, local learning models at district levels (referring to clusters of control devices in the same area) are combined into the global model over all districts within contentious distributed environment. It reduces the communication overhead, and in turn improves scalability of FL system.

Fleet-level aggregation of local models within a fleet—then sharing updates to that model globally across regions;

Benefits: HFL enjoys a reduced communication burden that scales well for training large transportation networks by restricting global updates.

4.3 Federated Learning with Only Asynchronous Updates

Asynchronous FL allows the devices on the edge (e.g., cars, traffic sensors) to share their model updates locally without waiting for central synchronization with all other parties.

Application:

This allows vehicles which are offline or temporarily out of communication range to catch up when they reconnect, ensuring that model updates can be delivered despite the latency introduced by connectivity failures.

For example, self-sufficient smart infrastructure such as traffic lights, parking sensors and road monitoring cameras can function on their own to receive updates whenever the network allows them which will in turn reduce latency of learning updates.

PROS: Asynchronous FL makes flash learning in distributed networks more feasible—for instance, when a device might be temporarily disconnected (e.g., moving vehicles; remote areas).

4.4 Communication Protocols—After Compiler Protocols Must Be Efficient

To work at the scale required in distributed environments, use efficient communication techniques such as sparsification, quantization and compression to minimize model update sizes Reducing Communication Overhead → Reduced message size & # messages sent.

Application:

The models of the vehicles can send condensed updates, which consist just from most important gradients and in that way reduce transferring data.

Traffic sensors and roadside units with limited bandwidth are those that can use these tricks to deliver smaller updates, so they do not delay real-time decision-making.

Techniques:

Juice runs only sends the top-k gradients to sparsify (i.e., represent a full model parameter set as sparse gradient vector and index).

Quantization: Decreases the precision of model updates (e.g., from 32-bit to 8 bit), that reduces data size.

Compression Algorithms: Compress updates further to minimize the amount of data that needs to be transferred between vehicles and servers.

4.5 V2X (Vehicle-To-Everything) Federated Learning

Approach: Federated learning could leverage Vehicle-to-Everything (V2X) networks, where vehicles and infrastructure communicate to work together on model training.

Application:

V2B (Vehicle-to-Vehicle): Vehicles share their model updates directly with each other allowing collective learning among nearby vehicle. An autonomous vehicle might use its learned model for pulling to the right or swerving left, an action that it can share with other AVs on some roads.

V2I (Vehicle-to-Infrastructure): Vehicles can provide software updates to intelligent infrastructure (such as traffic lights or smart roads), enabling a city's traffic control system deemed by vehicle data for optimization on a metropolitan scale.

V2N (Vehicle-to-Network): The vehicle and its sensor would update on a central server which lies in the cloud for aggregation. Asynchronous or periodic fetches.

Advantages—V2X supports high speed, low-latency communication and rapid model updates in real-time which will help respond faster and coordinate with other vehicles as well as infrastructures (Fig. 2).

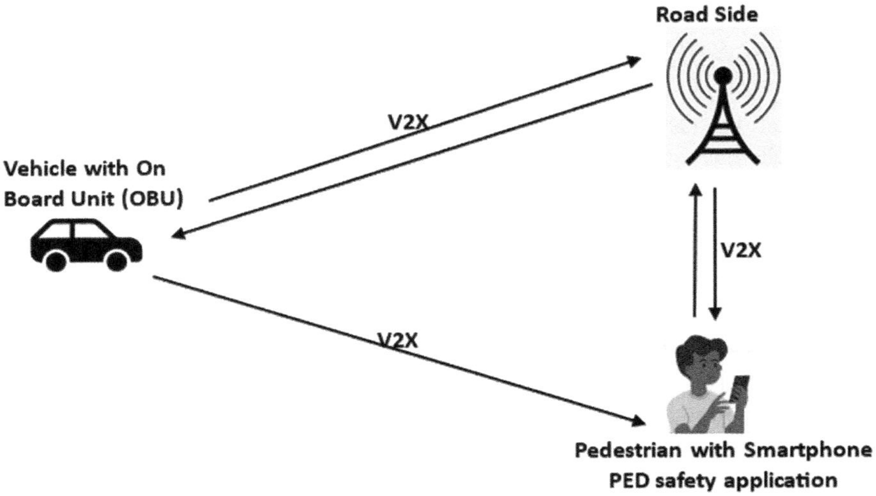

Fig. 2 V2X: vehicle to everything communication

4.6 How We Use Blockchain for Secured Aggregation of Model

Method: We propose integrating FL with blockchain technology to achieve decentralized secure and transparent model aggregation. This is where the Smart Contract enforces these guidelines for model updates and verifies the integrity of the information.

Application:

For multi-stakeholder networks (like city governments and private transportation companies) the permissioned blockchain is used to guarantee safe, verifiable, nonforgeable model updates. This is especially applicable in vehicle fleets or multi-agency traffic management systems.

Blockchain can be leveraged by autonomous vehicles to monitor and authenticate model updates across the network, so that no harmful update is introduced.

Benefits: The answer above has explained how blockchain can help improve the sekruiti, as well explain showed great about common fossil corporate society | collective in the learning of machine contains technology that helps to create transaction vision and unlimited aggregation process.

4.7 Cross-Client Federated Learning for Multi-agency Collaboration

Cross-silo FL: Similar to Horizontal FL but different organizations or agencies, e.g., multiple transport authorities and vehicle manufacturers collaboratively share updates of their models in a mutually protective way without sharing proprietary/ sensitive data [24].

Application:

Worldwide traffic models can be enhanced by the collaboration of municipal transit agencies without spreading raw data through jurisdictional limits.

By collaborating between manufacturers, it is possible to develop models for predictive maintenance alerts and driver behaviour analysis or autonomous driving systems whereas the sensitive vehicle data does not have to be shared among competitors.

Pros: Cross-silo FL rewards collaboration among many parties that own different parts of the data while preserving privacy and protecting IP.

4.8 Power-Constrained Device FL

Method: Distributed networks with less computationally and energetically loaded devices (e.g., traffic light sensor, parking meter) run the resource-aware FL techniques to train models economically at min potential resources.

Application:

Roadside units—devices installed along roadways to wirelessly transmit data between vehicles and networks—can use FL in which they train easier models or send just the portion of their updates due to less computational power, etc.

We show that electric vehicles (EVs) with limited battery life can decrease the rate and/or volume of updates to save power, but still benefit from participating in exchanging information tailored at collaborative model learning.

Techniques:

Adaptive FL—tunning update frequency and model size balancing with participating nodes' resources (e.g., battery, CPU power).

Energy-Accurate Learning Algorithm: Optimized for local model training on low-power devices with the objective of reducing energy consumption while it is being trained.

4.9 Privacy-Preserving Methodologies—Differential Privacy and Homomorphic Encryption

Approach: We exploit privacy-preserving solutions to ensure no sensitive data from the vehicles and infrastructure is compromised during FL training, as well as preservation of their security in aggregation.

Application:

For example, autonomous vehicles or traffic management systems are capable of sharing encrypted model updates without compromising the privacy about sensitive data e.g., location and driving behavior to other participants on network.

Differential Privacy: Injects noise to the model updates effectively making sure that each and every data point (car speed or location data) cannot be reversed from it using static models.

Homomorphically Encrypted: With this method, model updates are encrypted prior to sharing so that the central server can aggregate these private updates without seeing raw data.

Benefits: These methods protect the privacy of individual vehicle, passenger or infra data throughout collaborative learning.

4.10 Dynamic Participant Selection

Approach: FL systems should be able to change participating parties (vehicles, sensors, infrastructure) on the fly based upon data quality and sensor types, network conditions or computation resource availability.

Application:

Only vehicles with good enough connection and sensor data (a GPS position or camera image close to captured at the venue) are trained for a model update, while bad connectivity/shoddy quality data is temporarily excluded.

This means there is no need to hook up every piece of device unless we are in a real time optimization model where only high quality sensor traffic light (or the group thereof) and smart stabil network connectivity devices can be selected for critical decision making.

Pros: This feature increases model quality by invoking only the most reliable and high-performing participants in a FL task at each step.

5 Applications of Federated Learning in Smart Transportation

As it is the norm these days, our cities are increasingly fitted with sensors and hundreds of thousands of vehicles generate real-time data streams—a perfect match for Federated Learning (FL) which enables learning on edge devices as smart cars. FL approaches the challenges of smart transportation with keeping data decentralized and privacy-preserving techniques.

What makes FL valuable in smart transportation are illustrated with some of the prime applications [21–26].

1. **Autonomous Vehicles (AVs)**

Use Case: Autonomous Driving Model Co-Training

Issue: Autonomous vehicles (AVs) create vast sensor data streams—cameras, LiDAR and radar—that must be used to refine algorithms for obstacle detection, lane-keeping and navigation.

FL Solution: AVs can use FL to train models based on their driving experiences. These models can be centrally shared and aggregated on a central server without having to send raw data, thereby improving model performance across the fleet while still protecting driver privacy.

Example: Enable AVs to detect objects and pedestrians better, make real-time decisions without moving Protected Health Information (PHI), biometric information and other types of granular sensor data.

2. **Traffic Management and Prediction**

Use Case: Predicting traffic flow in real-time

Issue: Coordinating the transportation within cities is a problem because these already have real-time data from sensors (cameras, GPS and road infrastructure) apart stoplights. Putting this data into one place can cause latency and privacy problems.

FL Solution: Traffic sensors and smart traffic lights could train models to predict flow, congestion and optimal signal timing throughout the city without sharing raw data.

Example: Dynamic optimization of traffic light systems which reduces congestion, idle time and overall improves the efficiency of traffic.

3. **Vehicle-To-Vehicle (V2V) Communication Vehicle to Everything Communication.**

Utilization: Thereby enabling collaborative learning for better safety and navigation.

Challenge: Vehicles must exchange data to enable safety features, such as collision avoidance systems, but sharing raw sensor readings may compromise privacy and network capacity.

Vehicles near each other can use FL to collaboratively train common models for local navigation, threat velocities and vehicle-to-vehicle (V2V) communication. This allows vehicles to learn together without the need for sharing individual data, hence addressing privacy.

Example: Advanced assistance in accompanying autonomous and semi-autonomous vehicles over information on presence of impediments, street conditions or other vehicle conduct.

4. **Fleet Maintenance with Connected Fleets**

Use Case: Fleet Health Monitoring and Predictive Maintenance

Challenge: Fleet operators (e.g., buses, taxis, delivery vehicles) need to forecast the time when their vehicle is expected on road for maintenance in order not only prevent any breakdowns but also minimize downtime.

FL Solution: Every vehicle in the fleet can perform local data analysis from sensors (e.g., engine temperature, brake wear) and with FL communicate model updates to a fleet operator. For instance, aggregated models can predict when and where maintenance is needed throughout the entire fleet giving uninterrupted access to trucks in excellent condition for much longer periods of time.

5. **Transportation Optimization for Public**

Use case: Estimate passenger demand between different regions so that bus schedules can be optimized to decrease waiting times and increase service reliability.

Issue: Public TransportationPublic transportation systems (buses, trains) often run on siloed data sets that make it less efficient to determine optimized scheduling, routing and passenger load management.

FL Solution: FL allows transit systems to combine scheduling and route planning models without having sensitive real-time passenger counts or location data leave the boundaries.

6. **Ride-Sharing Optimization Dynamic**

Application: Better Ride-Hailing Algorithms.

Issue: Ride-sharing companies have to optimize the allocation of drivers and prices as well routes quickly according to demand, but this data is high privacy.

FL Solution: FL can be used to teach dynamic pricing and routing models using local ride data from drivers and passengers without compromising any sensitive information.

Example: Uber, LYFT or other ride-hailing platforms can use FL to enhance their surge pricing, driver location and customer waiting time algorithms preventing privacy violation.

7. **Smart Parking Systems**

Issue: Urban parking is not efficient. As a result, traffic congestion in urban areas has increased significantly later data structures keep the parking information in silos hindering the real time availability of optimization solutions.

FL Solution: Smart parking systems can predict the availability of free parking spaces by aggregating information with data from multiple numbers vehicle, allowing him to reduce his search time. Vehicles can learn from data gathered by parking sensors to predict when there will likely be open spots, while maintaining the privacy of model inputs.

Example: Application—Collaborative Parking Spot Prediction.

8. **Traffic Signal Control in Cooperation**

Use Case: Smart Traffic Light System.

Problem: Different traffic light systems require the capacity to be adjusted on-demand based on changes in vehicle flow, but vehicle data retention across a number of intersections may trigger privacy issues.

FL Solution: In this way FL allows the traffic signals and sensors at different intersections in order to train with their local traffic pattern data. These models for individual intersections can also be combined to help better coordinate the flow of traffic and optimize it, all without needlessly sharing raw traffic data.

Example: Improving traffic light timing and coordination across a city to reduce congestion, especially during rush hour while also improving the safety of roads.

9. **Smart Logistics and Freight Transport**

Use case: Freight Route Optimization with Offline Collaboration.

Problem: these optimisations require up-to-the-minute data on location and deliveries, which may make for a fair amount of privacy issues as well as potential industry-wide competitive advantages if this kind of thing were shared freely amongst companies.

FL Solution: FL enables multiple logistics companies to work together and train models that automatically optimize delivery routes according to traffic, weather, or road state. Local data can help every company, without disclosing proprietary delivery information.

Example: Optimization of route efficiency between different logistics providers (e.g. last-mile deliveries) to reduce delivery times and save fuel.

10. **Energy Management in Electric Vehicle (EV)**

Application: Shared use battery supervision and charging optimisation.

Issue: In order to manage the battery usage and charging schedules optimally, EVs must centralise this data for fleet management but in doing so certain privacy concerns might crop up.

FL Solution: EVs can locally train models used to predict battery degradation, energy consumption optimization and best charging schedules based on drive cycle patterns and available charged stationsiological characteristics.

Example: electric vehicle fleets (e.g., buses & taxis) can work together to enhance battery management and fleet-wide energy efficiency without sharing proprietary data.

11. **Smart City Infrastructure**

Urban Planning and Mobility Optimization Blockchains offer full traceability of actions, with potentially strong assurance features when used by permissioned actors, which makes them an excellent foundation for planning to help bring transparency to agents working in the smart city environment.

Problems: Smart cities use data streams from various sources—vehicles, public transport and road infrastructure are just a few examples—but centralizing all of this information violates privacy.

FL Solution: FL enables various departments of a city (transportation, utilities, public safety) to collaboratively train models for urban mobility optimization—optimizing traffic flow and coordinating public transit schedules and pedestrian safety without conventional data exchange or siloes.

Illustration: Using public transportation, ride-sharing and vehicle fleet data together to smart-up city-wide Transportation Systems whilst keeping the Data Privacy and Sovereignty bystander.

12. **Driver Behaviour Analysis and Safety**

Application: Driver Safety Improvement through Behavioral Analysis.

Issue: The collection of data on individual driving behavior (i.e. speed, braking, lane changes) includes privacy implications when used in monitoring or rating drivers.

FL Solution: FL enables models to be trained locally on individual vehicles as they show driver behaviour, so that risky behavior can then easily detect it and only model updates shared. Collaborative learning between otherwise isolated vehicles can then propagate useful knowledge to others, without leaking individual drivers' identity or driving history.

They could also reduce the costs and risks of accidents without resorting to invasive methods that would compromise driver privacy, all thanks to FL being used for evaluating driving behaviour.

6 Conclusion

Since Federated Learning permits decentralized data sources—like vehicles, traffic lights and infrastructure sensors—to learn from each other without sharing raw training data, the privacy is kept intact as fast model improvements can happen.

These examples and scenarios such as autonomous vehicle optimization, smart traffic management ridesharing requirement and predictive maintenance show how FL can improve transportation systems in general. AL not only enhances operational efficiency and safety, but also tackles more pressing environmental issues such as congestion, the use of scarce resources.

As transportation is going to be more and connected, data-driven FL has the potential of making our transportation networks smarter, faster reliable all this while being privacy preserving—a step towards future smart cities.

References

1. Wang, K., Zhang, D., Li, L., Guo, B.: Smart transportation: a comprehensive overview. IEEE Commun. Surv. Tutor. **19**(4), 1836–1857 (2017). https://doi.org/10.1109/COMST.2017.2706724
2. Xu, W., Meng, X., Wu, C., Li, M.: Smart transportation systems: concept, framework, and application. In: Proceedings of the International Conference on Big Data and Smart Computing, pp. 326–334 (2019). https://doi.org/10.1109/BigComp.2019.8679333
3. Li, S., Da Xu, L., Zhao, S.: The internet of things: a survey. Inf. Syst. Front. **20**, 577–590 (2018). https://doi.org/10.1007/s10796-017-9723-2
4. Alotaibi, S.R., Mehmood, R.: Smart transportation and smart cities in the Kingdom of Saudi Arabia. In: Smart Infrastructure and Applications, pp. 225–239. Springer (2018). https://doi.org/10.1007/978-3-319-70022-9_9
5. Pratama, A.Y., Mahmudah, A.R.: Smart transportation system: review of key enabling technologies. In: Proceedings of the 3rd International Conference on Smart Cities, Automation & Intelligent Computing Systems, pp. 45–51. Springer (2020). https://doi.org/10.1007/978-3-030-54042-9_6
6. Shaheen, S., Cohen, A., Zohdy, I.: Automated vehicles in the future of smart transportation. In: Impacts of Automated Vehicles on Public Transport, pp. 13–27. Springer, Cham (2016). https://doi.org/10.1007/978-3-319-46282-0_2
7. Li, Y., Zhao, R.: AI and machine learning for intelligent transportation systems. In: AI-Based Transportation Systems, pp. 91–108. Springer, Singapore (2020). https://doi.org/10.1007/978-981-15-2336-7_6
8. Qu, X., Li, L.: Artificial intelligence and autonomous vehicles: applications and trends. In: Smart Transportation, pp. 195–212. Springer, Singapore (2018). https://doi.org/10.1007/978-981-10-8639-7_12
9. Hussain, R., Zeadally, S.: Autonomous vehicles in smart transportation systems. In: Internet of Things for Smart Cities, pp. 85–104. Springer, Cham (2019). https://doi.org/10.1007/978-3-030-01795-2_5
10. Zhang, L., Wang, F.Y., Wang, X.: The role of artificial intelligence in intelligent transportation systems. In: Handbook of Intelligent Vehicles, pp. 379–397. Springer, London (2018). https://doi.org/10.1007/978-0-85729-085-4_39
11. Yang, Q., Liu, Y., Chen, T., Tong, Y.: Federated machine learning: concept and applications. ACM Trans. Intell. Syst. Technol. (TIST) **10**(2), 12 (2019). https://doi.org/10.1145/3298981

12. Kairouz, P., McMahan, H.B., Avent, B., Bellet, A., Bennis, M., Bhagoji, A.N., Zhao, S.: Advances and open problems in federated learning. Found. Trends Mach. Learn. **14**(1–2), 1–210 (2021). https://doi.org/10.1561/2200000083
13. Li, T., Sahu, A.K., Talwalkar, A., Smith, V.: Federated learning: challenges, methods, and future directions. IEEE Signal Process. Mag. **37**(3), 50–60 (2020). https://doi.org/10.1109/MSP.2020.2975749
14. Zhao, Y., Li, M., Lai, L., Sahu, A.K., Zaheer, M., Sanjabi, M., Mahajan, D.: Federated learning with non-IID data. In: Proceedings of the 4th Conference on Systems and Machine Learning, pp. 43–54. Springer (2020). https://doi.org/10.1007/978-3-030-53081-9_5
15. Bonawitz, K., Eichner, H., Grieskamp, W., Huba, D., Ingerman, A., Ivanov, V., van Overveldt, T.: Towards federated learning at scale: system design. In: Proceedings of the 2nd Conference on Machine Learning and Systems (MLSys), pp. 1–16. Springer (2019). https://doi.org/10.1007/978-981-15-6186-4_10
16. Chen, T., He, F., Yin, Y., Zhang, H.: A review on algorithms and techniques in intelligent transportation systems. In: Advances in Smart Vehicular Technology, Transportation, Communication and Applications, pp. 119–134. Springer, Singapore (2020). https://doi.org/10.1007/978-981-15-1728-1_9
17. Zheng, Z., Van Hentenryck, P.: Convolutional neural networks for large-scale transportation systems. Transp. Res. Part C: Emerg. Technol. **122**, 102861 (2021). https://doi.org/10.1016/j.trc.2021.102861
18. Xiong, Y., Zhang, F., Wu, H., Wang, Y.: A review on the application of machine learning and data mining in intelligent transportation systems. Neural Comput. Appl. **31**(6), 1555–1566 (2019). https://doi.org/10.1007/s00521-017-3212-3
19. Liu, R., Teng, F., Zhang, X., Zhang, L.: Deep reinforcement learning for traffic signal control. In: Artificial Intelligence in Transportation Systems, pp. 35–56. Springer, Cham (2022). https://doi.org/10.1007/978-3-030-93418-1_3
20. Xu, Z., He, J., Wang, S., Hu, S.: Evolutionary algorithms for optimizing urban traffic networks. In: Proceedings of the 16th International Conference on Natural Computation, Fuzzy Systems and Knowledge Discovery, pp. 847–856. Springer, Singapore (2020). https://doi.org/10.1007/978-981-15-8207-4_9
21. Yang, H., Wang, J., Guo, S.: Federated learning in intelligent transportation systems: a survey. IEEE Trans. Intell. Transp. Syst. (2021). https://doi.org/10.1109/TITS.2021.3076283
22. Chen, Y., Yang, J., Zhao, W., Xie, J.: Federated learning for intelligent transportation systems: a review. IEEE Commun. Surv. Tutor. **22**(3), 1930–1955 (2020). https://doi.org/10.1109/COMST.2020.2975935
23. Bonawitz, K., Hardjono, T., Steinwart, I.: Practical secure aggregation for privacy-preserving machine learning. In: Proceedings of the 2017 ACM SIGSAC Conference on Computer and Communications Security, pp. 1175–1191 (2017). https://doi.org/10.1145/3133956.3134001
24. Zhang, Y., Chen, S., Chen, Y., Li, Y.: A federated learning framework for connected vehicles: challenges and opportunities. IEEE Internet Things J. **8**(3), 1883–1897 (2021). https://doi.org/10.1109/JIOT.2020.2992299
25. Bennis, M., Mahajan, A., Debbah, M.: Federated learning for smart cities: a survey. IEEE Trans. Green Commun. Netw. **3**(2), 173–186 (2019). https://doi.org/10.1109/TGCN.2019.2891365
26. Li, H., Yang, Y., Liu, Y.: Federated learning: opportunities and challenges. IEEE Commun. Mag. **58**(4), 56–62 (2020). https://doi.org/10.1109/MCOM.2020.9039690

Federated Learning in Modern Education: Balancing Privacy, Scalability, and Effectiveness

Rita Ganatra, Monica Gahlawat, and Chetan R. Dudhagara

Abstract This chapter covers the crucial issues of data privacy, scalability, and efficacy while examining how federated learning might transform teaching methodologies. With FL, a decentralized machine learning technique, several academic institutions (clients) can work together to train a common model without jeopardizing the confidentiality of their private information. This chapter explores the complexities of choosing suitable frameworks for FL, guaranteeing strong security measures that adhere to data protection laws, and creating effective plans for deployment and training. It looks at collaboration protocols, secure communication channels, and real-world case studies of effective FL implementations in a variety of sectors. The chapter offers insightful analysis and recommended techniques for the possible integration of FL in educational contexts by examining these facets. In conclusion, the chapter demonstrates how FL enables academic institutions to take use of collaborative learning and decentralized data, opening up new avenues for improving a range of instructional settings. This innovative technology has promise for optimizing resource allocation, fostering fair access to high-quality education, and personalizing learning.

Keywords Federated learning · Artificial intelligence · Machine learning · Decentralization · Data privacy

---

R. Ganatra (✉) · M. Gahlawat
LJ Institute of Computer Applications, LJ University, Ahmedabad, Gujarat, India
e-mail: rita.gokani@ljku.edu.in

M. Gahlawat
e-mail: monica.gahlawat@ljku.edu.in

C. R. Dudhagara
International Agribusiness Management Institute, AAU, Anand, Gujarat, India
e-mail: drchetan32@gmail.com

R. Mohana et al. (eds.), *Federated Learning Applications in the Industrial Internet of Everything (IoE)*, Studies in Systems, Decision and Control 611,
https://doi.org/10.1007/978-3-031-99270-4_11

1 Introduction to Federated Learning: Exploring the Basics

In an era of data-driven innovations, the education sector is at a turning point. Large volumes of educational data have the potential to improve institutional effectiveness, tailor instruction, and allocate resources optimally, yet typical centralized approaches to data analysis pose serious privacy risks. Federated learning presents itself as a game-changing solution that permits cooperative model training without jeopardizing the confidentiality and security of private student information. The global model is then improved by the aggregator by combining these updates, and it is subsequently distributed back to the clients. Until the required degree of model accuracy is attained, this iterative procedure is carried out. For the education sector, this decentralized model offers numerous appealing advantages: Using Data Diversity, Increasing Data Security, Improving Data Privacy, Scalability, and Efficiency. FL was introduced by Google in 2016, which operates under the guidance of a central server, and several devices work together to jointly develop a machine learning model while maintaining the privacy of their personal information [1]. In Short, FL is a modern distributed AI model designed to facilitate collaborative machine learning that preserves privacy on a wide range of hardware and software platforms, even in situations where resources are limited.

1.1 Centralized Versus Distributed Versus Federated Learning

In classical machine learning, data exposure poses a privacy risk since all training data is collected and processed on a single central server. Federated learning, on the other hand, uses a distributed methodology that allows for model training on local data while maintaining participant anonymity by not disclosing sensitive information to the central server. This decentralized approach trains models using locally representative data, reducing communication overhead and promoting faster convergence. Federated learning proves to be a promising solution for large-scale applications by demonstrating superior convergence within limited communication rounds while protecting data privacy, as demonstrated by comparisons of convergence rates on publicly available datasets, such as logistic-regression-MNIST and image-classification-CIFAR-10 [2].

Figure 1 compares three distinct machine learning paradigms and emphasizes the information exchange that takes place during the learning process among clients and a central server. Several clients submit data to a master server for model training in centralized machine learning. Clients in the distributed machine learning system train their models independently, forwarding them to the master server for parameter averaging. The far-right illustration of federated learning shows how clients train models locally, upload them to the server for global model combining, and then get the parameters retrieved back to their clients. The goal of this procedure is to lessen the need to send big datasets while improving privacy.

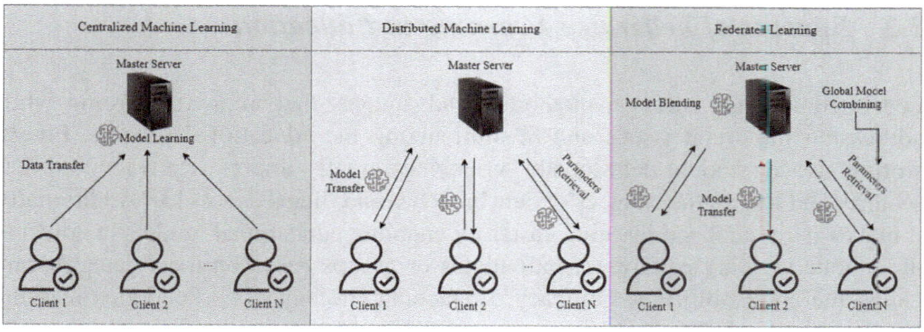

Fig. 1 Model classifier of centralized, distributed and federated learning

Centralized Machine Learning uses a traditional setup in which data is collected from several clients and transferred to a single master server. Clients 1, Client 2, …., Client N transfer their raw data straight to the master server in this setup [2]. Using the combined data from all clients, the master server handles all computational parts of training the model. Simplicity of having a single model that learns from the full dataset is the primary benefit of this strategy. However, it may raise concerns about privacy, as sensitive data is sent to and maintained on a master server and it needs a significant amount of bandwidth for transmitting potentially enormous datasets.

Distributed machine learning decentralizes the training process to some extent by allowing clients to perform computations on their local data [2]. Each client trains a model and transmits the updated or parameterized model to a master server. The global model is then updated by the master server aggregating these inputs, usually by averaging the parameters. After that, all clients receive a new distribution of the updated global model. Since just model parameters are exchanged, this approach minimizes the amount of raw data transferred; nonetheless, a strong mechanism is still needed to combine model updates efficiently and guarantee that the aggregated model functions properly on the distribution of data as a whole.

Federated learning further advances the concept of distributed learning with an emphasis on privacy and local computation [2]. In this setup, clients use their own data to train models, sending just the updated model parameters to a master server. These changes are combined by the server to produce a global model, which is subsequently transmitted back to the clients. Without exchanging raw data, this bidirectional communication enables ongoing model refinement. Federated learning reduces the danger of data exposure by keeping the data localized while still facilitating collaborative learning across many data sources, which is especially useful when data privacy is the utmost significance.

This chapter explores the core ideas of FL and considers how it could transform teaching methods while addressing privacy, scalability, and efficacy—three important issues. We will look at the core concepts, frameworks, and uses of FL in order to set the stage for a more thorough comprehension of its revolutionary influence on education going forward.

1.2 Benefits of Federated Learning in Education

Federated learning offers a plethora of advantages that improve learning while addressing important issues and revolutionizing the educational process. Firstly, storing critical student data locally on devices greatly improves privacy and data security and lowers the chances of data breaches and illegal access [3]. Additionally, it makes individualized learning easier by enabling educational models to adjust to the unique learning patterns of individuals or groups based on local data, guaranteeing the applicability and efficacy of teaching strategies [4]. Federated learning also promotes collaboration between academic institutions to enhance models while protecting privacy, resulting in strong models that may be used in a variety of learning environments [5]. By reducing data transfer and processing needs, it maximizes resource use, saving network resources and lowering costs [6]. Its scalability also makes it possible to run sizable learning networks without experiencing difficulties, facilitating system expansion and model upgrades gradually [7].

Furthermore, by enabling real-time changes to learning models based on continuing student interactions, federated learning promotes continuous learning and progress [8]. This iterative process informs evidence-based policymaking and instructional techniques while improving teaching effectiveness and offering insightful information about long-term educational trends.

Finally, by open access to excellent, individualized instruction, federated learning advances educational justice. Federated learning guarantees that learners obtain individualized educational experiences that represent the different needs and histories of the student population, regardless of the student's location or institutional affiliation. Federated learning promotes an inclusive and accessible educational environment by lessening bias and offering more equitable educational possibilities.

1.3 Types of Federated Learning

1.3.1 Vertical Federated Learning

Vertical federated learning is used for situations in which every device has a dataset containing distinct features derived from sample instances [1]. For instance, vertical FL can be used to create a shared machine learning model when two organizations have data on the same population but distinct feature sets. Vertical federated learning is also referred to as heterogeneous federated learning. When two data sets have different feature spaces but the same Instance ID space, they can be used using vertical federated learning, also known as feature-based federated learning as shown in Fig. 2.

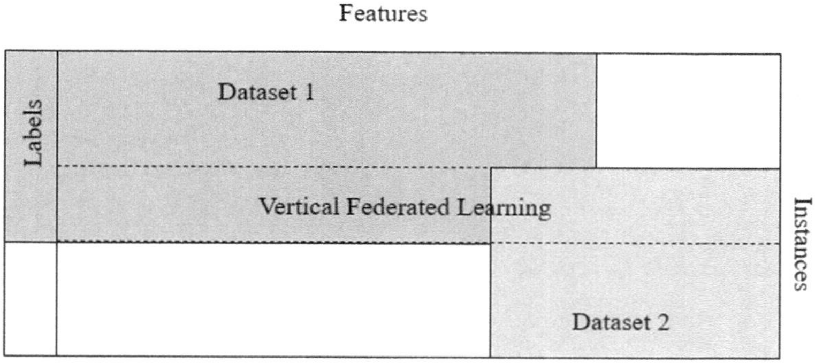

Fig. 2 Vertical federated learning

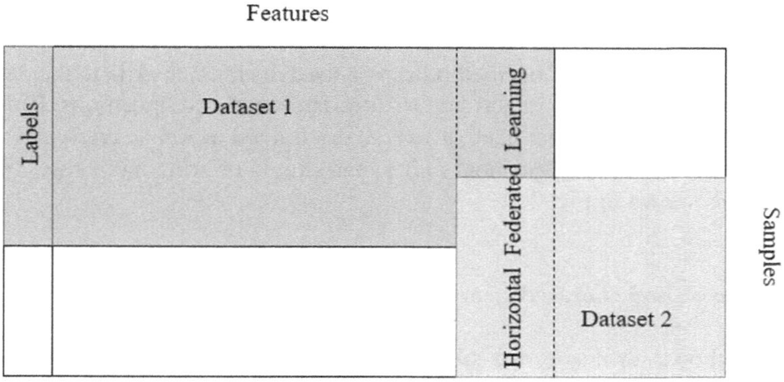

Fig. 3 Horizontal federated learning

1.3.2 Horizontal Federated Learning

Horizontal federated learning is used, when each device has datasets with the same feature space but different sample cases [1]. This kind of learning is used in the first FL use case, the Google keyboard, wherein the participating smartphones have different training data but the same functionalities. Horizontal federated learning, also known as sample-based federated learning, which is introduced in circumstances where data sets have different samples but share the same feature space as shown in Fig. 3.

1.3.3 Federated Transfer Learning

Federated transfer learning is improving a previously trained model to learn from new data that differs from the original training set in terms of features [1]. It enables the model to learn and perform outside of its original training domain. By extending

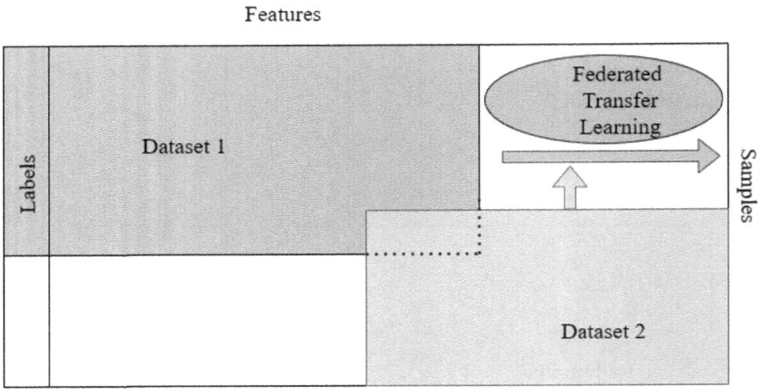

Fig. 4 Federated transfer learning

vertical federated learning, this method can be used in situations where the data sets have different feature types in addition to different sample compositions. Federated transfer learning, on the other hand, gives the pre-trained model an entire new set of features to acquire knowledge about rather than simply offering more cases for it to learn from as shown in Fig. 4.

1.3.4 Cross-Silo Federated Learning

Using a method known as "cross-silo federated learning," multiple organizations or "silos" collaborate to train an identical machine learning model while keeping their private information private [1]. Every organization distributes just model changes or gradients, and keeps its data private. Four distinct entities, 1, 2, 3 and 4, submit model updates to a central system, which combines the updates to create an enhanced, centralized model, as seen in Fig. 5. This example demonstrates the collaborative method of Federated Learning, in which several parties work together to create a machine learning model without directly exchanging data, protecting individual privacy.

1.3.5 Cross-Device Federated Learning

This type of learning occurs in situations where a lot of devices are involved. Selecting clients and creating incentive plans are two important strategies required to support this kind of FL [1] (Fig. 6).

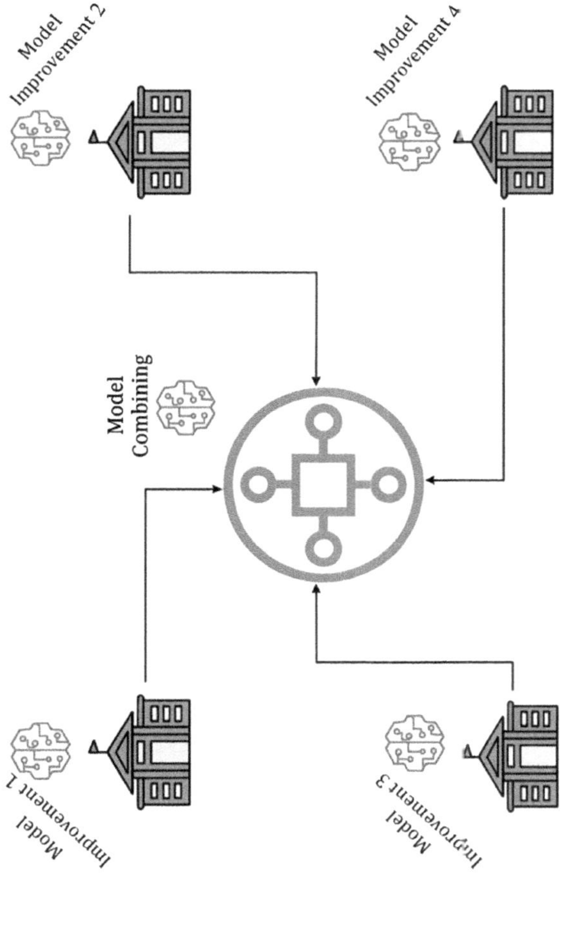

Fig. 5 Federated learning across organizations

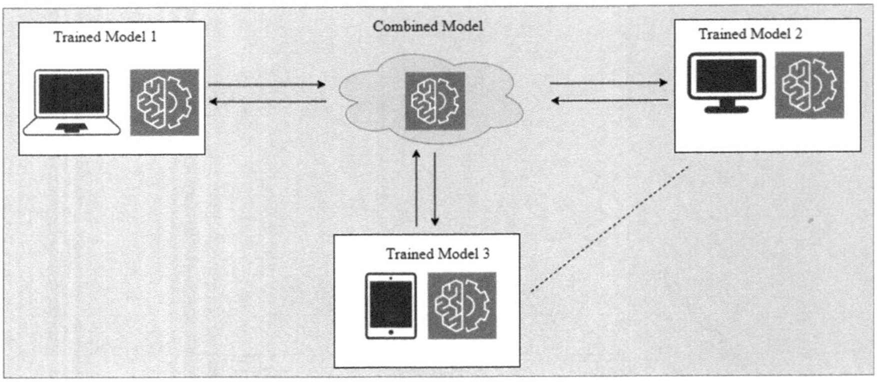

Fig. 6 Federated learning across devices

1.4 Architecture of Federated Learning

During the federated learning process, a central server gives the devices an initial model, which each device improves on its own using native data. These devices only transmit the modified information back to the central server, which compiles and refines the model.

As illustrated in Fig. 7, the cycle is repeated until the model achieves the necessary precision. By learning from a wide range of data points, this iterative approach strengthens the model's accuracy and reliability over time without compromising data privacy.

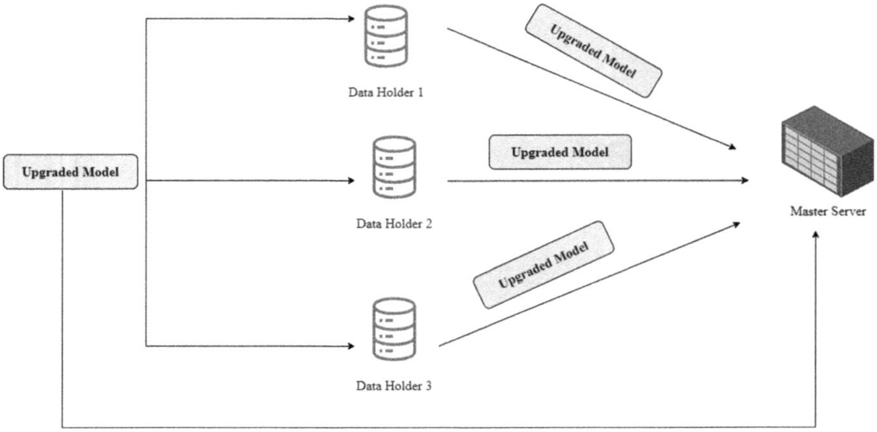

Fig. 7 General federated learning architecture

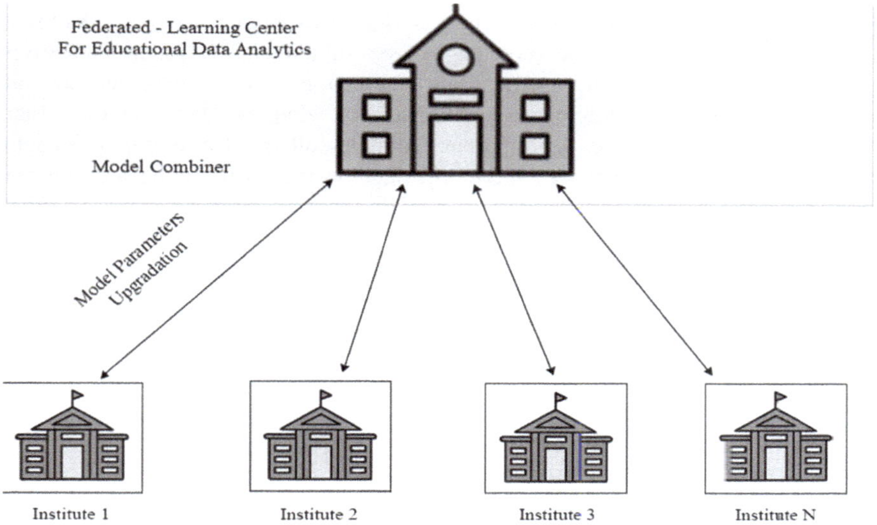

Fig. 8 General federated learning architecture for educational institutes

1.4.1 Federated Learning Architecture for Educational Institutes

The suggested cross-silo design for educational institutions is shown in Fig. 8 [4]. By evading the consequent consolidation of raw data, the federated learning framework allows a centralized evaluation of education system data. This strategy protects the privacy of instructors and students while enabling higher government organizations in charge of the educational system to perform data analytics using ML models. Instead of using vertical partitioning, these schooling cross-silo schemes use horizontal partitioning. In this instance, the various educational establishments utilize the same data structure, sharing the same student and course information among them.

1.5 Need of Federated Learning in an Education Setting

Federated Education gets access to data scattered over several platforms, locations, and companies. It makes it possible to train models on sensitive data—such as financial, medical, or educational records—securely and confidentially.

Federated learning is required in the educational system to improve educational assessments, optimize educational outcomes, handle data privacy issues, and facilitate collaborative learning. Federated learning may assist with personalized learning in educational institutions by delivering scalable, real-time precision while optimizing for individual variances [9]. Additionally, by maintaining confidential information about students confined inside every educational institution and only providing anonymized findings, it can help allay worries about data privacy. Additionally, by utilizing the combined knowledge of several universities, federated learning can enhance academic results and facilitate the creation of more reliable and accurate models. Finally, by encouraging cooperation and knowledge exchange among educational institutions, federated learning can facilitate collaborative learning. Students can gain from tailored learning experiences that consider their unique strengths and weaknesses by integrating federated learning into the educational system. Additionally, by guaranteeing that student personal information stays inside the school and is not shared outside, federated learning can safeguard student privacy. Overall, federated learning improves the effectiveness and efficiency of the educational system. However, the implementation of federated learning in education comes with its own set of challenges and considerations.

1.6 Federated Learning Frameworks for Educational Institutes

Federated Learning Frameworks that could be adapted for educational purposes are.

1.6.1 TensorFlow Federated (TFF)

Google created the open-source TensorFlow Federated framework to train machine learning models across several decentralized devices [10]. It works especially well in situations when privacy considerations prevent data from being centralized. Using data from several sources, educational institutions can jointly train models using TFF without exchanging sensitive information.

1.6.2 PySyft

A Python module named PySyft is used for private and secure machine learning [11, 12]. Popular deep learning frameworks such as PyTorch and TensorFlow can be integrated with it, enabling model training on distributed datasets without requiring the sharing of raw data. For educational institutions that wish to work together on model training while protecting data privacy, this approach is perfect.

1.6.3 Federated AI Technology Enabler (FATE)

The Chinese internet company Tencent started the open-source FATE project [10]. It offers a framework for federated learning that facilitates safe computing during model training and supports a range of machine learning methods. With FATE, educational institutions may create collaborative machine learning apps that protect student privacy and data protection.

1.6.4 Federated Learning by Microsoft Research (PyGrid)

Microsoft Research's Federated Learning architecture includes PyGrid [11, 12]. It interacts with multiple deep learning frameworks and allows for secure and privacy-preserving federated learning with PySyft. PyGrid can be used by educational institutions to enable collaborative training of machine learning models while safeguarding sensitive data.

1.6.5 Leaf

For decentralized and private machine learning, Leaf is an open-source federated learning system [13, 14]. It supports a number of communication protocols and machine learning algorithms and offers tools for creating and implementing federated learning systems. Leaf can be used by educational institutions to create federated learning programs that are customized to meet their unique requirements.

1.7 Objectives of the Chapter

The objectives of the chapter are:

- To establish the critical need for privacy-preserving machine learning in modern education and to position Federated Learning (FL) as a foundational framework for enabling data-driven pedagogical insights without compromising sensitive student information or violating regulatory compliance (e.g., FERPA);
- To explore the practical applications of Federated Learning across the educational landscape, including the development of personalized learning systems, early-warning models for student intervention, and collaborative institutional research, while maintaining data sovereignty at the individual school or district level;
- And, to critically evaluate the multifaceted challenges inherent in deploying FL in educational settings, focusing on the intricate balance between ensuring robust data privacy, achieving technical scalability across heterogeneous institutional infrastructures, and guaranteeing the pedagogical effectiveness and fairness of the resulting analytical models.

1.8 Organization of Chapter

The rest of the chapter is organized as: Sect. 2 explores federated learning applications. Section 3 enlightens literature review. Section 4 discusses diverse challenges of implementing federated learning in education. Section 5 focuses on various data privacy concerns in educational systems. Sections 6–9 illustrates overcoming

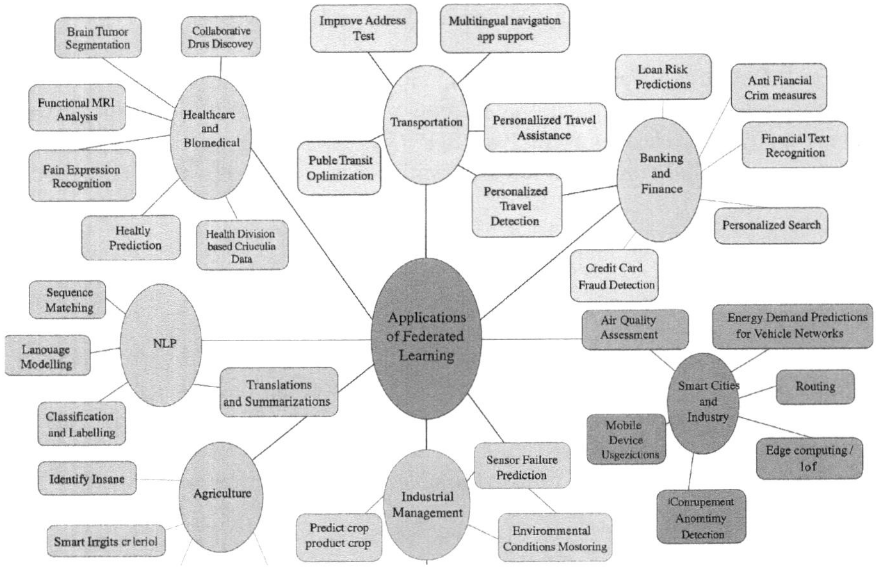

Fig. 9 Taxonomy for applications of federated learning across different domains

communication challenges in federated learning for education, personalized education through federated models, enhancing collaborative education with federated technologies and role of federated learning in remote education. And, finally, Sects. 10–12 Concludes the chapter with taxonomy of future direction and fure scope.

2 Exploring Federated Learning's Application

2.1 Federated Learning Applications in Different Domains

Federated Learning finds diverse applications across various domains. It's employed in healthcare and biomedical fields for tasks like functional MRI analysis and drug discovery. Industrial management benefits from it for optimizing processes, while recommender systems use it to personalize recommendations. In smart cities and autonomous industries, it aids in traffic flow prediction and AI object detection. Banking and finance utilize it for loan risk prediction and fraud detection. Additionally, NLP tasks and security monitoring also leverage Federated Learning for improved performance while maintaining privacy [15]. Overall, its versatility enables privacy-preserving, collaborative data analysis and model training across a wide range of sectors as shown in Fig. 9.

It improves addresses and geolocation predictive text in transportation and allows multilingual navigation app support [16]. Telecom gains from it by using data from several sources safely to enhance predictive maintenance and network efficiency [17]. Federated Learning is used in agriculture to predict crop production, identify insects, and analyze soil by combining data from several farms. This gives farmers control over their data while gaining access to pooled insights [18]. Moreover, in the field of education, it facilitates enhanced data privacy, improved remote learning options, collaborative model building, more individualized instruction, and adaptable learning environments [19].

2.2 *Federated Learning Applications in Education*

Some potential applications of federated learning in education include:

Federated learning allows for personalized learning, adaptive assessment, collaborative knowledge sharing, remote proctoring, predictive analytics, teacher professional development, educational research and policy analysis, student support and intervention, feedback systems, collaborative curriculum building and improvement, multilingual education systems, personalized educational robotics, and chatbots and virtual assistants for education as shown in Fig. 10. Personalized learning models can offer customized learning opportunities, materials, and advice, while adaptive assessment systems can dynamically modify test formats and difficulty levels based on student performance data.

Federated learning also ensures data privacy and security, allowing for collaborative insights and knowledge extraction. It can also help identify students at risk of struggling academically or dropping out, allowing for prompt interventions and support. Federated learning also facilitates cross-institutional collaboration between academics and policymakers, allowing for thorough analysis and well-informed decision-making in the field of education.

Federated learning enhances the educational process by enabling instantaneous feedback systems for instructors and students. It also facilitates collaborative curriculum building and improvement, enables multilingual education systems, and aids in personalized educational robotics. It also enables the creation of intelligent chatbots and virtual assistants for education, offering personalized help and tailored materials as shown in Fig. 10.

2.3 *Applying Federated Learning to Improve Educational Assessments*

In order to improve educational assessments, one approach that can be utilized is the application of federated learning. Federated learning optimizes learning while

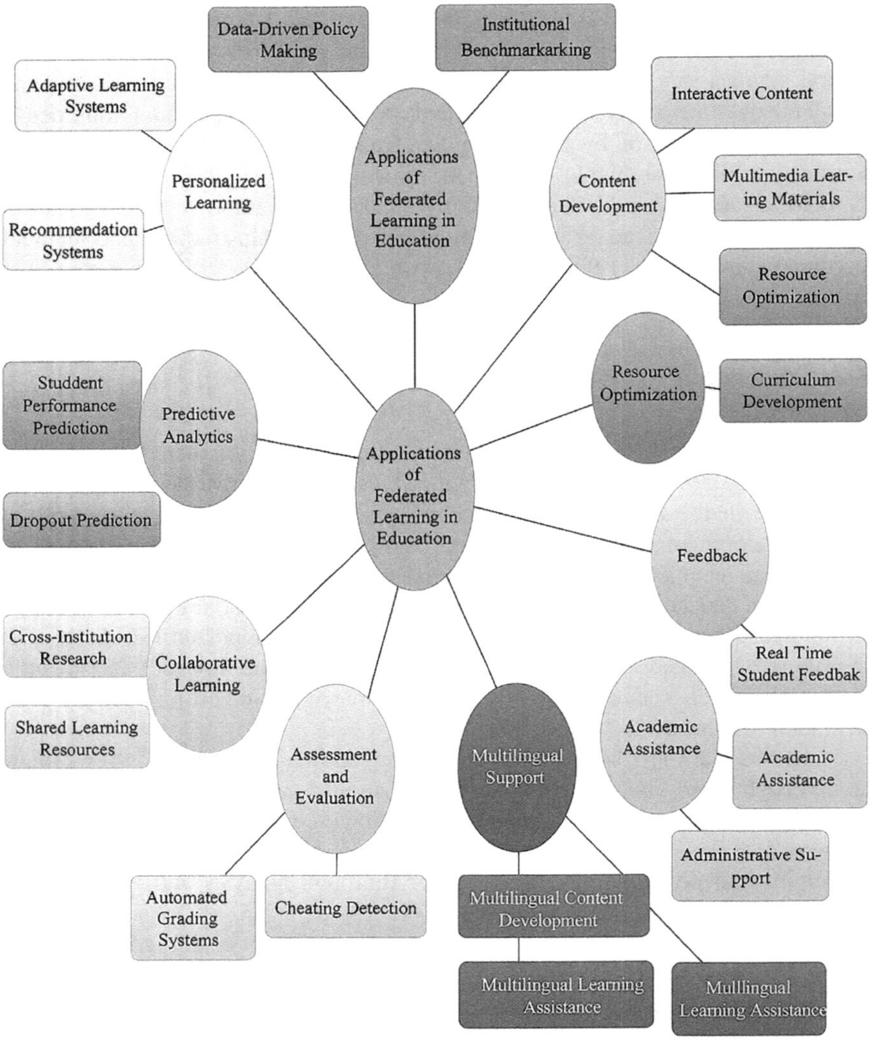

Fig. 10 Applications of federated learning in education

lowering the cost of delivering data to a centralized system by facilitating the use of data preserved across different devices. Multiple gains can be made in educational assessments by including federated learning.

Qin et al. [20], Woubie et al. [21] and Liu et al. [22] highlighted that federated learning provides a privacy preserving way. Student data security and privacy can be improved through federated learning as Federated learning allows data to remain on individual devices, providing better protection against potential breaches or misuse of sensitive information, as opposed to gathering and storing all student data in a single server.

Furthermore, Tan et al. [23] discussed personalized federated learning. By making use of distinct information which has been stored on every student's device, federated education encourages customized learning. This information can be used to customize instruction by identifying preferences, goals, and accomplishments.

Deng et al. [24] explored personalized federated learning. Federated learning aids in the creation of adaptive tests. Federated Learning is able to produce more precise and individualized evaluations for every student by exploring patterns and trends in the combined knowledge and data from various devices.

Ma et al. [8] and Li et al. [25] deliberated on real-time updates of federated learning system. Real-time feedback to teachers and students can also be obtained through the usage of federated learning in educational assessments. In response to this input, students can track their progress, identify areas for improvement, and adjust their learning strategies. Federated learning can also help teachers find gaps in their students' knowledge and modify their lesson plans to strengthen particular areas of difficulty.

Federated learning has the ability to improve privacy, support individualized learning, enable adaptive assessments, and offer real-time feedback in educational evaluations. These benefits will ultimately improve the educational process for both teachers and students. The correctness and dependability of eLearning tests administered without an instructor can be guaranteed by including federated learning. Since student data is stored on individual devices and is difficult to alter or cheat with, federated learning can help mitigate the problem of assessment malpractices.

3 Literature Review

S. no	Paper	Published date and year	Researcher(s)	References	Insights
1	Federated learning in the era of privacy preservation: a comprehensive review and future directions	16 Mar 2024	Kulshreshtha et al.	[47]	Federated learning enables collaborative model training without sharing data, ensuring privacy. Techniques like differential privacy and secure computation maintain data integrity. It revolutionizes predictive model construction in privacy-conscious environments
2	Integrated comprehensive analysis method for education quality with federated learning	10 Jan 2024	Wang et al.	[48]	The paper presents a method, CEQFL, utilizing federated learning to comprehensively assess education quality by aggregating data from universities for model enhancement, demonstrating strong assessment capabilities
3	FedEL: federated education learning for generating correlations between course outcomes and program outcomes for internet of education things	8 Jan 2024	Sengupta et al.	[49]	Federated Education Learning (FedEL) in the study utilizes federated learning for generating course outcomes (CO) and program outcomes (PO) correlations at the institution level, achieving 83.33% accuracy
4	Federated learning: advancements, applications, and future directions for collaborative machine learning in distributed environments	13 Apr 2024	Katyayani et al.	[50]	Federated Learning shows promise for education in 2024, enhancing privacy in decentralized settings. Challenges like non-IID data distribution and model heterogeneity are being addressed for improved efficacy
5	Revolutionizing AI-assisted education with federated learning: a pathway to distributed, privacy-preserving, and debiased learning ecosystems	20 May 2024	Hridi et al.	[51]	The paper discusses the potential of federated learning (FL) in education, highlighting benefits and challenges. It suggests future research directions for implementing FL in educational settings effectively
6	Federated learning-outcome prediction with multi-layer privacy protection	28 Dec 2023	Zhang et al.	[52]	The study introduces FecMap, a federated learning model for learning-outcome prediction, incorporating multi-layer privacy protection to enhance performance and preserve data privacy across institutions

(continued)

(continued)

S. no	Paper	Published date and year	Researcher(s)	References	Insights
7	A survey on blockchain-based federated learning and data privacy	29 Jun 2023	Chhetri et al.	[6]	The survey explores blockchain's role in enhancing data privacy in federated learning, comparing privacy mechanisms. It identifies research gaps and suggests future directions for blockchain-based federated learning
8	Accuracy degrading: toward participation-fair federated learning	15 Jun 2023	You et al.	[53]	The paper introduces the FedACC framework to address accuracy degradation in federated learning by ensuring fairness through controlling model accuracy based on participants' contributions
9	When decentralized optimization meets federated learning	04 Jun 2023	Gao et al.	[54]	Decentralized optimization enhances federated learning by addressing single-point failure risks, offering a robust alternative with improved privacy preservation and communication efficiency in distributed data settings
10	Migrating federated learning to centralized learning with the leverage of unlabeled data	04 May 2023	Wang et al.	[55]	The paper introduces a teacher-student framework using unlabeled data to enhance global model performance, combining federated learning's privacy preservation with centralized learning's accuracy
11	Stochastic clustered federated learning	02 Mar 2023	Zeng et al.	[56]	Federated learning in education allows leveraging private data on edge devices, enhancing model training without centralising data, ensuring privacy, and improving model performance through clustered federated learning
12	Federated learning for data analytics in education	20 Feb 2023	Fachola et al.	[4]	Federated learning in education offers privacy protection by training models on distributed data, enabling personalised learning trajectories, early dropout prediction, and comparable performance to centralised approaches without data centralization risks

(continued)

(continued)

S. no	Paper	Published date and year	Researcher(s)	References	Insights
13	Distributed sequential federated learning	31 Jan 2023	Wang et al.	[57]	Federated learning in education offers enhanced data security, minimises data transportation, and protects personal information, facilitating efficient information aggregation across multiple sites without centralising data
14	Federated learning for education data analytics	05 Jan 2023	Fachola et al.	[58]	Federated learning in education offers privacy protection by training models on distributed data, enabling analytics like dropout prediction and student classification without centralising sensitive information
15	Federated knowledge transfer for heterogeneous visual models	13 Dec 2022	Li et al.	[59]	Federated learning in education offers privacy-preserving collaborative training of heterogeneous visual models, enhancing generalisation performance and enabling wide applicability across various models
16	Federated learning scheme with dual security of identity authentication and verification	01 Dec 2022	Xu et al.	[60]	Federated Learning in education offers data privacy protection, decentralised storage, and collaborative model training, enhancing security and collaboration in educational settings
17	FedCL: federated multi-phase curriculum learning to synchronously correlate user heterogeneity	14 Nov 2022	Wang et al.	[46]	Federated Learning in education offers benefits like improved generalisation performance and handling user heterogeneity effectively, as shown in FedCL research
18	Federated learning	03 Nov 2022	Simeone	[61]	Federated learning in education offers enhanced privacy, scalability, and collaboration by enabling decentralised model training on local data without centralising sensitive information
19	Federated learning in multi-RIS-aided systems	15 Jun 2022	Ni et al.	[62]	Federated learning in education enables collaborative model training on edge devices without raw data transmission, improving accuracy, convergence speed, training loss, and inference accuracy significantly

(continued)

(continued)

S. no	Paper	Published date and year	Researcher(s)	References	Insights
20	Applied federated learning: architectural design for robust and efficient learning in privacy aware settings	01 Jun 2022	Stojkovic et al.	[3]	Federated learning in education offers enhanced privacy by keeping user data on devices, minimising centralization risks, and maintaining model performance with minimal degradation
21	Towards a secure and reliable federated learning using blockchain	26 Jan 2022	Moudoud et al.	[63]	Federated learning in education offers privacy preservation, communication efficiency, and resource conservation, enhancing collaborative model training while ensuring data security and reliability
22	Federated learning	01 Nov 2021	Ray et al.	[64]	Federated Learning in education allows decentralised devices to train models locally, preserving data privacy, enhancing scalability, and enabling collaborative learning without centralising data
23	A novel federated learning approach based on the confidence of federated Kalman filters	22 Aug 2021	Hu et al.	[65]	Federated learning in education offers enhanced privacy, addresses non-IID data challenges, and improves fault tolerance through FedCK with Federated Kalman Filters, enhancing robustness and accuracy
24	Federated Word2Vec: leveraging federated learning to encourage collaborative representation learning	19 Apr 2021	Garcia Bernal et al.	[66]	Federated Learning in education allows collaborative training of NLP models like Word2Vec across organisations' data without compromising privacy, ensuring quality results and efficient convergence
25	A survey on federated learning	15 Mar 2021	Zhang et al.	[67]	Federated learning in education offers decentralised training data, ensuring privacy, and reduces costs. It allows local training, model aggregation, and updates, enhancing data security and collaboration among multiple clients
26	Federated learning used for predicting outcomes in SARS-COV-2 patients	08 Jan 2021	Flores et al.	[68]	Federated Learning in healthcare showed improved model performance, data privacy, and collaboration across multiple institutes, suggesting similar benefits for education: enhanced model accuracy, privacy, and collaboration

(continued)

(continued)

S. no	Paper	Published date and year	Researcher(s)	References	Insights
27	From distributed machine learning to federated learning: in the view of data privacy and security	19 Oct 2020	Shen et al.	[69]	Federated learning in education offers enhanced data privacy and security by allowing local training on smart devices, reducing the risk of sensitive information leaving client devices
28	Pedagogical data analysis via federated learning toward education 4.0	04 Aug 2020	Guo et al.	[70]	Federated learning in education ensures student privacy, breaks data isolation, and maintains analysis quality without centralising data, as shown in the FEEDAN framework proposed in the paper
29	Active federated learning	27 Sep 2019	Goetz et al.	[71]	Advantages of federated learning in education include training models without centralising data, minimising transmission costs, and maximizing efficiency through Active Federated Learning's selective client sampling based on model and data

4 Challenges of Implementing Federated Learning in Education

There are few challenges with Federated Learning defined by author [26] in Table 1.

Rajbongshi et al. [26] highlighted challenges including device heterogeneity influencing performance, communication overhead from frequent updates, security vulnerabilities from possible attacks, non-IID data distribution affecting model training, and synchronization challenges resulting from intermittent as shown in Table 1.

In addition, the authors listed the difficulties in Table 2 that the Indian education system faced when implementing federated learning. It highlights issues like the necessity for FL to adjust to curricular and linguistic diversity, the necessity of teacher and student participation, the necessity of data privacy and security, the impact of resource constraints on scalability, and the requirement for regulatory compliance with data protection laws.

Different Authors have highlighted the challenges of implementing Federated Learning in Education such as:

Zhang et al. [26] and Rieke et al. [27] discussed the standardization challenge while implementing federated learning in education as there is absence of standardized procedures, structure, and interfaces, hindering resource sharing and collaboration.

Rajbongshi et al. [26], Rieke et al. [28], Al-Huthaifi et al. [29] and Mothukuri et al. [30] discussed the challenges of data privacy and security in federated learning, emphasizing the need for robust measures to protect sensitive student data, particularly in collaborative model training.

Rajbongshi et al. [26] and Zhang et al. [27] highlighted the concerns like limited resources and infrastructure which is crucial for implementing federated learning in education, especially for smaller centers with limited funding and technological resources. Addressing these constraints is essential for scalability and effectiveness.

Saini [31], Amidi and Giannopoulos [32] and Thummisetti and Atluri [33] discussed ethical considerations in federated learning, highlighting the need to adhere

Table 1 Challenges with federated learning [26]

Challenges	Description
Non-IID data distribution	Uneven distribution of data across nodes in federated learning setting can lead to suboptimal model training
Communication overhead	High communication costs between server and client devices due to frequent model updates and parameter exchanges
Security vulnerabilities	Potential threats such as model poisoning attacks, where adversarial actors
Device heterogeneity	Variations in computational capabilities and storage constraints among devices may result in consistent model performance and hinder collaborative learning
Synchronization issues	Challenges arising from intermittent connectivity, learning to synchronization issues and delays in model aggregation

Table 2 Challenges to implement FL in the Indian education system [26]

Challenges	Description
Linguistic and curricular diversity	FL must adapt to diverse language & curricula in Indian schools
Resource constraints	Limited resources hinder FL scalability in education institutions
Data privacy and security	Ensuring privacy of student data is crucial amidst FL implementations
Varying technological infrastructure	Disparities in tech infrastructure across institutes after FL integration
Teacher and student participation	Engaging teachers and students in FL process requires awareness and collaboration
Regulatory compliance	Compliance with data protection laws like personal data protection bill is essential for FL implementations in India

to legal and moral norms, including informed consent and data protection legislation, while maintaining morality and openness as shown in Fig. 11.

5 Overcoming Data Privacy Concerns in Educational Systems

As educational institutions manage sensitive student data, it is essential to protect its privacy and security. Federated learning provides a way around this problem by limiting the amount of data shared and keeping it confined inside each institution, protecting important student data only.

Choudhury et al. [35] and Gao et al. [36] proposed verifiable asynchronous federated learning for preserving privacy. By preserving data privacy, this strategy facilitates cooperation and coordination amongst educational institutions and makes it possible to create uniform guidelines and procedures for federated learning. Federated learning can address the issues with data privacy in educational systems by putting strong data anonymization techniques, differential privacy safeguards, and secure protocols for data exchange into practice. Aledhari et al. [9] discussed federated learning protocols for secure communication.

Liang et al. [34] discussed local model training in federated learning. By keeping the data localized inside every institution and only sharing aggregated insights that do not reveal personally identifiable information. Federated learning can assist address the issues of security and privacy of data in customized education, protecting the confidentiality and security of sensitive student information.

Nguyen et al. [37] said general data protection regulation (GDPR) makes FL an ideal solution for overcoming data privacy concern. Privacy concerns in educational systems that use federated learning can be addressed by following data privacy standards like GDPR and making sure that educational data protection laws are followed.

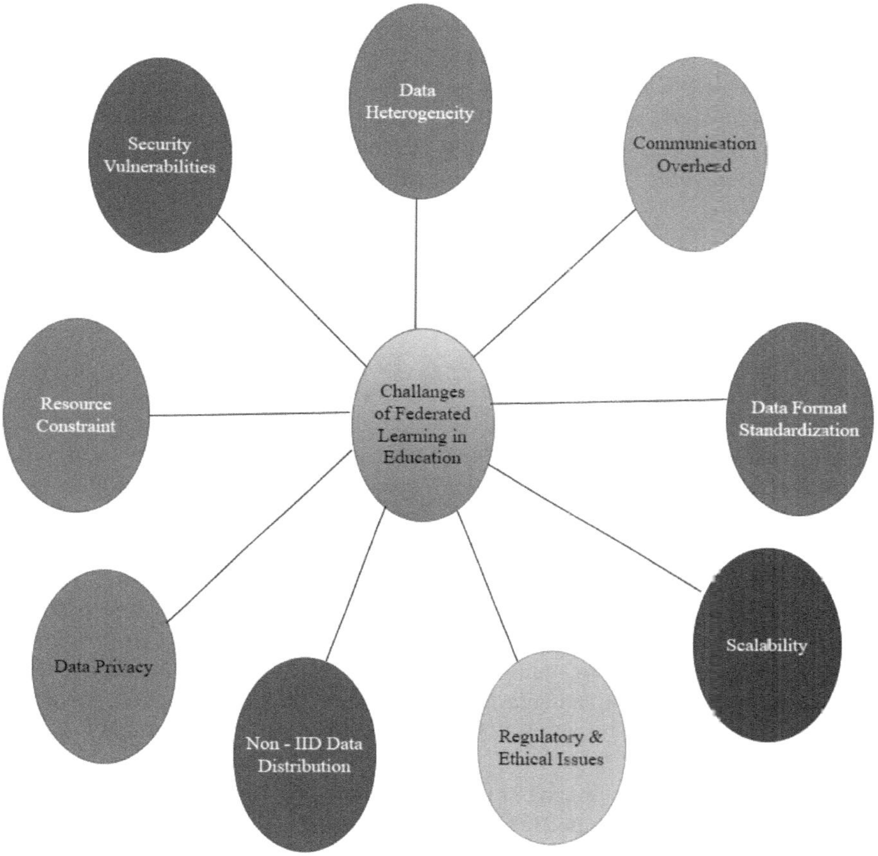

Fig. 11 Challenges of implementing federated learning in education

6 Overcoming Communication Challenges in Federated Learning for Education

Institutions might utilize many techniques to tackle communication obstacles in federated learning for educational purposes. First and foremost, it is imperative to set up explicit protocols for communication that include channels of agreement for exchanging updates and resolving problems during cooperative training. Second, using efficient and safe communication channels, including encrypted techniques, guarantees the privacy of student data is protected all along the way. Standardized reporting procedures can also make it easier to track developments across organizations, encouraging transparency and accountability. Furthermore, encouraging open communication, cooperation, and constructive criticism among participating institutions promotes information sharing, open conversation, and overall efficacy of cooperative model training. Last but not least, offering guidance and materials on effective communication gives participants the tools they need to overcome obstacles, such as resolving misunderstandings and collaborating across institutional borders.

By implementing these strategies into practice, educational institutions can successfully negotiate federated learning's communication challenges, promoting a cooperative model training environment and protecting student privacy and data ownership.

7 Personalized Education Through Federated Models

In the field of education, there is a growing need for personalized learning experiences that cater to individual student needs and preferences [4]. One-size-fits-all approaches are common in traditional educational paradigms, which may not adequately accommodate individuals' varied learning preferences and skill levels.

However, distributed models can now be used to produce individualized instruction as a result of advancements in federated learning approaches. Without jeopardizing the security and privacy of student data, educational institutions can use federated learning to train and develop machine learning models based on dispersed datasets across numerous devices. This method makes it possible to create user profiles, early dropout indicators, individualized learning trajectories, and other educational solutions that can significantly improve the efficacy and efficiency of the educational system. Tan et al. [23] discussed Personalized Federated Learning and its challenges.

Education establishments can reduce privacy issues and adhere to legal and ethical requirements by using federated learning to guarantee that data stays on local devices or separate data centers. Federated learning also makes it possible for academics and educational institutions to collaborate without exchanging raw data. Nguyen et al. [37] talked about implementing data privacy protection legislation.

In short, federated models of personalized education enable the creation of customized learning experiences that meet the demands of each individual student while honoring issues of security and privacy, scalability, and technical infrastructure.

8 Enhancing Collaborative Education with Federated Technologies

The goal of collaborative education is to improve students' learning experiences at educational institutions by combining in-person and online instruction. By encouraging student and teacher participation, knowledge sharing, and the development of new skills in virtual learning environments, federated technologies can further improve collaborative education. Federated learning approaches, which enable the training and development of machine learning models based on dispersed datasets across numerous devices, can be used to achieve this.

Federated learning also makes it possible to use quantitative methods to educational data, including learning analytics. These methods can be used to create user profiles, customize learning paths, and spot warning signs and early dropouts. In conclusion, by guaranteeing privacy, promoting information sharing, and permitting customized learning trajectories, the incorporation of federated technologies in collaborative education has the potential to completely transform the learning process. The importance of reliable weather forecasts in the quickly evolving world of today cannot be emphasized. To optimize the advantages for both educators and students in the context of collaborative education, federated technologies must be utilized. Students and teachers can take part in virtual classrooms, exchange knowledge, and develop new skills by utilizing federated technologies in collaborative education [38]. This may result in more successful and interesting learning opportunities as well as better student outcomes.

In summary, we can improve the learning process through personalized learning trajectories, data protection and privacy, knowledge sharing, and active engagement between teachers and students in virtual learning environments by utilizing federated technologies in collaborative education.

9 The Role of Federated Learning in Remote Education

Federated learning plays a crucial role in remote education by addressing key challenges and optimizing educational outcomes. It makes it possible for many educational institutions to work together, pooling their resources and knowledge to advance the creation of machine learning models and increase instructional strategies. Educational institutions can overcome the drawbacks of traditional centralized approaches—where collaboration is frequently impeded by worries about data privacy and security—by utilizing federated learning.

Through federated learning, academic institutions can train machine learning models on their localized data as a group without exchanging private information. This encourages the availability of different data for more thorough and accurate models while simultaneously guaranteeing privacy protection. All things considered, federated learning in education shows promise for revolutionizing personalized learning, facilitating distributed collaboration among institutions, resolving privacy issues, and optimizing educational outcomes through the use of decentralized data processing and collective intelligence.

Federated learning, which addresses issues like privacy, scalability, and technical infrastructure, has the potential to completely transform personalized learning in education by fusing the advantages of data-driven insights with tailored instruction. Federated learning also gives students the ability to take an active role in their own education. Federated learning encourages students to take an active role in their education by enabling them to contribute their data to the training of tailored models without requiring their personal data to escape their devices.

10 Taxonomy of Future Direction and Opportunities for Federated Learning in Education

Future goals and opportunities for federated learning in education involve dealing with ethical and legal dilemmas surrounding the handling of sensitive data for purposes like unsupervised student classification and dropout prediction [4, 39]. Furthermore, improvements in federated editable learning (FEL) can ease privacy problems, augment knowledge, and aid in knowledge editing [4, 40]. Early dropout indications, user profile determination, and individualized learning trajectories can all result from the application of quantitative methodologies based on educational data [41]. Federated learning can open the door for more effective and confidentiality-preserving learning analytics solutions by safeguarding student privacy, boosting educational outcomes, and strengthening data security in educational environments.

1. Enhancing individualized learning, predicting dropouts early, and protecting data privacy are some of the future routes that federated learning in education will take. Opportunities exist for enhancing academic performance while protecting private data [4].
2. Federated Editable Learning (FEL), improving privacy issues, enabling individualized learning, and creating effective federated unlearning methods—which offers prospects for fresh knowledge creation and educational advancements—are some of the future directions in federated learning for education [39].
3. Promising prospects for educational data analytics are presented by the next steps for federated learning in education, which include increasing predictive analytics for student performance, promoting individualized learning, and strengthening privacy protection [4].
4. Optimizing resource allocation, data protection, and individualized learning are some of the future directions for federated learning in education. FL provides chances for safe, cooperative educational progress [40].
5. Promising prospects for educational applications can be found in the following future areas of federated learning for education: algorithm improvements, tailored learning, and resolving security and hardware limitations, enabling continuous learning, managing nonstandard data, cost reduction, and regulatory compliance [41].
6. Education using federated learning will go in new ways in the areas of collaborative model training, individualized learning, and student data protection. There are opportunities to use decentralized data utilization to improve educational outcomes [42].
7. In the future, federated learning in education will focus on tackling issues like data security and model aggregation while improving privacy in student data sharing, individualized learning models, and collaborative research opportunities [43].

8. Personalized learning models, data privacy safeguards, and cooperative research projects to improve educational results using decentralized machine learning systems are some of the future directions for federated learning in education [42].
9. Personalized online FL for customized learning experiences, resolving issues with sporadic data access and device-specific features in IoT/CPS applications, and facilitating real-time learning from various data sources in educational settings are some of the possible future recommendations for federated learning in education [44].
10. Prospective avenues for Over-The-Air Federated Learning (OTA-FL) in the education sector could encompass improving system efficiency, dependability, and credibility, while tackling issues such as model imbalance, accessibility, and distortion [45].
11. Improving user heterogeneity handling, boosting generalization performance, and investigating dynamic multi-phase curricula to synchronously improve learning speed are some of the future topics in federated learning for education [46].

11 Future Direction and Opportunities for Federated Learning in Education

Future development of Federated Learning in educational institutes should focus on enhancing scalability and efficiency in collaborative model training, focusing on optimizing communication protocols and minimizing network latency. Advanced security measures, such as encryption techniques and differential privacy mechanisms, are also essential for safeguarding student data. Specialized tools and resources are needed to facilitate the adoption of Federated Learning in educational settings. Additionally, research should explore advanced applications beyond traditional machine learning models, such as natural language processing and computer vision, to revolutionize educational data utilization and protection.

12 Conclusion

Federated learning offers a compelling fresh vision for education, one that maintains the critical value of student privacy while simultaneously enabling collaborative data analysis. The potential of FL to transform educational practices has been highlighted in this chapter; however, its successful integration depends on resolving important issues and pursuing a research agenda that is focused on the future.

In order to suit a variety of educational contexts, future development must prioritize improving the efficiency and scalability of collaborative model training. This will involve maximizing communication protocols and decreasing network latency. Securing sensitive student data requires strong security measures that are non-negotiable, such as differential privacy methods and powerful encryption techniques. Specialized tools and resources designed to meet the unique requirements of educational institutions will be necessary to promote greater use. Furthermore, studies should investigate FL's enormous potential beyond conventional machine learning models, branching out into fields like computer vision and natural language processing to transform the use and security of educational data.

Federated learning can enable a paradigm change in educational technology by adopting these future paths, guaranteeing the ethical and responsible use of data while promoting fair access to high-quality education.

References

1. Mammen, P.M.: Federated learning: opportunities and challenges (2021). arXiv preprint arXiv: 2101.05428
2. Asad, M., Moustafa, A., Ito, T.: Federated learning versus classical machine learning: a convergence comparison (2021). arXiv preprint arXiv:2107.10976. arXiv:2101.05428
3. Stojkovic, B., Woodbridge, J., Fang, Z., Cai, J., Petrov, A., Iyer, S., Huang, D., Yau, P., Kumar, A.S., Jawa, H., Guha, A.: Applied federated learning: Architectural design for robust and efficient learning in privacy aware settings (2022). arXiv preprint arXiv:2206.00807
4. Fachola, C., Tornaría, A., Bermolen, P., Capdehourat, G., Etcheverry, L., Fariello, M.I.: Federated learning for data analytics in education. Data **8**(2), 43 (2023)
5. Bernal, D.G., Giaretta, L., Girdzijauskas, S., Sahlgren, M.: Federated word2vec: leveraging federated learning to encourage collaborative representation learning (2021). arXiv preprint arXiv:2105.00831
6. Chhetri, B., Gopali, S., Olapojoye, R., Dehbashi, S., Namin, A.S.: a survey on blockchain-based federated learning and data privacy. In: 2023 IEEE 47th Annual Computers, Software, and Applications Conference (COMPSAC), pp. 1311–1318. IEEE 2023
7. Ray, N.K., Puthal, D., Ghai, D.: Federated learning. IEEE Consum. Electron. Mag. **10**(6), 106–107 (2021)
8. Ma, X., Wen, C., Wen, T.: An asynchronous and real-time update paradigm of federated learning for fault diagnosis. IEEE Trans. Indus. Inf. **17**(12), 8531–8540 (2021)
9. Aledhari, M., Razzak, R., Parizi, R.M., Saeed, F.: Federated learning: a survey on enabling technologies, protocols, and applications. IEEE Access **8**, 140699–140725 (2020)
10. Kholod, I., Yanaki, E., Fomichev, D., Shalugin, E., Novikova, E., Filippov, E., Nordlund, M.: Open-source federated learning frameworks for IoT: a comparative review and analysis. Sensors **21**(1), 167 (2020)
11. Ziller, A., Trask, A., Lopardo, A., Szymkow, B., Wagner, B., Bluemke, E., Nounahon, J.M., Passerat-Palmbach, J., Prakash, K., Rose, N., Ryffel, T.: Pysyft: a library for easy federated learning. Federated Learning Systems: Towards Next-Generation AI, pp. 111–139 (2021)
12. Hilberger, H., Hanke, S., Bödenler, M.: Federated learning with dynamic model exchange. Electronics **11**(10), 1530 (2022)
13. Caldas, S., Duddu, S.M.K., Wu, P., Li, T., Konečný, J., McMahan, H.B., Smith, V., Talwalkar, A.: Leaf: a benchmark for federated settings (2018). arXiv preprint arXiv:1812.01097
14. Patel, N.P., Parekh, R., Amin, S.A., Gupta, R., Tanwar, S., Kumar, N., Iqbal, R., Sharma, R.: LEAF: a federated learning-aware privacy preserving framework for healthcare ecosystem. IEEE Trans. Netw. Serv. Manage. (2023)
15. Shaheen, M., Farooq, M.S., Umer, T., Kim, B.S.: Applications of federated learning; taxonomy, challenges, and research trends. Electronics **11**(4), 670 (2022)

16. Rahman, K.J., Ahmed, F., Akhter, N., Hasan, M., Amin, R., Aziz, K.E., Islam, A.M., Mukta, M.S.H., Islam, A.N.: Challenges, applications and design aspects of federated learning: a survey. IEEE Access **9**, 124682–124700 (2021)
17. Mangla, U.: Application of federated learning in telecommunications and edge computing. In: Federated Learning: A Comprehensive Overview of Methods and Applications, pp. 523–534. Springer International Publishing, Cham (2022)
18. Manoj, T., Makkithaya, K., Narendra, V.G.: A federated learning-based crop yield prediction for agricultural production risk management. In: 2022 IEEE Delhi Section Conference (DELCON), pp. 1–7. IEEE (2022)
19. Guo, S., Zeng, D., Dong, S.: Pedagogical data analysis via federated learning toward Education 4.0. Am. J. Educ. Inform. Technol. **4**(2), 56–65 (2020)
20. Qin, Y., Li, M., Zhu, J.: Privacy-preserving federated learning framework in multimedia courses recommendation. Wireless Netw. **29**(4), 1535–1544 (2023)
21. Woubie, A., Solomon, E., Attieh, J.: Maintaining privacy in face recognition using federated learning method. IEEE Access (2024)
22. Liu, X., Li, H., Xu, G., Chen, Z., Huang, X., Lu, R.: Privacy-enhanced federated learning against poisoning adversaries. IEEE Trans. Inf. Forensics Secur. **16**, 4574–4588 (2021)
23. Tan, A.Z., Yu, H., Cui, L., Yang, Q.: Towards personalized federated learning. IEEE Trans. Neural Netw. Learn. Syst. (2022)
24. Deng, Y., Kamani, M.M., Mahdavi, M.: Adaptive personalized federated learning (2020). arXiv preprint arXiv:2003.13461
25. Li, L., Xiong, H., Guo, Z., Wang, J., Xu, C.Z.: SmartPC: Hierarchical pace control in real-time federated learning system. In: 2019 IEEE Real-Time Systems Symposium (RTSS), pp. 406–418. IEEE (2019)
26. Rajbongshi, S.K., Sarmah, K., Sarmah, S.: Unveiling the tapestry: federated learning challenges and opportunities in the Indian educational landscape. Math. Stat. Eng. Appl. **71**(2), 747–762 (2022)
27. Zhang, T., Gao, L., He, C., Zhang, M., Krishnamachari B., Avestimehr, A.S.: Federated learning for the internet of things: applications, challenges, and opportunities. IEEE Internet Things Mag. **5**(1), 24–29 (2022)
28. Rieke, N., Hancox, J., Li, W., Milletari, F., Roth, H.R., Albarqouni, S., Bakas, S., Galtier, M.N., Landman, B.A., Maier-Hein, K., Ourselin, S.: The future of digital health with federated learning. NPJ Digit. Med. **3**(1), 1–7 (2020)
29. Al-Huthaifi, R., Li, T., Huang, W., Gu, J., Li, C.: Federated learning in smart cities: privacy and security survey. Inf. Sci. **632**, 833–857 (2023)
30. Mothukuri, V., Parizi, R.M., Pouriyeh, S., Huang, Y., Dehghantanha, A., Srivastava, G.: A survey on security and privacy of federated learning. Futur. Gener. Comput. Syst. **115**, 619–640 (2021)
31. Saini, T.: Re: Open issues in federated learning (2023). Retrieved from: https://www.researchgate.net/post/Open_issues_in_federated_learning/64e9c3bf2d3d87971607eb57/citation/download
32. Amidi, A., Giannopoulos, A., Trakadas, P.: Federated Learning Techniques: From Ethics to Regulation (2023). Available at SSRN 4588972
33. Thummisetti, B.S.P., Atluri, H.: Advancing healthcare informatics for empowering privacy and security through federated learning paradigms. Int. J. Sustain. Dev. Comput. Sci. **1**(1), 1–16 (2024)
34. Liang, P.P., Liu, T., Ziyin, L., Allen, N.B., Auerbach, R.P., Brent, D., Salakhutdinov, R., Morency, L.P.: Think locally, act globally: federated learning with local and global representations (2020). arXiv preprint arXiv:2001.01523
35. Choudhury, O., Gkoulalas-Divanis, A., Salonidis, T., Sylla, I., Park, Y., Hsu, G., Das, A.: Anonymizing data for privacy-preserving federated learning (2020). arXiv preprint arXiv: 2002.09096
36. Gao, Y., Wang, L., Zhang, L.: Privacy-preserving verifiable asynchronous federated learning. In: Proceedings of the 2021 3rd International Conference on Software Engineering and Development, pp. 29–35 (2021)

37. Nguyen, D.C., Ding, M., Pathirana, P.N., Seneviratne, A., Li, J., Poor, H.V.: Federated learning for internet of things: a comprehensive survey. IEEE Commun. Surv. Tutor. **23**(3), 1622–1658 (2021)
38. Alvarez, C., Alarcon, R., Nussbaum, M.: Implementing collaborative learning activities in the classroom supported by one-to-one mobile computing: a design-based process. J. Syst. Softw. **84**(11), 1961–1976 (2011)
39. Wu, L., Guo, S., Wang, J., Hong, Z., Zhang, J., Zhou, J.: On knowledge editing in federated learning: perspectives, challenges, and future directions (2023). arXiv preprint arXiv:2306.01431
40. Banabilah, S., Aloqaily, M., Alsayed, E., Malik, N., Jararweh, Y.: Federated learning review: fundamentals, enabling technologies, and future applications. Inf. Process. Manage. **59**(6), 103061 (2022)
41. Ding, J., Tramel, E., Sahu, A.K., Wu, S., Avestimehr, S., Zhang, T.: Federated learning challenges and opportunities: an outlook. In: ICASSP 2022–2022 IEEE International Conference on Acoustics, Speech and Signal Processing (ICASSP), pp. 8752–8756. IEEE (2022)
42. Bharati, S., Mondal, M., Podder, P., Prasath, V.B.: Federated learning: applications, challenges and future directions. Int. J. Hybrid Intell. Syst. **18**(1–2), 19–35 (2022)
43. Zhu, J., Cao, J., Saxena, D., Jiang, S., Ferradi, H.: Blockchain-empowered federated learning: challenges, solutions, and future directions. ACM Comput. Surv. 55(11), 1–31 (2023)
44. Gogineni, V.C., Werner, S., Gauthier, F., Huang, Y.F., Kuh, A.: Personalized online federated learning for IoT/CPS: challenges and future directions. IEEE Internet Things Mag. **5**(4), 78–84 (2022)
45. Xiao, B., Yu, X., Ni, W., Wang, X., Poor, H.V.: Over-the-air federated learning: Status quo, open challenges, and future directions. Fundam. Res. (2024)
46. Wang, M., Guo, J., Jia, W.: Federated multi-phase curriculum learning to synchronously correlate user heterogeneity. IEEE Trans. Artif. Intell. (2023)
47. Kulshreshtha, K., Goel, B., Kanyal, H.S., Singhal, V.: Federated learning in the era of privacy preservation: a comprehensive review and future directions. In: 2024 2nd International Conference on Disruptive Technologies (ICDT), pp. 763–768. IEEE (2024)
48. Wang, R., Chen, T., Wang, J., Lai, J., Li, J., Zhang, M., Chen, X.: Integrated comprehensive analysis method for education quality with federated learning. In: International Conference on Computer Science and Educational Informatization, pp. 439–449. Springer Nature Singapore, Singapore (2023)
49. Sengupta, D., Khan, S.S., Das, S., De, D.: FedEL: federated education learning for generating correlations between course outcomes and program outcomes for internet of education things. Internet of Things **25**, 101056 (2024)
50. Katyayani, M., Keshamoni, K., Murthy, A.S.R.C., Rani, K.U., Reddy, S.L., Alapati, Y.K.: Federated learning: advancements, applications, and future directions for collaborative machine learning in distributed environments. J. Electr. Syst. **20**(5s), 165–171 (2024)
51. Hridi, A.P., Sahay, R., Hosseinalipour, S., Akram, B.: Revolutionizing AI-assisted education with federated learning: a pathway to distributed, privacy-preserving, and debiased learning ecosystems. In: Proceedings of the AAAI Symposium Series, vol. 3, No. 1, pp. 297–303 (2024)
52. Zhang, Y., Li, Y., Wang, Y., Wei, S., Xu, Y., Shang, X.: Federated learning-outcome prediction with multi-layer privacy protection. Front. Comp. Sci. 18(6), 186604 (2024)
53. You, X., Liu, X., Lin, X., Cai, J., Chen, S.: Accuracy degrading: toward participation-fair federated learning. IEEE Internet Things J. **10**(12), 10291–10306 (2023)
54. Gao, H., Thai, M.T., Wu, J.: When decentralized optimization meets federated learning. IEEE Netw. **37**(5), 233–239 (2023)
55. Wang, X., Zhu, T., Ren, W., Zhang, D., Xiong, P.: Migrating federated learning to centralized learning with the leverage of unlabeled data. Knowl. Inf. Syst. **65**(9), 3725–3752 (2023)
56. Zeng, D., Hu, X., Liu, S., Yu, Y., Wang, Q., Xu, Z.: Stochastic clustered federated learning (2023). arXiv preprint arXiv:2303.00897
57. Wang, Z.F., Zhang, X.Y., Chang, Y.I.: Distributed sequential federated learning (2023). arXiv preprint arXiv:2302.00107

58. Fachola, C., Tornaria, A., Bermolen, P., Capdehourat, G., Etcheverry, L., Fariello, M.I.: Federated learning for education data analytics (2023)
59. Li, W., Zhu, Z., Huang, T., Sun, L., Yuan, C.: Federated knowledge transfer for heterogeneous visual models. In: Proceedings of the 4th ACM International Conference on Multimedia in Asia, pp. 1–7 (2022)
60. Xu, Y., Su, M., Hou, J., Nie, C.: Federated learning scheme with dual security of identity authentication and verification. In: 2022 IEEE International Conference on Trust, Security and Privacy in Computing and Communications (TrustCom), pp. 695–702. IEEE (2022)
61. Simeone, O.: Federated learning. In: Machine Learning for Engineers, pp. 541–562. Cambridge University Press, Cambridge (2022)
62. Ni, W., Liu, Y., Yang, Z., Tian, H., Shen, X.: Federated learning in multi-RIS-aided systems. IEEE Internet Things J. **9**(12), 9608–9624 (2021)
63. Moudoud, H., Cherkaoui, S., Khoukhi, L.: Towards a secure and reliable federated learning using blockchain. In: 2021 IEEE Global Communications Conference (GLOBECOM), pp. 01–06. IEEE (2021)
64. Ray, N.K., Puthal, D., Ghai, D.: Federated learning. IEEE Consum. Electron. Mag. **10**(6), 106–107 (2021)
65. Hu, K., Wu, J., Weng, L., Zhang, Y., Zheng, F., Pang, Z., Xia, M.: A novel federated learning approach based on the confidence of federated Kalman filters. Int. J. Mach. Learn. Cybern. **12**, 3607–3627 (2021)
66. Garcia Bernal, D., Giaretta, L., Girdzijauskas, S., Sahlgren, M.: Federated Word2Vec: leveraging federated learning to encourage collaborative representation learning (2021)
67. Zhang, C., Xie, Y., Bai, H., Yu, B., Li, W., Gao, Y.: A survey on federated learning. Knowl.-Based Syst. **216**, 106775 (2021)
68. Flores, M., Dayan, I., Roth, H., Zhong, A., Harouni, A., Gentili, A., Abidin, A., Liu, A., Costa, A., Wood, B., Tsai, C.S.: Federated learning used for predicting outcomes in SARS-COV-2 patients. Res. Square (2021)
69. Shen, S., Zhu, T., Wu, D., Wang, W., Zhou, W.: From distributed machine learning to federated learning: in the view of data privacy and security. Concurr. Comput. Pract. Exp. **34**(16), e6002 (2022)
70. Guo, S., Zeng, D., Dong, S.: Pedagogical data analysis via federated learning toward Education 4.0. Am. J. Educ. Inform. Technol. **10**(2), 56–65 (2020)
71. Goetz, J., Malik, K., Bui, D., Moon, S., Liu, H., Kumar, A.: Active federated learning (2019). arXiv preprint arXiv:1909.12641

GPSR Compliance

The European Union's (EU) General Product Safety Regulation (GPSR) is a set of rules that requires consumer products to be safe and our obligations to ensure this.

If you have any concerns about our products, you can contact us on ProductSafety@springernature.com

In case Publisher is established outside the EU, the EU authorized representative is:

Springer Nature Customer Service Center GmbH
Europaplatz 3
69115 Heidelberg, Germany

Batch number: 09409918

Printed by Printforce, the Netherlands